Smart Energy Management for Smart Grid

Smart Energy Management for Smart Grid

Editor

Miguel Jiménez Carrizosa

MDPI

Basel • Beijing • Wuhan • Barcelona • Belgrade • Novi Sad • Cluj • Manchester

Editor
Miguel Jiménez Carrizosa
Universidad Politécnica de
Madrid, Madrid, Spain

Editorial Office
MDPI
St. Alban-Anlage 66
4052 Basel, Switzerland

This is a reprint of articles from the Special Issue published online in the open access journal *Energies* (ISSN 1996-1073) (available at: https://www.mdpi.com/journal/energies/special_issues/Smart_Energy_Management_for_Smart_Grid).

For citation purposes, cite each article independently as indicated on the article page online and as indicated below:

Lastname, A.A.; Lastname, B.B. Article Title. *Journal Name* **Year**, *Volume Number*, Page Range.

ISBN 978-3-0365-8272-6 (Hbk)
ISBN 978-3-0365-8273-3 (PDF)
doi.org/10.3390/books978-3-0365-8273-3

Contents

About the Editor

Miguel Jiménez Carrizosa

Miguel Jiménez Carrizosa (PhD) received an M.S. Industrial Engineering degree in Electrical Engineering from the Universidad Politécnica de Madrid (UPM), Madrid, Spain, in 2011, an M.S. degree in Energy Systems from Supélec, Gif-sur-Yvette, France, in 2011, and a Ph.D. degree in Electrical Engineering and Automatic Control from Paris-Sud XI University, Orsay, France, in 2015.

He is currently an Assistant Professor at the "Departamento de Energía y Combustibles", Escuela Técnica Superior de Ingenieros de Minas y Energía, UPM, and a researcher with the Centro de Electrónica Industrial (CEI-UPM). His research interests include smart grids, HVDC systems, the control of power converters, transmission and distribution systems, the integration of renewable energies and optimal power flow methods.

energies

MDPI

Review

A Review of the Conceptualization and Operational Management of Seaport Microgrids on the Shore and Seaside

Nur Najihah Abu Bakar [1,2], Josep M. Guerrero [1,*], Juan C. Vasquez [1], Najmeh Bazmohammadi [1], Yun Yu [1], Abdullah Abusorrah [3] and Yusuf A. Al-Turki [3]

1 Center for Research on Microgrids (CROM), AAU Energy, Aalborg University, 9220 Aalborg East, Denmark; nurnbab@energy.aau.dk (N.N.A.B.); juq@energy.aau.dk (J.C.V.); naj@energy.aau.dk (N.B.); yyu@energy.aau.dk (Y.Y.)
2 Faculty of Electrical Engineering Technology, Universiti Malaysia Perlis (UniMAP), Kampus Pauh Putra, Arau 02600, Perlis, Malaysia
3 Center of Research Excellence in Renewable Energy and Power Systems, Department of Electrical and Computer Engineering, Faculty of Engineering, K. A. CARE Energy Research and Innovation Center, King Abdulaziz University, Jeddah 21589, Saudi Arabia; aabusorrah@kau.edu.sa (A.A.); yaturki@kau.edu.sa (Y.A.A.-T.)
* Correspondence: joz@energy.aau.dk

Abstract: Seaports are well known as the medium that has evolved into the central link between sea and land for complex marine activities. The growth in maritime logistics particularly necessitates a large volume of energy supply in order to maintain the operation of sea trade, resulting in an imbalance between generation and demand sides. Future projections for three major concerns show an increase in load demand, cost of operation, and environmental issues. In order to overcome these problems, integrating microgrids as an innovative technology in the seaport power system appears to be a vital strategy. It is believed that microgrids enhance seaport operation by providing sustainable, environmentally friendly, and cost-effective energy. Although microgrids are well established and widely used in a variety of operations on land, their incorporation into the seaport is still limited. The involvement of a variety of heavy loads such as all-electric ships, cranes, cold ironing, and buildings infrastructure renders it a complicated arrangement task in several aspects, which necessitates further research and leaves space for improvement. In this paper, an overview of the seaport microgrids in terms of their concepts and operation management is presented. It provides the perspectives for integrating the microgrid concept into a seaport from both shore side and seaside as a smart initiative for the green port's vision. Future research directions are discussed towards the development of a more efficient marine power system.

Keywords: cold ironing; electrification; operation management; renewable energy source; seaport microgrids; shipboard microgrid; maritime

Citation: Bakar, N.N.A.; Guerrero, J.M.; Vasquez, J.C.; Bazmohammadi, N.; Yu, Y.; Abusorrah, A.; Al-Turki, Y.A. A Review of the Conceptualization and Operational Management of Seaport Microgrids on the Shore and Seaside. *Energies* **2021**, *14*, 7941. https://doi.org/10.3390/en14237941

Received: 19 October 2021
Accepted: 17 November 2021
Published: 26 November 2021

1. Introduction

Recent decades have witnessed the rapid expansion of the global economy. This development was motivated by an even faster escalation in international trade. Seaport plays a significant role in this economic liberalization as sea trading is in high demand due to its cheap prices and capability to support goods transportation at a large volume through cargo services. Currently, the marine transport network accounts for more than 90 percent of global trade, and according to the International Maritime Organization (IMO) annual report, it is predicted to be tripled by 2050 [1]. Seaports have become complex hubs due to the multifunctional operations that need to comply with various factors such as marine regulations, technologies, transportation, operational, and policy requirements. Hence, high investment is required to expand ports facilities and infrastructure and to enhance their operation management and maintenance in order to boost all marine operations.

The increasing number of ships, cranes, and trucks for port transportation requires high-energy consumption. Moreover, the size of ships has become much larger nowadays, with ships measuring hundreds of meters long and kilotons in terms of weight. All of these factors consequently result in a huge volume of fossil fuel burning in order to fulfill the entire load at the port. Overuse of energy always results in unbalance energy issues, which makes the energy system inefficient and prone to frequent power outages and thereby huge economic losses.

Research in developing more efficient power systems for port operations has attracted special interest in recent years. Incorporating renewable energy sources (RESs) and energy storage system (ESS) technologies through microgrid systems has been viewed as an essential route. By utilizing microgrids, power generation is not limited to the conventional fossil-based resources anymore, and the flexibility to integrate multiple RESs such as wind turbines (WTs), photovoltaic (PV) systems, biomass, ocean energy, hydrogen, and geothermal sources all contribute to sustainable energy when introduced efficiently. Deploying ESS at seaport will enhance better energy distribution by providing backup energy during emergencies and providing the capability of storing excess energy generation.

The promising benefits of microgrids to the marine sector can be categorized into four groups, namely energy efficiency, economic, environmental, and security benefits. However, the variation of heavy loads energy demand at the port both at the shore and seaside results in complex control for seaport microgrids. Although microgrids coordination is an undeniably established technology in land-based applications that currently provides tremendous economic and environmental benefits, there are still several issues that need to be addressed before incorporating them in maritime applications. Moreover, the literature in this field is far from conclusive. Many aspects need further exploration in order to ensure the optimal benefits of integrating the microgrid concept into seaports.

In this perspective, this paper presents an overview of the seaport microgrids concept and its operational measures for seaport applications on both shore and seaside (shipboard microgrid). Moreover, from a macro perspective to most co-occurrence research studies, VOSviewer visualization analysis is provided to illustrate the trending and probable future research directions of microgrid applications in the maritime sector.

The rest of this paper is organized as follows. Section 2 is dedicated to introducing the variation of load demand and current major issues at ports as well as energy evolution in seaports over time. In Section 3, the conceptualization of microgrids on shore and seaside is presented, while topology and operation management of the seaport microgrids are also discussed. In addition, a few constraints and potential aspects for further research investigation are highlighted in Section 4. Finally, the significant findings of the paper are summarized in Section 5.

2. Seaport

Seaports are well known as significant interfaces for both sea and land transportations, which is supported by their strategic geographical locations between the sea and land, as shown in Figure 1. Ports are among the largest parts of the industrial sector for economic and social development across the world. However, the growing demand for logistics globally increases the volume of sea traffic.

Port functions have aggressively changed along with global development compared with the conventional ports as shown in Figure 2, where the first generation of ports in the early 1960s only focused on the modes of transportation without any trading and commercial activities. It then experienced more advances in terms of technology, networking, international trading, and logistics. In the fifth generation, ports moved toward smart ports by employing automation, advanced technologies, hybrid and intelligent infrastructure, and efficient energy management systems.

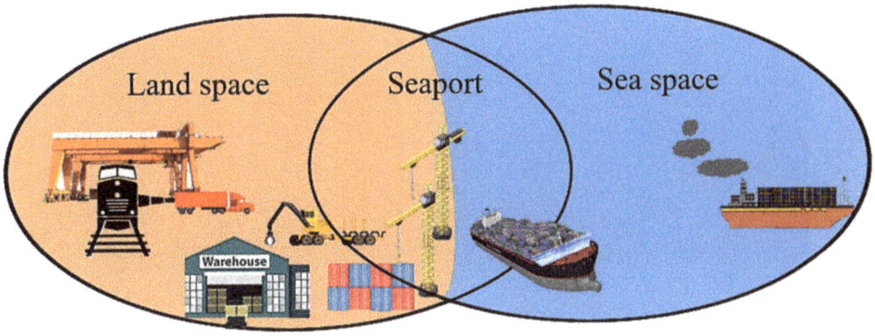

Figure 1. Seaport intersection between land and sea space [2].

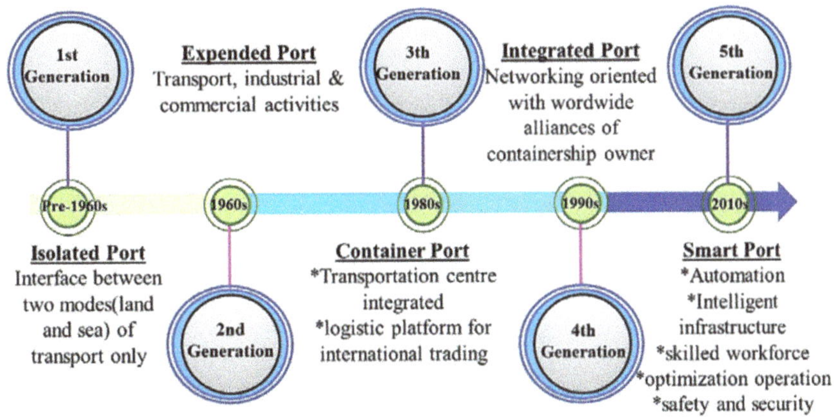

Figure 2. The evolution of the seaport functions over time [2–5].

2.1. Port Activities and Power Consumers

Port engagement with respect to energy can be classified into energy generation and power consumption mainly in the form of electricity and fuel. It is crucial to comprehend the load components and monitor energy-related activities taking place within the port before initiating any energy planning. The purpose is to analyze the amount of energy demand and ensure that there is enough power supply to prevent power shortages. However, load profile measures at seaports vary according to several factors such as the type of port and activities conducted at that particular port.

Seaports by definition can be viewed as centers of economic activities associated with any kind of arrival (tourist or goods), service of ships, and cargo [6]. The most common ports can be categorized into commercial/industrial ports [2], container terminal ports [7], and intermodal ports [8,9]. In another classification provided in [10], ports are classified into three types, such as local ports, national ports, and international ports. These classifications are based on the ports' characteristics, region of the port, the volume of loads it serves, vessel type, port's operation and services, the annual number of passengers, and the annual number of ships berthing in and out from the port. Different types of ports and their characteristics are summarized in Table 1.

Table 1. Classification of ports [10].

Port Type	Characteristics
Local port	Serves for local needs Limited space and capacity Small size No logistic activity handling Do not support cruise ships Boats, vessels, yachts, and small-sized ships < 500 passengers
National port	Serves country needs Medium-sized (larger than local port) Cover all ships type with small logistic and cruise activities Medium-sized ships < 2500 passengers, cargo (packages), and logistics (only trucks)
International port	Serves international needs Largest sized Provide huge logistic infrastructure Cruise ships > 2500 passengers, cargo, containerships, and RTG cranes

Geographically, a seaport is the point of interaction between land and sea, which involves two modes of transportation (water and road traffic) with different characteristics. From this perspective, the seaport area can be sorted into the seaside, shore side, and intermediate space. Each space is utilized for a different type of activity, transportation, and application, all of which have an impact on the amount of energy demand. Figure 3 illustrates the most common activities running in the seaport area and its load components.

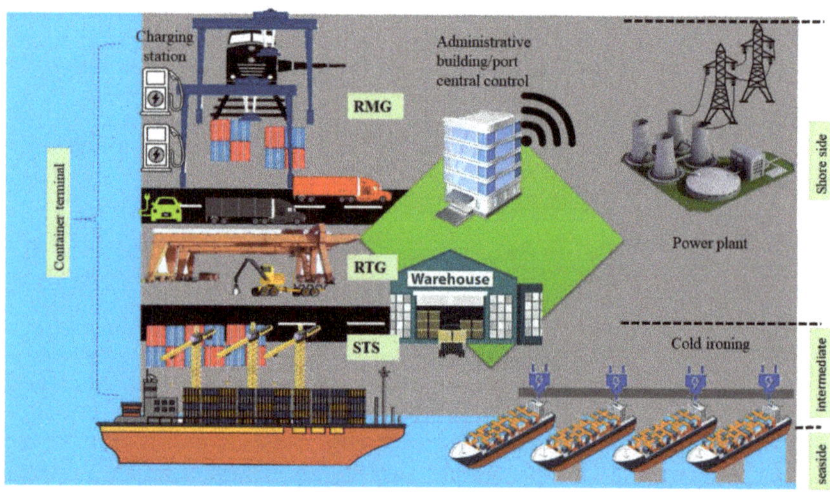

Figure 3. Seaport activities and load components (abbreviations: rail mounted gantry (RMG), rubber tyred gantry (RTG), and ship to shore (STS)).

Loads on the shore-side: On the shore side, administrative buildings and a custom facility, as well as a warehouse for goods, are built. In this infrastructure, electricity is consumed mostly for lighting; Heating, Ventilation, and Air Conditioning (HVAC) system; and equipment [11]. Many factors influence the building's energy consumption, including weather conditions, building materials, occupant behavior, work durations, equipment, and electrical load used [12].

In terms of transportation, different land vehicles such as trucks, cranes, yard tractors, and trains are powered by diesel fuel, and electricity is needed for electric vehicle (EV) charging stations. Typically, diesel is used as the primary source for motor movement in the cranes. However, recent research studies on electrical cranes show a lot of interest in energy storage systems for storing potential energy regenerated from lowering and lifting cranes operations [13].

Loads on the seaside: The seaside is includes a marine vessel or any watercraft transportation voyages across the ocean that serves the purpose of carrying passengers or delivering cargo. Ships will make more voyages, consume more fuel, and increase the volume of water traffic as global trade expands. Normally heavy fuel oil (HFO) is consumed for big vessels and bulk carriers. General cargo vessels consume the most HFO, followed by oil tankers and cruise ships accounting for 66,000 t, 43,000 t, and 25,000 t of HFO annual consumption, respectively [14]. The spill of fossil-based marine fuel into the water and its combustion into the air can become a major threat to the environment [15]. Despite its undesired environmental impacts, HFO continues to be the preferred fuel of the maritime transport industry due to its relatively low cost, widely available resources, and the ability to suit engines that were originally designed for HFO [14].

In order to prevent heavy utilization of HFO and to be in line with port development, ship technology is embracing the electrification concept by implementing an Integrated Power System (IPS) and storage system known as All Electric Ship (AES) [16]. Energy usage for a vessel is hard to measure as it depends on numerous aspects such as the size of the vessel, onboard loads, vessel speed, sea waves, and weather conditions.

During berthing for transit or transferring goods, the auxiliary engines of the ships are kept operating in order to supply the energy for onboard loads [17]. To overcome continuous fossil fuel burning, cold-ironing facilities are provided at the intermediate area between the sea and shore side. Power requirement varies from 300 kW to 7 MW depending on the type of ships and berthing duration [18]. In addition to the onshore power supply, the port must accommodate shore charging facilities. Some of the vessels with storage components such as hybrid vessels and AES will need a charging station in the port area for recharging their batteries and supplying onboard loads.

Loads in the intermediate area: In this region, there are cold ironing, charging stations, and container terminal facilities. The container terminal is an important part of international logistics and requires a large amount of energy, both fuel and electricity [19], for loading and unloading cargo as well as the transshipping goods to the next mode of transportation [20]. Container terminals serve three primary functions: yard side, quayside, and landside [21]. Each side has its operation and transportation needs. Cold storage facilities among the terminal operation consume a lot of energy, as it is a temperature-controlled storage solution for perishable goods. The Port of Wilmington built a 101,000 square foot refrigerated warehouse to refrigerate food, pharmaceutical, floral, and other items that require maintaining specific temperatures for storage [22]. It allows distribution companies to deliver their goods locally, nationally, and globally in good conditions.

In the container terminal, there are various types of cranes in use, from ship to shore (STS)/quay cranes (QC) to automated guided vehicles (AGVs)/automated straddle carriers (ASCs), and, finally, rail-mounted gantry (RMG) cranes, before they are transferred into truck/trains [23]. Considering peak power demand, cranes need about 72% of total energy (STS cranes about 37%, ASC cranes 32%, and RMG around 3%) [24]. The breakdown of the cranes' power demand is shown in the pie chart of Figure 4.

From the appliances that are used and activities conducted within the marine port, it can be observed that the seaport sector has a large energy demand that makes the energy-handling problem a complex task. Hence, a reliable power system is required that can provide sufficient energy supply to all distribution loads. Any shortcoming in energy delivery will cause a big disruption to the seaport's operations.

Therefore, it is very important to distinguish between critical and non-priority loads so that the critical loads will receive priority during emergencies in order to obtain energy supplies [25]. Table 2 summarizes the findings regarding seaport-related services and their load.

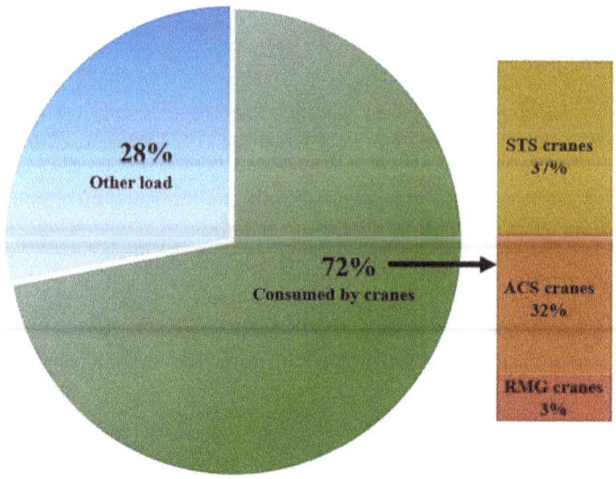

Figure 4. Breakdown of cranes power demand during demand peak intervals.

Table 2. Seaport activities and loads variation [2,26].

Seaport's Services	Load	Factor Influence Energy Consumption	Load Classifications	Form of Energy
Vessel	Passenger ships (cruise, ferry), container ships, electric ships, tugs, gliders, bunkers, boats, tankers, hovercraft, sailboats, submarines, yachts	Size of the ship, activity conduct on the ship, time of operation, weather, wave, speed	(1) Peak load (2) Critical load (3) Non-critical load (4) Variable load (5) Constant load	(1) Electricity (2) Fossil fuel
Goods handling	Cargo, container, quay, logistic, freight forwarder, customs warehouse, storage, security, loading-unloading	Number of cranes, amount of cargo, hours of operation		
Administration	Management and administrative building, planning, service solution, IT, monitoring	Type of electrical equipment, weather, building material, hours of operation, occupant behavior		
Transportation	Electric vehicles, cranes, trucks, yard tractors, trains	Number of transportation, hours of consumption		
Electric Facility	Cold ironing, charging station for electric vehicles	Time of berthing, number of ships per berthing, size, and ship's load		
Maintenance	Repair and maintenance	Type of the maintenance		

2.2. Port Critical Concerns and Green Maritime Policy

The urgent need for a more efficient maritime system demands a great effort to increase the performance of every subsystem linked to it by implementing an energy efficiency program. It is critical to identify the root of the problem and core issues in the port in order to achieve a good outcome from planning.

The European Sea Ports Organisation (ESPO) in their latest report (October 2020) highlighted the top ten priorities of the port sector from 1996 to 2020 [27]. Figure 5 shows the top three port issues from 2009 until 2020. As observed in this figure, most of the issues listed have remained the same while their relative positions have changed over time. However, air quality and energy consumption remained mostly the highest priorities. These data are important as it identifies the vital concerns that port managing sectors are working on.

According to the top three priority issues highlighted in Figure 5, air quality has persisted as number one from 2013 to 2020 followed by energy consumption. The problem arises from the fact that air quality, noise, port waste, and climate change are strongly related to environmental concerns. This indicates that environmental issues are big problems in the port sector and require immediate action. Without fast action, the emissions from maritime transportations are predicted to rise by 250% in 2050 from the amount released in 2012 [1]. Heavy utilization of fossil fuel in marine transportation and other areas contributes to serious air pollution from Carbon Dioxide (CO_2) and greenhouse gas (GHG) emissions. This is because the fossil sources encompass hazardous contents, which in burning it will emit chemical and dangerous gases such as particulate matter (PM); (CO_2); sulphur dioxide (SO^2); nitrogen oxides (NOx); and black carbon (BC) to the air [28]. Consequently, this emission results in acid rain and serious climate changes. It is also capable to reaching 400 km [29] of land and produces a negative effect on human health such as asthma, tuberculosis, impact on children's lung growth, cardiovascular disease, and lung cancer.

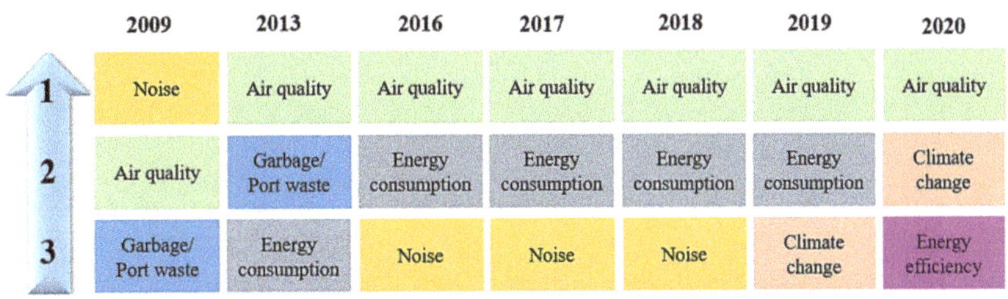

Figure 5. Top three issues in the port from 1996 to 2020 [27].

Meanwhile, ports' energy consumption has increased due to various reasons. Development and expanded port functionalities over time raise the energy demand from many facilities and loads with low and heavy consumption. Continuous heavy use of energy will cause fast energy resources depletion. The inequality between energy production and demand results in frequent unplanned power outages. Poor power quality, lack of energy monitoring, and old instruments are also among the factors that cause the high energy consumption problem resulting in additional energy ***costs*** in ports' daily operation. The need for improvement both in infrastructure and port power system will acquire a huge amount of investment. Lacking proper planning and a solid development framework will cause great losses. Based on these scenarios, three major issues related to seaports include energy, environment, and cost, as shown in Figure 6.

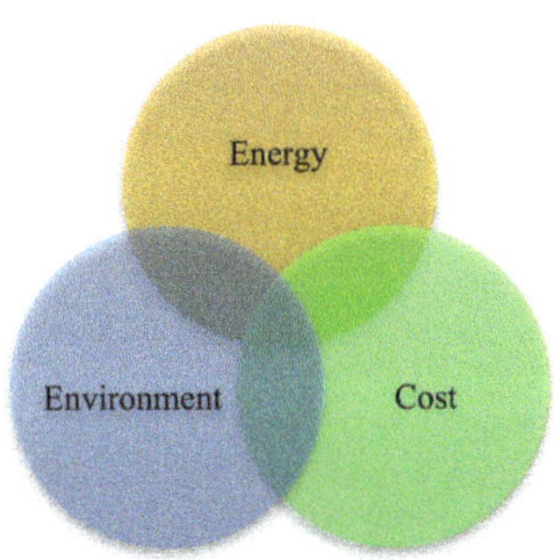

Figure 6. The three vital issues in the maritime sector.

Due to the serious environmental impacts caused by pollution from marine logistics, ports are moving toward a greener industry by implementing various alternatives. Particularly, in recent years, the arising awareness on environmental issues has made this target one of the compulsory goals in achieving high-energy efficiency levels. Authorities have formulated port policies consistently with tight regulation in order to ensure minimum GHG emissions. In this regard, the International Maritime Organization (IMO) in the latest regulation limits the sulfur content in fuel to 0.5% m/m, known as 'IMO 2020' [30]. The change in sulfur limitation from 2000 to 2020 is shown in Table 3. The five most important outcomes from IMO 2020 are as follows: (1) cleaner air, (2) higher quality fuel, (3) positive impact on human health, (4) ship operator role, and (5) changes for enforcement authorities.

Table 3. Change in fuel sulfur limit. Source: Marpol 2018, Marpol Annex VI.

<table>
<tr><th rowspan="2">Date</th><th colspan="2">Sulfur Limit in Fuel (% m/m)</th></tr>
<tr><th>SO_x ECA</th><th>Global</th></tr>
<tr><td>2000</td><td>1.5</td><td rowspan="2">4.5</td></tr>
<tr><td>2010</td><td rowspan="2">1.0</td></tr>
<tr><td>2012</td><td rowspan="2">3.5</td></tr>
<tr><td>2015</td><td rowspan="2">0.1</td></tr>
<tr><td>2020</td><td>0.5</td></tr>
</table>

In line with the green port objective, the Port Authority of Genoa (GPA) is developing a plan integrating renewable energy in their marine sector known as the Port Energy Environmental Plan (PEEP). The ultimate goal of the PEEP is to reduce 20,000 t of CO_2 annually by utilizing 12 plug cold ironing facilities, wind turbines, and photovoltaic power stations with an overall investment of 60 million Euros [31].

In order to move towards a healthier environmental space, ports must plan and manage their operations and future potential expansion in a sustainable manner. Saeyeon Roh et al. [32] in their research stated that the majority of the existing literature's emphasis is on the environmental aspects for sustainable development, but they fail to clarify what factors influence this process. Several authors studied the causes that contribute towards air pollution and environmental damage from the operation of the harbors. Bunkering from the vessels generates the risk of oil spill with potentially disastrous impacts on the food chains of beaches [33]. Matishov and Selifonova [34] pointed out that the source of water resources damage comes from a high density of ship transportation via waterborne traffic. Meanwhile, Brigitte Behrends and Gerd Liebezeit [35] addressed that the two leading destructive factors generated by shipping movement are atmospheric and seawater contaminations. In an attempt to safeguard nature and waterways, new legislation for future growth of ports and their construction at both international and domestic levels are released from time to time, aiming to handle environmental issues based on strict standards for core ports' strategies. For instance, legislations in a few countries are listed in Table 4 below [33];

Table 4. List of legislations in a few countries.

Country	Legislation
EU	Classification Societies—Regulation (EC) No 391/2009; Ship-Source Pollution—Directive 2000/59/EC; Marine Equipment—Directive 96/98/EC and Directive 2014/90/EU
Australia	Environmental Protection Act 1986 (WA)
New Zealand	Resource Management (Marine Pollution) Regulations
USA	Diesel Emission Reduction Act (DERA)
Singapore	Environmental Protection and Management Act (Cap.94A)

Without a doubt, all of these environmental legislations offer vast advantages in terms of health, clean air, economy, energy, and potential for new technology if they are properly considered in port planning. Nevertheless, the great challenge is that all the planning steps must be compatible with the rules, from the initial step of research and development, collecting real-time data, collaboration among the parties, implementing, monitoring, and analysis. All of these stages are time consuming and involve high capital investment where payback period analysis is necessary. Cost is one of the main drivers to run all strategic planning. Due to this risk, the relevant parties, especially the ports themselves, are reluctant to implement any environmental program.

2.3. Seaport Energy Revolution

2.3.1. Conventional System

In the beginning phases of the maritime system, there has been no concern for energy because it was solely utilized for shipping transportation. When the industrial revolution arrived, energy demand began to raise. Electricity and fuel are the two main forms of energy used in this sector. Depending on the availability and variation of the primary sources in a particular country, both electricity and fuel are produced from fossil-based sources (coal, oil, and natural gas). Electricity is not freely available in nature, so it must be produced by transforming primary energy to electricity through a process. Figure 7 displays the entire power station network, from generation to the seaport.

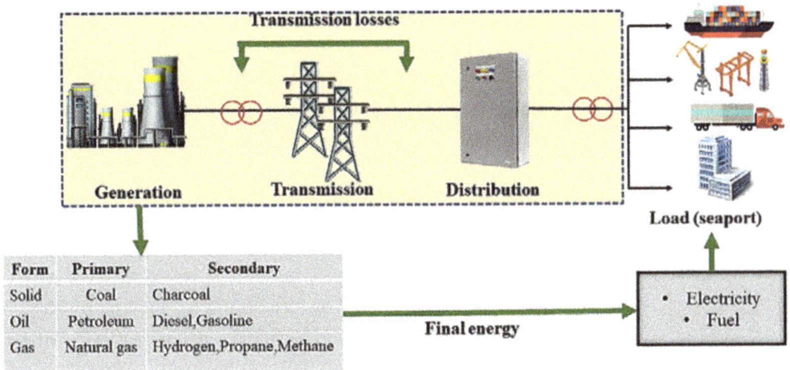

Figure 7. The overall framework of the conventional structure.

In the 1880s, coal was first used to generate energy and became a fundamental driver of steamships [36]. Coal is a combustible black rock that is formed over millions of years by the decay of land vegetation and generates energy under high pressure. Coal is convenient to use because it is easily combustible and cheap. Apart from electricity, another fundamental source of energy, particularly for maritime transportation, is fossil fuel, which contains hydrogen and carbon in its particles.

Previously, many vessels, particularly those with bunker fuel, used HFO to allow the ship's propeller or alternator to rotate. This can be accomplished by either burning fuel in the engine combustion chamber or producing steam in the boiler. Due to its relatively higher density and lower cost, HFO is a preferable marine fuel source. HFO is characterized by its heavy molecules containing long-chain hydrocarbons, density greater than 900 kg/m^3 at 15 °C, and kinematic viscosity greater than 180 mm^2/s at 50 °C [37]. The HFO comes in three varieties, which are listed in Table 5.

Table 5. Three variants of HFO and their sulfur content [30].

No	Variant	Sulfur Content
1	High Sulfur Fuel Oil (HSFO)	3.5%
2	Low Sulfur Fuel Oil (LSFO)	1.0%
3	Ultra-Low Sulfur Fuel Oil (ULSO)	0.1%

Sulfur content is a key differentiator between these three types of HFO. However, from 1 January 2020, HFO can only be used by vessels equipped with exhaust gas cleaning systems (EGCS), and IMO has set new regulations banning vessels from using fuels that have sulfur substances above 0.5% [30]. Due to the tight prohibitions, HSFO and LSFO are no longer practical options for marine transport. The only choice to comply with this limit is ULSO, which has a sulfur content of 0.1%. Unfortunately, the desulfurization process of HFO requires very high costs and is not economical anymore.

Even though electricity and fuel are greatly used in the seaport sector since their early stages of development, due to the arising issues, there is a great interest in alternative solutions. This alternative includes considering electrification of vehicles, cold ironing systems, alternative fuels (i.e., LNG, biofuels, methanol, hydrogen, and low-sulfur fuels), and implementation of both renewable energy and energy storage by implementing microgrid power systems.

2.3.2. Ports and Ships Electrification

From the seaport perspective, electrification can be viewed as the replacement of fossil fuel-based energy with new advanced technologies through the use of electricity. Marine transportation contributes significantly to air pollution and climate change mainly for two reasons: (1) The port is the area where emission sources are least regulated and (2) high reliance on fossil fuels. With the increasing number of ships, the ports' energy management is required to provide sufficient power to supply the ship during berthing. Large ships have a power range between 1 MW and 6 MW [38]. While the main engines are normally switched off during berthing, the vessel's auxiliary engines are switched on to supply power to all loads inside the vessel such as lighting, ventilation, cooling, and other onboard equipment. A study on different types of vessels in Reference [18] shows that the average berthing time per vessel is between 21 and 52 h. Burning diesel oil during berthing for such a long time injects severe GHG emission.

In order to resolve this problem, shore-side electricity practice known as cold ironing is introduced to mitigate undesirable environmental impacts at seaports. Employment of this technique allows a ship to shut off its auxiliary engines when docking, and it receives electricity supplies from the shore to maintain the entire load in the ship. It provides substantial advantages by reducing the dependency on fossil fuels and lessening GHG emissions. The energy required for cold ironing can be supplied directly from the utility grid, RESs, and ESS. Alexander Innes and Jason Manios [39] in their article highlighted a few ports around the world that already utilized cold ironing. Despite the promising benefits of a cold ironing system, many ports, particularly small ports, do not install it due to a few constraints. In order to overcome this, alternative power sources are formed by using shipboard microgrid (SMG), where this method acts as mobile cold ironing facilities by sharing power from multiple ships in the port [16,40]. Another innovative solution with the same concept is known as vehicle-to-grid (V2G) and boat-to-grid (B2G) paradigms [41–43].

However, with the high energy requirement for the entire port operation, reliance on power from the grid alone is not enough. Thus, the port industry has started to explore clean energy resources including renewable energy. A few ports even built their renewable energy power plant. Currently, Jurong Port is the world's largest port that has installed a solar power generation facility, which is estimated to generate more than 12 mil kWh of solar energy per year and covers 60% of the port power demand [44]. In addition, it also does help to reduce 5200 tones carbon emissions in a year. Meanwhile, Aalborg Port, Denmark, not only constructed solar cell systems with an annual production of 80,000 kWh but also included a 2 MW wind turbine in the plan [45].

By utilizing multiple RESs and paying attention to the need for a storage solution, smart technologies such as a microgrid system, are being introduced. Microgrid's alternative is great at solving power supply problems, and it has been gaining more attention lately. The interest in microgrids is increasing due to their promising advantages in providing sustainable, reliable, efficient, and environmentally friendly power supply [46]. Market acceptability of microgrid technologies due to their reliability and cost-efficient power supply makes them a practical and effective solution in power delivery. Although microgrids have been widely developed around the world, their application in harbor areas remained limited. Due to this situation, the development of a microgrid in a port presents significant challenges. Diversity of loads in the seaports (cranes, ship, container, cold ironing, building, etc.) induces difficulties in load forecasting and accurate measurements of power requirement. In order to avoid prohibitive costs and inefficient systems, a comprehensive study on ports microgrids planning, energy management, regulation, and other aspects is required.

Along with the industrial developments and innovations in power systems, ship transportation is also emerging toward the All-Electric Ship (AES) concept [47,48]. Conventionally, ships used a steam turbine as a prime mover, but it switched to steam piston engines in 1850 [49]. The ship's electrification began with an IPS that uses electric propulsion. It gradually evolved into a hybrid power system (HPS) by incorporating energy storage elements. S. fang et al. [50] in their research study modeled the next version of AES by using photovoltaic sources. The goal is to provide better energy production through a hybrid concept that combines diesel generation with renewable sources.

3. Seaport Microgrids

Implementing a microgrid system is advantageous to the seaport because the port's geographical location can provide a strong base for RES production. A port is an area with a large flat surface that is suitable for solar panel installation, such as on the rooftop of a warehouse, a storage area, or a flat roof from a building. However, such infrastructure may not always be appropriate for large-scale solar energy utilization. In addition to wind and PV, some ports utilized other forms of energy such as waves (e.g., Port Kembla in Australia), tide differentials (Port of Digby, Nova Scotia), and geothermal energy (Hamburg) [31]. Thus, seaport microgrids appear to be a feasible option for future power systems in the harbor area.

Thereby, it is important to understand the significant aspects of this research field. VOSviewer is a visualization tool that is useful for mapping large co-occurrence keywords from research sources, especially from Web of Science and Scopus databases [51]. With the assistance of such visualization tools, evolutionary patterns of seaport power systems can be easily interpreted to identify trends and potential future research directions from a macro perspective to the most current research. It will enhance the understanding of the research field in the seaport microgrids and provide an intuitive overview. Figure 8 shows the co-occurrence analysis of the maritime field with the search keywords of (microgrid* AND (seaport* OR "Maritime" OR marine OR ("Ship Harbor"))) from the Scopus database. The larger nodes indicate that more articles have been published in that area, showing that the research field is trending. Small nodes in the network show low co-occurrence of keywords, which might be because of the fact that the topic is still new, providing the opportunities for further research. In this case, larger nodes in Figure 8 represent 'microgrid/microgrids/ship' keywords. It indicates that microgrid power systems are increasingly popular in the maritime sector. Meanwhile, the keywords from the small nodes reflect that researchers from all over the world are engaged in marine power systems, including energy management, optimization, energy efficiency, and other power-related subject areas. Since small nodes are caused by low co-occurrence, this allows more research opportunities for the improvement of marine power systems.

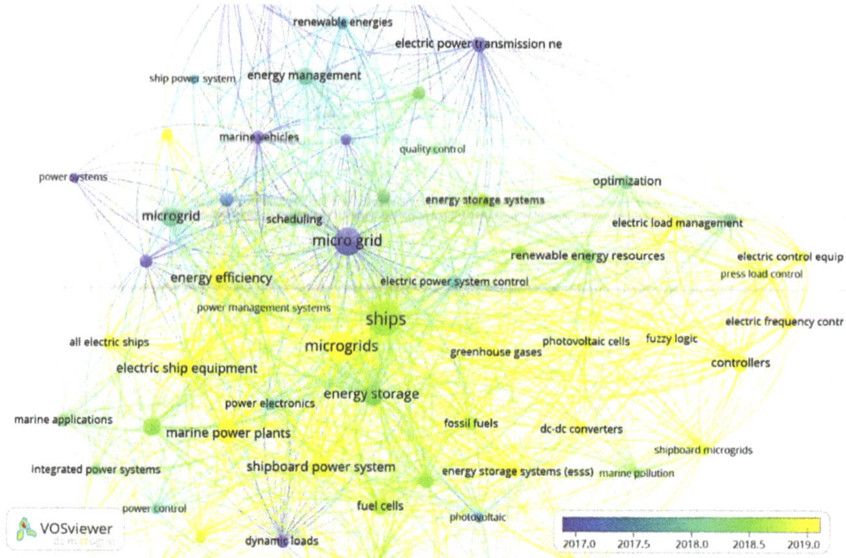

Figure 8. VOSviewer visualization for co-occurrence keywords from recent Scopus research publications on harbor power systems (Accessed on 20 July 2021).

3.1. Seaport Microgrid Topology

A microgrid is a local aggregation of distributed energy resources (DERs) including generators or RES, ESS, power converters, and loads [52]. It links the low voltage (LV) or medium voltage (MV) network to the utility grid at a single point known as point of common coupling (PCC) [53]. Microgrids mainly work in two modes of operation, namely (1) grid-connected and (2) island mode [54,55]. In the grid-connected mode with the main grid fully functional, the microgrid acts as an extension part of the generator by supplying power to the utility grid. However, it can also absorb power from the grid. The inconsistency of power generation from RESs can sometimes produce an abundance of power that exceeds the microgrid's energy demand, thereby providing an opportunity to sell energy back to the main grid and to make a profit [56]. Huang et al. [57] considered 'market-based time-of-use (TOU) pricing' strategies for a microgrid in order to reduce the cost of electricity. A microgrid can also operate independently as an autonomous power system in the island mode [58]. Microgrids help in power restoration after the occurrence of a failure on the utility side, ensuring that end consumers have continuous access to energy, particularly for critical loads. Utility failures can occur as a result of an unplanned shutdown due to the fault or when power demand surpasses power generation [59,60].

Basically, microgrids frameworks can be classified into three topologies including (1) AC microgrids, (2) DC microgrids, and (3) hybrid AC/DC microgrids [53,59,61]. In AC microgrids, the common bus is the AC bus where DC-to-AC inverters are needed for ESSs and DC generating units such as PV and fuel cells. Meanwhile, AC-to-DC rectifiers are normally used for supplying DC loads. However, for DC microgrids, the common bus is in DC and requires AC-to-DC rectifiers for AC generating components, and AC loads are supplied via DC-to-AC inverters. In the case of hybrid microgrids, both types of buses exist in the same network. It allows the flexible connection of DC and AC components to their bus accordingly and reduces unnecessary converters. Figure 9 shows all three topologies and their basic network connections.

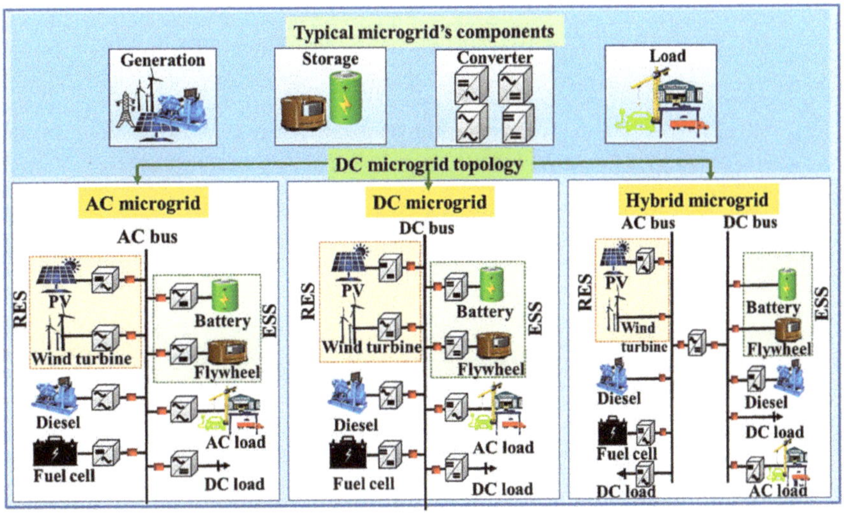

Figure 9. Microgrid topology options for a seaport.

The AC topology is the most commonly used topology that has become the standard choice from the early phase of microgrid invention due to its ability to be synchronized with the AC distribution network, its simple structure, and its economic viability [54]. Normally, when a microgrid has a connection with the main utility, AC microgrids become an ideal choice. It has the capability to transmit power over a long distance, adjust the AC voltage is easy to adjust into different levels for various applications, and it is applicable for induction motors. Despite all of these benefits, long-distance transmission lines are one of the barriers that make it cost-ineffective and impractical in some cases.

However, if the microgrid system is completely isolated without connection to the grid, DC microgrids happen to be the preferred option. The advancements of power electronics technology have resulted in arising the number of DC loads, indicating that modernization of the existing energy system is inevitable [62]. Continuously using AC configuration will require a lot of converters for DC applications and reduces the microgrids' efficiency [63]. In addition, encouragement toward storage systems and clean energy resources such as PVs in power systems, which are in DC, oblige better coordination for the distribution network. As a result, DC microgrids have drawn a lot of interest in the research community in examining possible direct connections with DERs, ESSs, and DC loads through the DC bus and decreasing the impact on the AC networks. DC distribution systems are now gradually used for various applications such as aircraft, automotive, marine, and manufacturing industries [64,65]. Z. Jin et al. [66] investigated the concept of DC microgrids for the onboard power systems of AES. In the maritime sector, the adoption of DC topology brings an enormous range of advantages to the onboard power system by eliminating frequency constraints, allowing the utilization of high-speed generators and providing systematic management [66]. R. Prenc et al. [67] stated that DC ship power systems in the maritime sector prevailed over the AC systems due to the following reasons: (1) improvements in prime mover performance and fuel cost savings; (2) reduction in weight and space; (3) unity power factor for generators; (4) low transmission losses; (5) faster and easier parallel connection of generators; (6) flexible and simple implementation of ESSs.

Both AC and DC microgrids, however, rely on the converter when they interact with the opposite network source that attaches to the buses, resulting in unavoidable power losses during the conversion process [61]. Accordingly, a hybrid AC/DC microgrid appears as a flexible solution for integrating AC or DC-based components while reducing reliance on converters. The reduction in conversion equipment improves overall system efficiency and reliability whilst also lowering costs [68]. X. Liu et al. [69] looked into different operating modes of hybrid microgrids and applied coordinated control schemes to produce maximum power from RES, to reduce power transfer between AC and DC networks, and to enhance stabilization in operation with various networks of DER, ESS, and load components.

By the presence of two types of buses in a hybrid microgrid, control, operation, and management become more complex compared to individual AC or DC microgrids. Furthermore, the exploration of hybrid microgrids is still in its formative development stages and requires extensive research. Table 6 presents the reference of research publications that apply three kinds of microgrid topology into seaport applications.

Table 6. Seaport application with different topologies.

Topology	Seaport Application	References
AC	Shipboard microgrid	[70–72]
	Cold ironing	[73]
DC	Ship-based seaport microgrid	[16,40]
	Shipboard microgrid	[70,74–77]
	Cold ironing	[78]
	Electric ship	[79]
	Offshore application	[80]
Hybrid AC/DC	Electric ferry shipboard	[81]
	Shipboard microgrid	[82]
	Cranes	[22]
	Cold ironing	[83]

3.2. *Conceptual Seaport Microgrids in Shore Side and Seaside*

3.2.1. Shore Side

Conventionally, the relationship between a seaport and ship mainly corresponds to logistic activities. The electrification innovation in the maritime sector is represented by the following: (1) seaport microgrid and (2) all-electric ship (AES), which is the promising solution toward achieving zero carbon footprint in future seaports [51]. Apart from the logistic side, the rising electrification trends in this sector extended the connection of seaport-ships on the electric side. With the broadened infrastructure and multiple functions conducted in the port area, their energy demand continuously grows. Hence, the maritime sector is experiencing considerable power shortage as well as environmental pollution from the burning of fossil fuels. These constraints encourage the incorporation of RES and ESS into the seaport power system, further complicating its control and management. Therefore, the concept of seaport microgrid is introduced in order to provide better coordination for multiple energy resources, storages, and variation in seaport loads [84]. Figure 10 illustrates the concept of microgrid integration on the shore side and seaside.

German maritime sector is an example for manifesting the integration of seaport microgrid, where Hamburg port draws a power of 24.5 MW from renewable energy by installing more than 20 units wind turbines [85]. Moreover, their warehouse rooftops are filled with solar panels with projections to generate 500 MWh of electricity per year [85]. German ports emphasize the importance of renewable energy as they are also considering other sources of energy such as tidal power generation, wave energy, and geothermal energy [86]. Another positive step taken by the Hamburg Port is towards electrification through the installation of cold ironing [39]. These measures are being developed to increase power supply while simultaneously assisting the port in achieving its green maritime goals.

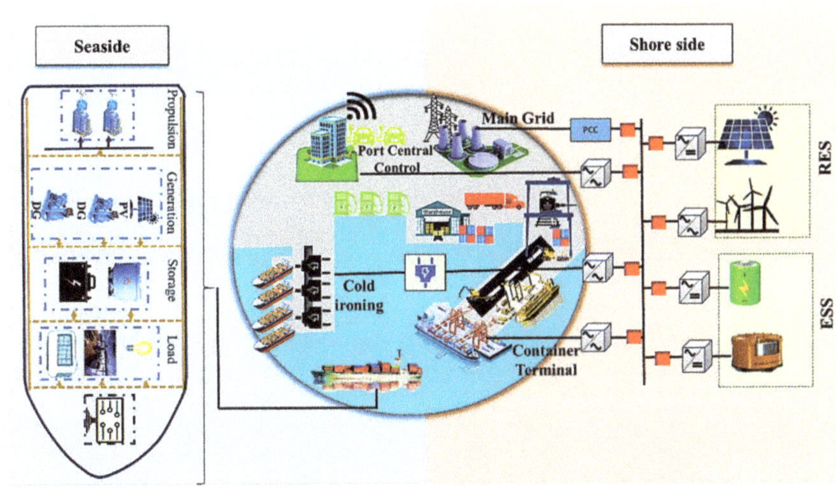

Figure 10. The concept of seaport microgrid on the shore and seaside.

On the shore side, microgrid integration will enhance more electrification and automation for port applications such as heavy lifting devices and transport machinery and enables the transition from diesel to cold ironing during berthing [56]. On the other hand, all of these practices are currently limited as the conventional grid cannot bear the heavy load.

By definition, a seaport microgrid refers to a port that employs microgrid technologies to support its power system and distributed loads. The goal is to improve the port's operation efficiency, maximize renewable energy penetration, provide flexibility for storage installation and allow for more energy sales to the market through the main grid [50]. Basically, seaport microgrids share similarities with land-based microgrids, but they have slight differences in some aspects.

Commonalities: The framework of seaport microgrid has the same components as land-based microgrids, for instance, RES, ESS, converter, and load components. Both microgrids can operate in grid-connected and island modes. In this regard, their basic control and operation frameworks may be similar [50].

Differences: The significant difference between these two types of microgrids stems from their application. Land-based microgrids normally support the power necessary for building electrical appliances, and the most common electricity demand is for lighting and thermal and HVAC systems [11]. Meanwhile, seaport microgrids must consider both the logistic and the electric sides [50]. On the logistic side, a lot of factors need to be considered such as berth allocation [87], port crane scheduling [88], route and voyage scheduling [89], and various cargo transportation. On the electrical side, each application needs a different level of energy, and the energy is volatile depending on many factors.

In addition, cold ironing applications on seaport microgrids will cause frequent berth in and berth out, which will influence the microgrid's performance. Even more difficult, the port is also restricted by the strict policies that are changing over time [89]. Many involved parties such as port entities, port stakeholders, policymakers, port authorities, and the government add difficulty in any decision in the planning of ports. All of these circumstances render seaport microgrids more complex compared to land-based microgrids. Further exploration of the system requirements and modifications for better seaport microgrids is necessary to have compatible and efficient solutions for port electrification.

3.2.2. Seaside (Shipboard Microgrid)

The invention of electric propulsion has resulted in the total electrification of shipboard power systems, referred to as AES. The innovation brings numerous advantages to t maritime ships by reducing manpower, minimizing maintenance workload, improving fuel consumption, fast start-up of ships, and reducing emissions from diesel combustion [48]. AES is considered as the maritime microgrid because of the integration of microgrid technology on shipboard power systems. AES operates in both grid-connected and automation modes. It has similar components as a typical microgrid, whereas generators and ESS deliver power via an energy network in order to supply propulsion and the ship's load. In future shipboards, renewable energy implementations such as photovoltaic systems become trendy. However, the limited space in the ship makes it one big constraint. Ships work in two modes of operation.

Berthed in mode: During berthing, AES is connected to the cold ironing system, receiving electricity so that its auxiliary engine can be switched off. The cold ironing system normally obtains supply from the main grid. Hence, the same concept as the grid-connected mode of operation for land-based microgrids is applied.

Berthed out mode: In the situation where the ship becomes physically independent from the seaport, the electric connection between these two entities (port and ship) no longer exists. The AES voyage at sea is viewed as a mobile microgrid and in island mode of operation. It moves and supplies its entire onboard load with its onboard power system.

3.2.3. Operation Management and Energy Planning of Seaport Microgrid

The complexity of the coordination of various power resources in a microgrid, load management, synchronization with the main grid, meeting policy obligations, and environmental criteria recognizes the importance of the power/energy management system (PMS/EMS) in the seaport microgrids. Energy management systems (EMS) in ships send the signal to the particular component through the communication network after determining the optimal outputs for generators and batteries [50]. P. Xie et al. [90] categorized the PMS and EMS of shipboard power systems into rule-based and optimization-based techniques, where rule-based techniques are highly dependent on human expertise, preconfigured strategies, and priorities. Meanwhile, optimization-based techniques are becoming more trendy as they are capable in providing a better solution by using analytical strategies or numerical optimization algorithms [91]. The diverse assortment of the PMS/EMS strategies in a microgrid system entails managing each component and sub-component by hierarchical control schemes, including primary, secondary, tertiary, and upper-level control systems with different functionalities and time scales [71].

In this section, the importance of the operation management system for better energy planning within the seaport microgrid is highlighted. The demand for energy in the maritime sector keeps increasing over time because of the expanding infrastructure, increasing size and number of seaport transportations, need for handling multi functions, and increase in global demand for logistics. The load will continue to rise due to the above-mentioned factors, but there are several operational practices for controlling energy demands, including load scheduling, load forecasting analysis, improving load factor, peak shaving, and enhancement of ESS utilization. Moreover, price and tax incentives also play a vital role in the operation management of seaport microgrids.

Shipboard/Seaport Microgrid Power Management and Load Scheduling

In the maritime sector, seaport controls on the shore side and seaside (shipboard power system) normally have different administrators that seek different goals [50]. The coordination between these two administrative bodies is necessary. For instance, the vessel could choose a berth-in time when the electricity price is low, and the seaport can make more electricity price savings by adjusting the berth allocation during off-peak hours. Load scheduling at the time of the minimum electricity price is beneficial to energy consumers with respect to minimizing total electricity cost while at the same time meeting environmental requirements.

Pricing policy is one of the vital considerations in load scheduling as the price of electricity fluctuates over time. Different pricing policies apply to different applications in industries. Similarly, in the marine sector, in order to achieve energy-saving and to optimize the cost of energy, pricing policy plays a significant role. Sun and Li [92] described two pricing policies, namely Time-Of-Use (TOU) pricing and Critical Peak Pricing (CPP). TOU pricing is a dynamic pricing strategy where electricity is charged at several price levels for off-peak, mid-peak, and on-peak intervals during the day [93]. This pricing strategy allows energy consumers to shift their loads from peak load intervals to off-peak periods and to avoid high electricity prices [94]. In the peak event situation where production is skyrocketing due to very high demand, CPP is the most effective scheme, which selects one price for critical periods [95]. Kyaw Hein et al. [1] proposed robust coordinated operational scheduling for grid-connected seaport microgrids. In that framework, ship-to-shore (STS) power demand is scheduled by using day-ahead and hour-ahead scheduling with different time horizons. The aim is to reduce emissions and minimize the cost of port operation.

In addition, this scheduling technique is widely used in shipboard microgrids. Unlike land-based microgrids, AES at the sea operation can be regarded as mobile microgrids. Hence, power consumption varies with cruise speed and voyage distance. With the help of electric propulsion motors, AESs are capable of adjusting cruising speeds in order to achieve more economical operations. AESs operate at different speeds during different operating modes including docking mode (when ship approaching or leaving the port), cruising mode, and berthed mode. Yuqing Huang et al. [96] could reduce the operation cost and GHG emissions by 17.4% and 23.6%, respectively, by implementing voyage generation scheduling methods in the AES. In [97], the optimal scheduling for a ferry is achieved by employing a rule-based algorithm considering three scheduling models of different DG units and ESS. Simulation results show that by optimally scheduling the onboard power sources, poor low-efficiency situations can be prevented. Appropriate DG selection and load scheduling for the ship can maximize fuel-saving. Srinivasa Rao K et al. [98] minimized fuel consumption in an offshore support vessel with a dynamic positioning system by scheduling generation resources. The Genetic algorithm is applied for the optimal load sharing due to the nonlinear specific fuel consumption curves of diesel engines.

Load Factor Improvement

The load factor is expressed as the fraction of average load during a specified period to peak load in that timeframe [99]. This technique is useful for determining the efficiency of microgrid operations. A low load factor signifies that an electricity system is being operated inefficiently and is economically poor [100]. Thus, a high load factor is desirable in order to ensure that the seaport microgrid is economically viable by utilizing the total plant capacity for the longest possible period [101]. As a result, the overall cost of providing electric energy will be reduced. F.Robert et al. [102] analyzed the impact of different load factors of 0.2, 0.3, and 0.4 on the energy cost at the design stage of a microgrid. The analysis shows that the highest value of load factor can reduce energy costs up to 48% approximately for a solar-based microgrid [102].

In the harbor, instead of running all big operations such as heavy loading/unloading cargo and multiple ship berthing at the same time, which causes peak demand in the specific time, scheduling the time of ship and cargo arrivals is a wise action. When all the high-demand operations are not running simultaneously, peak demand can be distributed over off-peak hours and can increase the average load as well as the load factor.

At the same time, it will result in reducing the total energy purchase costs. Thus, improving the load factor by reducing or shifting the peak load is necessary in order to ensure the profitability of microgrid power systems [103]. In [97], power scheduling techniques are applied to a hybrid ferry microgrid. The results show that the right arrangement of DG and ESS is capable for improving the load factor.

Peak Shaving

Electrification technologies in ports, such as AES, the transition from diesel-based cranes to electric cranes, cold ironing, electric vehicles, electric trains, and electric trucks for cargo transportation, change the port's dynamic behavior. Peak shaving is a favorable method to control the operational load at ports. Figure 11 illustrates three peak shaving techniques to scale down peak load profiles and to distribute the load over low-demand time intervals.

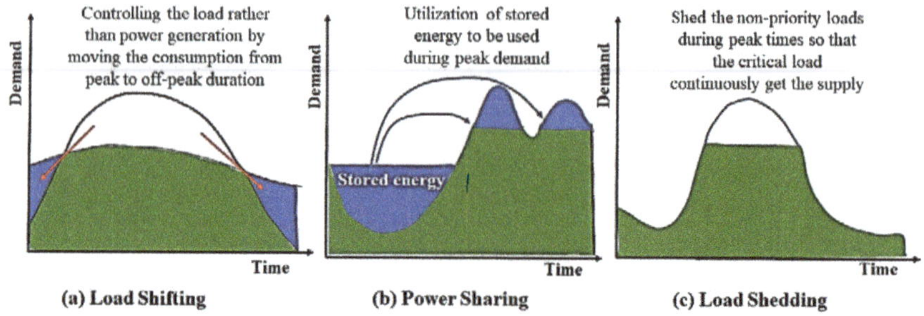

Figure 11. Three type of peak shaving methods where (**a**) is load shifting approach, (**b**) is power sharing approach, and (**c**) is load shedding approach [20,22,52,104,105].

Peak shaving is based on the incorporation of energy storage systems. K. Mostafa et al. [23] coordinated the duty cycles for the STS cranes through full automation of the crane terminal and integrating peak shaving methods through the deployment of ultra-capacitor (UC). The outcome shows that the higher amount of the peak load absorbed by the UC results in more operational earning. Harry Geerlings et al. [106] investigated the effectiveness of peak shaving at the container terminal (CT) by applying 'new rules of operation' with the aim of cost saving while analyzing its effect on the handling time of containerships. They highlighted that the peak shaving method not only reduces the handling costs at CT but also enhances power stability. However, because the number of terminal equipment and processes operating at the same time may be limited, this may result in additional handling time for containerships [107]. This load-leveling technique also applies to shipboard microgrids. In [76], in order to organize the operation in shipboard, a specialized hierarchical control scheme is proposed. Load sharing levels in primary control aim to distribute the load from peak to off-peak hours. Peak shaving in the shipboard helps to absorb the load variation in the system so that engines can continue to operate at their most efficient operating point. Kiyoune Kwon et al. [77] proposed a load frequency-based approach to manage shipboard power and load sharing in DC hybrid electric ships. Simulation results indicate that load frequency-based power management offers effective load distribution and is capable of protecting the generator from sudden load variations, which results in poor power quality and causes damage in mechanical systems.

Load Forecasting

On the shore side, dynamic load from multiple operations of logistics (container terminal, cargo transportation, vessel, and charging station) and administration infrastructure (communication center, warehouse building, and maintenance) might result in unexpected power consumption peaks. In this situation, a substantial amount of power supply is required to prevent disruption in port operation due to power shortages. Thus, load forecasting in ports is vital for optimizing resource utilization, scheduling the load, and managing alternative energy from RES and ESS. Furthermore, understanding load behavior brings advantages for complex decision making in energy management. However, several uncertainty parameters influence the dynamic load of port operation in seaport microgrids, demanding the use of innovative load forecasting techniques. Various techniques can be used to implement load forecasting ranging from parametric-based methods to the evolutional artificial intelligent techniques listed in Table 7.

Table 7. Load forecasting methods [108–112].

<table>
<tr><th>Load Forecasting Approach</th><th>Technique</th><th>Parameter Requirement</th><th>Load Forecasting Time Horizon</th></tr>
<tr><td>Traditional parametric techniques</td><td>(1) Gray dynamic methods
(2) Regression methods
(3) Time-series prediction methods (autoregressive (AR), moving average (MA), autoregressive moving average (ARMA), autoregressive integrated moving average (ARIMA), seasonal autoregressive integrated moving average (SARIMA), autoregressive moving average exogenous (ARMAX))</td><td rowspan="2">– Historical data of load profile
– Influential factors: weather, air temperature, humidity, wind speed, calendar seasonal information, economical events, and geographical information</td><td rowspan="2">– Short-term load forecast (STLF)-typically from one hour to one week for economic dispatch
– Medium-term load forecast (MTLF)—typically from one week to one year
– Long-term load forecast (LTLF)-typically longer than one year suitable for capacity expansion</td></tr>
<tr><td>Artificial intelligence-based techniques</td><td>(1) Artificial neural network (ANN)
(2) Fuzzy logic
(3) Genetic algorithm (GA)
(4) Support Vector Machine (SVM)</td></tr>
</table>

One of the electrification innovations in ports is related to the transition from a diesel-powered rubber-tyred gantry (RTG) to an electric RTG [113]. To understand the energy behavior of electric RTG, Feras Alasali et al. [114] implemented three different methods of load forecasting, namely autoregressive integrated moving average with exogenous (ARIMAX), artificial neural network (ANN), and autoregressive moving average exogenous-support vector machine (ARMAX-SVM). The purpose is to find the most accurate forecasting method so that peak demand can be reduced effectively. Several simulation analyses are conducted by implementing STLF techniques (24 h) on the load profile of an electric RTG crane from the Port of Felixstowe [114]. The results show that ANN outperforms ARIMAX and ARIMAX-SVN forecasting techniques with minimum errors during a prediction interval.

Another vital seaport application, onshore power system (OPS) or cold ironing, involves the interaction between ship and shore-side power facilities. Accordingly, the dynamic load of OPS fluctuates with various parameters from both sides. The power of the ships varies depending on the load they carry and the size of the ship. Moreover, the arrival and departure times of ships affect the power consumption of OPS. In addition, ship traffic at the port is also inconsistent. Thus, these factors have an impact on OPS load behavior. However, in a few forecasting techniques conducted for OPS load forecasting, the mentioned factors are either neglected or only one of them is considered. Hence, research studies on how these parameters affect the OPS load profile are still lacking. The authors in [115] conducted OPS load forecasting by considering the traffic volume of ship berthing at two different seasons, winter and summer. The goal is to observe how the number of ship berthing varies in different seasons. Still, it is better to conduct a forecasting method for all seasons including spring and autumn in order to obtain the overall pattern of ship berthing during all year. D'Agostino et al. [116] performed OPS load forecasting by using two methods, namely the Environmental and Protection Agency (EPA)-based method and Monitoring, Reporting, and Verification (MRV)-based method. These techniques measure the ship's load demand based on four operating modes: sailing at sea, maneuvering, cargo handling in port, and idle mode. The results show that peak demand varies depending on the ships' categories and operating mode.

The amount of shore power demanded by ships changes randomly depending on the number of berthed ships and their auxiliary power requirement. In order to determine OPS power capacity, Yun Peng et al. [117] used the stochastic nature of arriving ships. The goal is to observe the manner in which ships' arriving patterns affect OPS power consumption. In another research study [118], OPS power requirement is determined by utilizing the predicted energy consumption of auxiliary engines and by obtaining hourly energy consumption for different kinds of ships by using the Monte Carlo procedure. The power consumption prediction data are necessary for energy management. In [119], forecasting data are utilized for the day ahead optimization of a hybrid ferry with OPS. Two conflicting objectives of minimizing the operation cost and degradation of energy storage are considered.

Regarding the seaside microgrid (isolated shipboard), the dynamic positioning (DP) system, which is used in marine vessels to keep the ship position from displacement, is important for determining how thrusters should act to stabilize the position and heading of vessels. However, the uncertainty during the voyage that stems from sea waves and weather conditions affects DP, vessel's speed, and ship power. Hence, DP load forecasting by considering the uncertainty is essential for better shipboard power consumption management. M. Mehrzadi et al. [120] utilized the deep learning method for DP load forecasting by using the Levenberg–Marquardt algorithm based on a nonlinear recurrent neural network to predict thrusters' power consumption in different sea states. The case study is divided into several clusters of sea conditions including calm, moderate, rough, and high states. As DP power consumption considerably changes based on the sea states, wave height and wind speed are significant factors to be taken into account. The sea state parameters change the vessel's motion and affect the thrusting power demand, thereby resulting in a dynamic load pattern. Meanwhile, Kyaw Hein et al. [121], with the same consideration of sea states parameters, provide the simulation results in which sea states have a big influence on voyage path and ship's velocity. The volatility of the sea wave and wind speed uncertainties influence propulsion power requirements. This DP forecasting practice greatly aids the ship's operation management and power system planning in maintaining the stability of the ships' DP handling. However, there is no analysis based on real-time data, which is very important for observing the effectiveness of the proposed load forecasting technique.

Storage Management

The incorporation of a RES component in a seaport microgrid necessitates the use of an ESS to store excess energy and absorb RES power fluctuations. Due to the weather-dependency of the RES power generation and uncontrolled weather conditions, the generated energy will fluctuate, which highlights the importance of ESS integration. Chun Sing Lai et al. [122] reviewed two types of energy storage systems for storing low carbon energy, namely generation-integrated energy storage (GIES) and non-GIES. GIES stores a substantial amount of energy along with the transformation from primary energy to electricity. GIESs typically consists of power generating technologies and are more efficient for RES generation sources such as wind and solar thermal energy. Non-GIES is a common type of ESS that converts primary energy directly into electricity for storage. Meanwhile, Amirante et al. [123] made an overview of the three types of ESS:

(1) Mechanical-compressed air energy storage (CAES), pumped storage hydropower (PSH), and flywheel;
(2) Electrical-supercapacitors and superconducting magnetic energy storage (SMES);
(3) Electrochemical-lead-acid batteries, lithium-ion batteries, and flow batteries;
(4) Hydrogen.

In maritime transportations, ESS is beneficial for strengthening the system in order to avoid instability, which is caused by the engines' delay in responding to load demand [124]. It will also serve as an additional power reserve in the event of a generator failure, reducing the risk of a blackout. In [96], a virtual ESS between shipboard thermal storage and thermal load is used in an AES for mutual support between voyage scheduling and economic dispatch. The optimization model is formulated in order to effectively coordinate voyage scheduling and power generation considering different load conditions. The results show that operational cost and emission can be reduced by 17.4% and 23.6%, respectively.

However, in some operating conditions, a low load factor of the parallel generator in a ship might cause a detrimental effect on the fuel consumption rate. In order to solve this issue, M. Othman et al. [97] utilized a battery for optimally arranging power generation in a hybrid shipboard microgrid. Three case studies with different numbers of diesel generators with and without batteries are performed. Optimization results illustrate that fuel consumption can be reduced significantly, and the load factor is improved with battery application compared to solely using a diesel generator. However, coordination between several generators and battery needs to be carefully planned in order to achieve better results.

M. Mutarraf et al. [17] signified the importance of battery banks in the DC bus seaport microgrid in providing a mobile cold ironing facility at the harbor. Due to the highly dynamic behavior of the port's load, wise distribution of RES and ESS during peak and off-peak duration help greatly in balancing demand and supply. For instance, during a sharp rising in energy demand from cold ironing due to high traffic of ships berthing, where generation is insufficient, ESS can come to the scene.

In another case, the storage component provides a solution to store energy generated from ports' applications that generates energy. In [13], a flywheel storage system is used to harvest energy from the harbor's electrical cranes. This idea comes from the problem of conventional cranes operation that ignores the regenerated energy from cranes when the container is lowered. As a result, most of this energy is dissipated as heat in resistor banks. Thus, developing this flywheel storage system will avoid energy loss when the cranes are lowered. In this manner, a significant amount of energy can be stored and reused during peak hours.

From the emissions point of view, ESS is a promising solution relative to the port for reducing pollution. Kyunghwa Kim et al. [125] proposed a hybrid storage system by integrating supercapacitors (SC) and lithium-ion batteries (LIB) by targeting a bulk carrier with four deck cranes. The capacity of SC and LIB is selected based on the load consumption from the cranes in loading and unloading modes. Simulation results show that emissions were reduced by roughly 77%, 93%, and 99% for CO2, SOX, and NOX, respectively, for this specific case study. Emission reductions might vary in different marine applications.

Several different methods for reducing peak demand have been proposed in the literature. Some countries offer rewarding incentives so that energy consumers willingly shift their electricity consumption to off-peak periods. Storage technology has merit over this rewarding program in allowing customers to have their normal daily lives while lowering their peak demand charge. The cost-benefit analysis is based on the battery's lifespan, state of health, and discharge time [100]. With the increase in storage capacity, more loads can be scheduled at the minimum cost duration, thereby reducing the potential for higher electricity charges [126]. The trade-off between financial gain and operational efficiencies is quantified. In order to ensure the high efficiency of ESS integration, suitable type storage systems must be carefully adapted by considering the type and scale of applications. Nowadays, in some complex maritime applications, storage technology is growing toward hybrid energy storage in order to satisfy demand from the ports' dynamic loads.

Price and Tax Incentives

Demand response is an efficient strategy to moderate electricity consumption in response to the market incentives relative to either price or tax reduction. There are two types of demand response programs: time-based rate (TBR) programs and incentive-based programs (IBPs) [127]. TBR programs provide consumers with time-varying rates based on the price of electricity in different periods (real-time pricing (RTP), TOU, and CPP), motivating users to adjust their consumption patterns by changing price signals [128]. On the other hand, IBP is the form of incentive that provides advantages in the time of stress. Considering environmental issues at ports, port authorities from U.S and EU offer shipping rebates to reward ship operators that satisfy environmental requirements [129]. In practice, the Port of Long Beach implemented the Green Ship incentive program to reduce NOx emissions from shipping [28]. Another alternative is the introduction of various environmental indexes such as the Environmental Ship Index (ESI), Green Award (GA), the Clean Shipping Index (CSI), and Blue Angel (BA) [28]. These programs give special discounts on port dues if the port score satisfies the baseline index. For instance, in Bremen, ships with an ESI score of 30 to 40 receive a 5% discount on port dues, while ships with an ESI score of more than 40 receive a 10% discount. These incentives will encourage port entities to take the necessary actions to improve their port so that they can achieve more discounts on port dues. J. Sanz et al. [130] reviewed four types of incentives for microgrids, including (1) feed in tariffs, (2) market premium, (3) green certificates, and (4) tenders. Each incentive is initiated for different conditions and rewards but with the same goal toward energy saving and zero footprint ports.

4. Seaport Microgrid Challenges and Future Trends

4.1. Challenges in Developing Microgrid Systems at Seaports

The increasing number of publications from researchers around the world indicates a growing trend towards maritime microgrids. However, real implementation of seaport microgrids in the harbor is scarce due to several issues and barriers that exist from various perspectives. Anthony Roy et al. [43] discussed a few aspects of seaport microgrids, namely technical and managerial. Security and regulatory aspects are also among the important perspectives. Some of these challenges are addressed below.

4.1.1. Technical Challenge

Relying solely on the utility grid is inefficient when considering heavy appliances at the harbor. The grid must be close to the port or the cables will be very expensive. Thus, harnessing energy from RES is more economical due to local generation advantages. However, ports have limited space for RES installation [81]. Undeniably, solar panels can be installed on the flat rooftop of the buildings or warehouse but with restricted capacity, and it is not applicable for large-scale exploitation. Shipboard microgrids suffer from the same problem. Many of the commercial AES use diesel generators and batteries for propulsion and supplying onboard loads [131].

For the next AES evolution, the implementation of RES such as photovoltaic systems is discussed in [50] to support the power required by the shipboard microgrid. However, there is no discussion on a compatible unit of solar panels that can be placed on the ship. It is well known that ships have very limited space due to existing bulky equipment on board, especially in cargo ships, which usually use the space to place the cargo. Furthermore, coordination of the RES will be complicated because ships are always on the move and isolated when they are at sea [81].

For the wind turbines, installation can take up a significant portion of land and deforestation to set up a wind farm that treats wildlife such as birds and bats. Moreover, it does not only require a high upfront investment but is also prone to noise disturbance and undesirable visual impacts.

Another challenge for RES installation is uncertainty in weather conditions. RESs are unable to produce consistent power, and their power production depends on several factors such as ambient temperature, wind speed, irradiation, and time of the season, among others. Hence, storage elements are always integrated with the RES.

4.1.2. Managerial

According to the literature, a few ports, such as the Port of Hamburg and Port of Genoa, have already taken the initiative for microgrid deployment [31]. In comparison to the massive use of land-based microgrids, port-based microgrid implementation is considered scarce. This situation creates an opportunity to bring microgrids technology into seaports, but it is also difficult to access references in terms of their needs, design, operation, and maintenance requirements. Particularly during the design phase, it is a complicated task to find the optimal design of the seaport microgrid with the compatible configuration and the right size of its components. The research conducted on testing and analysis of seaport microgrids lacks real-life data. This can be explained by many of the studies that are conducted based on the simulated data sets. Thus, it is hard to know if the simulation result is compatible with real implementation or not.

In addition, considering that the seaport microgrid is a relatively new topic, the required manpower that has the expertise and enough experience is limited. Due to this, port entities find that new power system technology is difficult to handle, leading them to decide not to take any risks by denying microgrid implementations. Not only that, the planning for the implementation takes a long time before it can be executed. Revolutionizing the ports to microgrid technology may appear great on paper; however, one of the biggest hurdles that have to be overcome is the human factor with different mentalities. Moreover, the requirement to make an agreement and synchronization between all parties involved such as port entities, stakeholders, port authorities, policymakers, and government makes it more difficult. This explains the reason why the majority of the ports are more comfortable with traditional power systems.

4.1.3. Security and Regulation

Operation management and control of seaport microgrids are very different from conventional power systems due to the characteristics of power electronic converters and highly dynamic load behavior. A port is the site that has highly volatile loads due to the frequent arrival and departure of ships. Microgrids must achieve a balance between supply and demand in order to maintain voltage/frequency stability. Particularly during isolated operations, system instability is a major concern. In order to effectively manage these issues, hierarchical control techniques have been proposed [21]. On top of this, shore-side and shipboard microgrids (seaside) have different administration and rules of conduct.

In terms of security, the lack of an efficient monitoring system makes microgrids extremely vulnerable to cyber threats [132]. Malicious cyber assaults are a threat to power network operations, causing massive damages from different technical, economic, social, and control points of view. According to statistics from the energy sector, more than 150 cyber-attacks occurred in 2013 and 79 in 2014 [133]. Accordingly, the cost of power outages in the United States is estimated to be around $ 80 billion per year [133].

Above all of the aforementioned concerns, one of the biggest challenges of the maritime sector is the implementation of strict policies for marine operations both on the shore and seaside. Policy regulations are in place either to restrict emissions or to stimulate the use of improved technologies at ports. These guidelines must be followed in any future port planning. It is becoming more difficult for existing appliances that require immediate replacement. In accordance with a current IMO rule limiting sulfur content in fuel to 0.5% m/m, ships must either find alternative clean fuels or integrate a fully electric ship.

On the positive side, certain legislation exists that provides incentives and subsidies to stimulate the use of new technologies. Taxes, for example, are proposed as a strategy by increasing taxes on older systems while lowering taxes on newer systems. It is important to formulate a legislative scheme that favors microgrids as a power system at ports, as this legislative framework will result in switching to microgrids that are more cost-effective than staying with conventional fossil-fuel-based systems. Adam Hirsch et al. [94] attracted the attention to two key questions about the rules that require clarification, where the answer eventually has a significant influence on microgrids. The first question is whether a microgrid is considered as electrical distribution utilities and, thus, whether it should be subject to state regulatory control or not. The second clarification is, in the case that microgrids are excluded from state regulation as utilities, do microgrids fit into existing legal frameworks governing energy sale and purchase, as well as rights to generate and distribute electricity. Microgrids require a rightful legal identity and regulatory certainty in order to ensure that their implementation is profitable; otherwise, the upfront cost is too high, and the profit is too uncertain with respect to rationalizing the investment of time and money.

4.2. Potential Future Research Directions

The ongoing challenges of seaport microgrids including the barriers mentioned in the previous subsection and other technical aspects require more investigation to find effective solutions to improve the ports' performance. The know-how to manage all of these will benefit the ports in a variety of ways, including energy efficiency, cost-saving, and environmental issues. Below are some of the identified potential issues on the seaport microgrids that can be considered for future studies.

4.2.1. Mobile Cold Ironing

Currently, cold ironing in the harbors mostly uses direct energy from the main grid. There is a chance that a few large vessels with high power requirements will arrive at the same time. Thereby, the reliance on grid supply alone is insufficient to keep all parts of ports operational and might result in power shortages. Furthermore, most of the smaller seaports are not equipped with cold ironing stations, resulting in continuous pollution from auxiliary engines during berthing [17]. In order to address these issues, mobile cold ironing by forming a power-sharing network between nearby ships called a 'moveable shipboard microgrid' appears to be a potential solution for providing temporary power. Further research on this technique can be pursued with respect to putting it into practice.

4.2.2. Optimal Port Planning

Considering that seaport microgrids involve different kinds of load, various objectives, and constraints, it is critical to plan efficient coordination among the components in the microgrid. There is limited literature that addresses compatible configurations and appropriate size/capacity for each component in a seaport microgrid. Outstanding coordination is needed in addition to determining the right type of RES/storage component/converter installation and suitable topology. This is to ensure that there is no lacking or oversizing in microgrid implementation, allowing power to be distributed efficiently and avoiding unnecessary investments.

4.2.3. Cluster Seaport Microgrids

From the available literature, most of the publications on seaport microgrids contribute to the overall performance of the microgrid. However, the port itself deals with many kinds of large energy applications such as container terminals, cold ironing, cargo transportation, vessel, and administrative buildings for port handling. Thus, it is a good opportunity to implement a cluster seaport microgrid by grouping the loads into different load clusters.

4.2.4. Optimization

Due to the current firm policy for the port operation to control prolonged toxic pollution, an optimization algorithm that will benefit both the legislative body and the port entity is required. The analysis will help port entities in future planning so that it is compatible with the port policy.

4.2.5. Economical Analysis

All the alternatives carried out in the marine sector from renewable energy harnessing to electrification innovation undeniably bring a plentiful amount of energy-saving and emission reduction benefits. Most previous research has focused on the technical aspects, and a comprehensive economic analysis is still lacking. For a stakeholder, financial management and revenue generation from investment is vital before executing any planning process. Further economic analysis on the seaport considering the market price, the tax charged, energy price, and return on investment would be necessary. The authors in [134] provide net present value for the OPS system at ports. However, this OPS installation involves retrofitting on both sides of ships and shore facilities. It would be insightful if economic analysis can be provided in terms of investment needed and potential savings made from the point of view of ships' owners and port entities. In addition, in order to observe the effectiveness of the return on investment from OPS implementation, a comprehensive comparison between energy production cost from diesel auxiliary engines and electricity purchase cost from OPS is necessary. If it is proven that this strategy can bring economic advantages, it will encourage more OPS implementation in seaports worldwide.

Another alternative for future study is to develop an economic model that benefits all parties in the port. This can be explained by the fact that national electricity rates vary per country. Some countries offer electricity at a high cost, while others obtain lower costs. In a country where electricity is expensive, the maritime industry has a big potential to maintain using diesel energy regardless of whether it would harm the environment. A good economic model that is capable of attracting all port stakeholders not only because of environmental concerns but also because of economic benefits can be implemented. For instance, by reducing energy tax for those shifting to electrification applications, more savings are obtained compared to the traditional paradigm of port operation. The port policymaker, port entities, and ship owner should all be taken into account.

5. Conclusions

Ports are sites with major contributions to pollution as large vessels keep their engines running even when berthed, and heavy lifting work is operated by diesel-powered cranes. Dynamic behavior and the diversity of the port applications emphasize the promising role of microgrid technology in supporting the utility grid. This paper addressed the concept of the seaport microgrid and its integration into seaport from both shore and seaside (shipboard microgrid) points of view as well as its operational management. After the thorough literature review, the following points are some of the significant conclusions drawn in this paper:

- Three major concerns at the ports include energy, environment, and cost. Future port planning should be geared toward addressing these issues.
- A microgrid is a promising power system for the marine sector that is capable of supporting the industry's heavy loads. It enables the diversification of alternative energy resources, such as harnessing power from RES, rather than being limited to only fossil-based energy. A port will manage to achieve a substantial amount of cost-saving since electricity is generated locally by RES in the harbor area. It will reduce the investment costs in both utility grid expansion and long distribution cables.
- Furthermore, ESS components in the microgrids help in improving port performance and serve as a useful tool for demand-side management. The good practice from the operation management in seaport microgrids enhances better operation at a lower price.
- With a seaport microgrid, it is possible to bring more electrification and automation into the ports than compared to the conventional grid, which cannot support factors such as large-scale cold ironing, full electrification of cranes, improved charging stations, and electrification of other modes of transportation.

Electrification movement in the ports implies that the connections between seaport and ships are no longer on the logistic side only but are also on the electric side. It requires an efficient transportation system and power system, rendering the seaport microgrid more complicated than the traditional land-based microgrids. Thus, further actions for improvement in ports are needed in order to synchronize both the logistic and energy side in a more coordinated manner.

Author Contributions: Conceptualization, N.N.A.B.; software, N.N.A.B. and Y.Y.; validation, N.B.; formal analysis, N.N.A.B., visualization, N.N.A.B. and Y.Y., writing—original draft preparation, N.N.A.B.; writing—review and editing, N.N.A.B. and N.B.; supervision, J.M.G. and J.C.V.; project administration, J.M.G.; funding acquisition, A.A. and Y.A.A.-T. All authors have read and agreed to the published version of the manuscript.

Funding: The Deanship of Scientific Research (DSR) at King Abdulaziz University, Jeddah, Saudi Arabia has funded this project, under grant no. (RG-49-135-40). The research work was also funded by a Villum Investigator grant (no. 25920) from The Villum Fonden.

Institutional Review Board Statement: Not applicable.

Informed Consent Statement: Not applicable.

Data Availability Statement: Not applicable.

Acknowledgments: This research work was supported by a Villum Investigator grant (no. 25920) from The Villum Fonden, University Malaysia Perlis, and Ministry of Education Malaysia. The Deanship of Scientific Research (DSR) at King Abdulaziz University, Jeddah, Saudi Arabia, has funded this project under grant no. (RG-49-135-40).

Conflicts of Interest: The authors declare no conflict of interest.

References

1. Hein, K.; Xu, Y.; Gary, W.; Gupta, A.K. Robustly coordinated operational scheduling of a grid-connected seaport microgrid under uncertainties. *IET Gener. Transm. Distrib.* **2021**, *15*, 347–358. [CrossRef]
2. Hlali, A.; Hammami, S. Seaport concept and services characteristics: Theoretical test. *Open Transp. J.* **2018**, *11*, 120–129. [CrossRef]
3. Molavi, A.; Lim, G.J.; Race, B. A framework for building a smart port and smart port index. *Int. J. Sustain. Transp.* **2020**, *14*, 686–700. [CrossRef]
4. Kaliszewski, A. Fifth and sixth generation ports (5Gp, 6Gp)—Evolution of economic and social roles of ports, translated from Polish: "Porty piątej oraz szóstej generacji (5GP, 6GP)—Ewolucja ekonomicznej i społecznej roli portów". *Studia I Materiały Instytutu Tra.* **2017**, *32*. [CrossRef]
5. Montwiłł, A. The role of seaports as logistics centers in the modelling of the sustainable system for distribution of goods in urban areas. *Procedia-Soc. Behav. Sci.* **2014**, *151*, 257–265. [CrossRef]
6. De Langen, P.W. *Governance in Seaport Clusters*; Palgrave Macmillan: London, UK, 2015; pp. 138–154.
7. Versteegt, G. *Berthing Loads in Strcutural Design*; Delft University of Technology: South Holland, The Netherlands, 2013.
8. Bintiahamad, N.B. Integration of Microgrid Technologies in Future Seaports. 2019. Available online: https://www.forskningsdatabasen.dk/en/catalog/2471559831 (accessed on 20 July 2021).
9. De Langen, P.W.; Sharypova, K. Intermodal connectivity as a port performance indicator. *Res. Transp. Bus. Manag.* **2013**, *8*, 97–102. [CrossRef]
10. Sifakis, N.; Tsoutsos, T. Planning zero-emissions ports through the nearly zero energy port concept. *J. Clean. Prod.* **2021**, *286*, 125448. [CrossRef]
11. Abu Bakar, N.N.; Hassan, M.Y.; Abdullah, H.; Rahman, H.A.; Abdullah, M.P.; Hussin, F.; Bandi, M. Energy efficiency index as an indicator for measuring building energy performance: A review. *Renew. Sustain. Energy Rev.* **2015**, *44*, 1–11. [CrossRef]
12. Zhao, H.X.; Magoulès, F. A review on the prediction of building energy consumption. *Renew. Sustain. Energy Rev.* **2012**, *16*, 3586–3592. [CrossRef]
13. Binti Ahamad, N.B.; Su, C.L.; Zhaoxia, X.; Vasquez, J.C.; Guerrero, J.M. Modeling and controls of flywheel energy storage systems for energy harvesting from harbor electrical cranes. In Proceeding of the 2018 IEEE Industry Applications Society Annual Meeting (IAS), Portland, OR, USA, 23–27 September 2018; pp. 1–8. [CrossRef]
14. Comer, B.; Olmer, N.; Mao, X.; Roy, B.; Rutherford, D. *Prevalence of Heavy Fuel Oil and Black Carbon in Arctic Shipping, 2015 to 2025*; International Council on Clean Transportation (ICCT): Washington, DC, USA, 2017; p. 68. Available online: https://www.theicct.org/publications/prevalence-heavy-fuel-oil-and-black-carbon-arctic-shipping-2015-2025 (accessed on 25 June 2021).
15. Yiğit, K.; Kökkülünk, G.; Parlak, A.; Karakaş, A. Energy cost assessment of shoreside power supply considering the smart grid concept: A case study for a bulk carrier ship. *Marit. Policy Manag.* **2016**, *43*, 469–482. [CrossRef]
16. Thongam, J.S.; Tarbouchi, M.; Okou, A.F.; Bouchard, D.; Beguenane, R. All-electric ships—A review of the present state of the art. In Proceeding of the 2013 Eighth International Conference and Exhibition on Ecological Vehicles and Renewable Energies (EVER), Monte Carlo, Monaco, 27–30 March 2013; pp. 1–8. [CrossRef]
17. Mutarraf, M.U.; Terriche, Y.; Nasir, M.; Guan, Y.; Su, C.L.; Vasquez, J.C.; Guerrero, J.M. A Decentralized control scheme for adaptive power-sharing in ships based seaport microgrid. In Proceeding of the IECON 2020 the 46th Annual Conference of the IEEE Industrial Electronics Society, Singapore, 18–21 October 2020; pp. 3126–3131. [CrossRef]
18. Ahamad, N.B.B.; Guerrero, J.M.; Su, C.L.; Vasquez, J.C.; Zhaoxia, X. Microgrids technologies in future seaports. In Proceeding of the 2018 IEEE International Conference on Environment and Electrical Engineering and 2018 IEEE Industrial and Commercial Power Systems Europe (EEEIC/I&CPS Europe), Palermo, Italy, 12–15 June 2018; pp. 1–6. [CrossRef]
19. Boile, M.; Theofanis, S.; Sdoukopoulos, E.; Plytas, N. Developing a port energy management plan: Issues, challenges, and prospects. *Transp. Res. Rec.* **2016**, *2549*, 19–28. [CrossRef]

20. Kusuma, L.T.W.N.; Tseng, F.S. Analysis of the impact of the "sea toll" program for seaports: Resilience and competitiveness. *Appl. Sci.* **2019**, *9*, 3407. [CrossRef]
21. Sadiq, M.; Ali, S.W.; Terriche, Y.; Mutarraf, M.U.; Hassan, M.A.; Hamid, K.; Ali, Z.; Sze, J.Y.; Su, C.-L.; Guerrero, J.M. Future greener seaports: A review of new infrastructure, challenges, and energy efficiency measures. *IEEE Access* **2021**, *9*, 75568–75587. [CrossRef]
22. Port of Wilmington Cold Storage Becomes First in-Port Cold Storage Facility in North Carolina. Available online: https://www.provisioneronline.com/articles/104257-port-of-wilmington-cold-storage-becomes-first-in-port-cold-storage-facility-in-north-carolina (accessed on 23 July 2021).
23. Mostafa, K.; Giuseppe, P.; Ben, C.; Luigi, M. Ultracapacitors for port crane applications: Sizing and techno-economic analysis. *Energies* **2020**, *13*, 2091.
24. Kermani, M.; Parise, G.; Martirano, L.; Parise, L.; Chavdarian, B. Optimization of peak load shaving in STS group cranes based on pso algorithm. In Proceeding of the 2018 IEEE International Conference on Environment and Electrical Engineering and 2018 IEEE Industrial and Commercial Power Systems Europe (EEEIC/I&CPS Europe), Palermo, Italy, 12–15 June 2018. [CrossRef]
25. Hariharan, R.; Usha Rani, P.; Muthu Kannan, P. Sustain the critical load in blackout using virtual instrumentation. In *Intelligent and Efficient Electrical Systems*; Springer: Singapore, 2018; pp. 77–88.
26. Pei, R.; Xie, J.; Zhang, H.; Sun, K.; Wu, Z. Robust multi-layer energy management and control methodologies for reefer container park in port terminal. *Energies* **2021**, *14*, 4456. [CrossRef]
27. Rosa Mari Darbra, M.P.; Wooldridge, C. ESPO Environmental Report 2020. 2020. Available online: https://www.espo.be/media/EnvironmentalReport-WEB-FINAL.pdf (accessed on 23 July 2021).
28. Zhang, X. Analysis of the Incentives in Environmental Strategies Implementation in Chinese Ports. Master's Thesis, Erasmus University Rotterdam, Rotterdam, The Netherlands, 2016; pp. 1–76.
29. Endresen, Ø.; Sørgård, E.; Sundet, J.K.; Dalsøren, S.B.; Isaksen, I.S.A.; Berglen, T.F.; Gravir, G. Emission from international sea transportation and environmental impact. *J. Geophys. Res. Atmos.* **2003**, *108*, 14–21. [CrossRef]
30. IMO. IMO 2020—Cutting Sulphur Oxide Emissions. Available online: https://www.imo.org/en/MediaCentre/HotTopics/Pages/Sulphur-2020.aspx (accessed on 8 August 2021).
31. Acciaro, M.; Ghiara, H.; Inés, M.; Cusano, M.I. Energy management in seaports: A new role for port authorities. *Energy Policy* **2014**, *71*, 4–12. [CrossRef]
32. Roh, S.; Thai, V.V.; Wong, Y.D. Towards sustainable asean port development: Challenges and opportunities for vietnamese ports. *Asian J. Shipp. Logist.* **2016**, *32*, 107–118. [CrossRef]
33. Edoho, F.M. Oil transnational corporations: Corporate social responsibility and environmental sustainability. *Corp. Soc. Responsib. Environ. Manag.* **2008**, *15*, 210–222. [CrossRef]
34. Matishov, G.G.; Selifonova, Z.P. New scientifically based methods for controlling ship ballast in ports. *Dokl. Biol. Sci.* **2008**, *422*, 342. [CrossRef]
35. Behrends, B.; Liebezeit, P.G.; Gregory, D. *Reducing SO2 and NOX Emissions from Ships by a Seawater Scrubber*; BP Marine Report; Research Centre Terramare: Wilhelmshaven, Germany, 2003; pp. 11–64.
36. Fouquet, R. Energy Research & social science historical energy transitions: Speed, prices and system transformation. *Chem. Phys. Lett.* **2016**, *22*, 7–12. [CrossRef]
37. Heavy Fuel Oil. 2015. Available online: https://www.oiltanking.com/en/news-info/glossary/details/term/heavy-fuel-oil-hfo.html (accessed on 15 August 2021).
38. Papoutsoglou, T.G. A Cold Ironing Study o Modern Ports, Implementation and Benefits Thriving for Worldwide Ports. Bachelor's Thesis, Marine Engineering National Technical University of Athens, Zografou, Greece, 2012.
39. Innes, A.; Monios, J. Identifying the unique challenges of installing cold ironing at small and medium ports—The case of aberdeen. *Transp. Res. Part D Transp. Environ.* **2018**, *62*, 298–313. [CrossRef]
40. Mutarraf, M.U.; Terriche, Y.; Nasir, M.; Guan, Y.; Su, C.-L.; Vasquez, J.C.; Guerrero, J.M. A communication-less multi-mode control approach for adaptive power-sharing in ships-based seaport microgrid. *IEEE Trans. Transp. Electrif.* **2021**, *7*, 3070–3082. [CrossRef]
41. Mocci, S.; Porru, M.; Serpi, A.; Soma, G.G. The poseidon project: Microgrid in port areas to improve energy efficiency by the integration of res, flexible loads and smart mobility. In Proceeding of the 2019 1st International Conference on Energy Transition in the Mediterranean Area (SyNERGY MED), Cagliari, Italy, 28–30 May 2019; pp. 1–5. [CrossRef]
42. Salleh, N.A.S.; Muda, W.M.W.; Abdullah, S.S. Feasibility study of optimization and economic analysis for grid-connected renewable energy electric boat charging station in Kuala Terengganu. In Proceeding of the 2015 IEEE Conference on Energy Conversion (CENCON), Johor Bahru, Malaysia, 19–20 October 2015; pp. 510–515. [CrossRef]
43. Roy, A.; Auger, F.; Olivier, J.; Schae, E.; Auvity, B. Design, Sizing, and Energy Management of Microgrids in Harbor Areas: A Review. *Energies* **2020**, *13*, 5314. [CrossRef]
44. Soh, A. Jurong Port Starts World's Largest Port-Based Solar Facility. 2016. Available online: https://www.businesstimes.com.sg/energy-commodities/jurong-port-starts-worlds-largest-port-based-solar-facility (accessed on 23 August 2021).
45. Port of Aalborg Becomes the First co2 Neutral Port in Denmark. Available online: https://safety4sea.com/port-of-aalborg-becomes-the-first-co2-neutral-port-in-denmark/ (accessed on 23 August 2021).
46. Shahidehpour, M.; Clair, J.F. A functional microgrid for enhancing reliability, sustainability, and energy efficiency. *Electr. J.* **2012**, *25*, 21–28. [CrossRef]

47. Vicenzutti, A.; Bosich, D.; Giadrossi, G.; Sulligoi, G. The role of voltage controls in modern all electrical ships toward the all-electric ship. *IEEE Electrif. Mag.* **2015**, *3*, 49–65. [CrossRef]
48. McCoy, T.J. Electric ships past, present, and future [technology leaders]. *IEEE Electrif. Mag.* **2015**, *3*, 4–11. [CrossRef]
49. Murphy, J.F.H.M.; May, J.; Riksheim, R. International Cooperation on Marine Engineering Systems/Electric Propulsion—State-of-the-art and Trends in Electric Power Generation, Distribution, and Propulsion, and Their Associated Control Systems Report from Technical Committee B: Electric Pr. 2000; pp. 1–36.
50. Fang, S.; Wang, Y.; Gou, B.; Xu, Y. Toward future green maritime transportation: An overview of seaport microgrids and all-electric ships. *IEEE Trans. Veh. Technol.* **2020**, *69*, 207–219. [CrossRef]
51. Ding, X.; Yang, Z. Knowledge mapping of platform research: A visual analysis using VOSviewer and CiteSpace. *Electron. Commer. Res.* **2020**, 1–23. [CrossRef]
52. Bakar, N.N.A.; Hassan, M.Y.; Sulaima, M.F.; Mohd Nasir, M.N.; Khamis, A. Microgrid and load shedding scheme during islanded mode: A review. *Renew. Sustain. Energy Rev.* **2017**, *71*, 161–169. [CrossRef]
53. Rajesh, K.S.; Dash, S.S.; Rajagopal, R.; Sridhar, R. A review on control of ac microgrid. *Renew. Sustain. Energy Rev.* **2017**, *71*, 814–819. [CrossRef]
54. Mousavi, S.Y.M.; Jalilian, A.; Savaghebi, M.; Guerrero, J.M. Flexible compensation of voltage and current unbalance and harmonics in microgrids. *Energies* **2017**, *10*, 1568. [CrossRef]
55. Farrok, O.; Sheikh, M.R.I.; Islam, M.R. An advanced controller to improve the power quality of microgrid connected converter. In Proceeding of the 2015 International Conference on Electrical & Electronic Engineering (ICEEE), Rajshahi, Bangladesh, 10 March 2016; pp. 185–188. [CrossRef]
56. British Columbia Ministry of Agriculture. *Market Opportunity Report: JAPAN*; British Columbia Ministry of Agriculture: Abbotsford, BC, Canada, December 2014. [CrossRef]
57. Huang, Z.; Zhu, T.; Gu, Y.; Irwin, D.; Mishra, A.; Shenoy, P. Minimizing electricity costs by sharing energy in sustainable microgrids. In Proceedings of The 12th ACM Conference on Embedded Network Sensor Systems, Memphis, TN, USA, 3–6 November 2014; pp. 120–129. [CrossRef]
58. Haidar, A.M.A.; Fakhar, A.; Helwig, A. Sustainable energy planning for cost minimization of autonomous hybrid microgrid using combined multi-objective optimization algorithm. *Sustain. Cities Soc.* **2020**, *62*, 102391. [CrossRef]
59. Lotfi, H.; Khodaei, A. AC versus DC microgrid planning. *IEEE Trans. Smart Grid* **2017**, *8*, 296–304. [CrossRef]
60. Lu, X.; McElroy, M.B.; Nielsen, C.P.; Chen, X.; Huang, J. Optimal integration of offshore wind power for a steadier, environmentally friendlier, supply of electricity in China. *Energy Policy* **2013**, *62*, 131–138. [CrossRef]
61. Wu, X.; Wang, Z.; Ding, T.; Li, Z. Hybrid AC/DC microgrid planning with optimal placement of DC feeders. *Energies* **2019**, *12*, 1751. [CrossRef]
62. Justo, J.J.; Mwasilu, F.; Lee, J.; Jung, J.W. AC-microgrids versus DC-microgrids with distributed energy resources: A review. *Renew. Sustain. Energy Rev.* **2013**, *24*, 387–405. [CrossRef]
63. Gao, L.; Liu, Y.; Ren, H.; Guerrero, J.M. A DC microgrid coordinated control strategy based on integrator current-sharing. *Energies* **2017**, *10*, 1116. [CrossRef]
64. Lago, J.; Heldwein, M.L. Operation and control-oriented modeling of a power converter for current balancing and stability improvement of DC active distribution networks. *IEEE Trans. Power Electron.* **2011**, *26*, 877–885. [CrossRef]
65. Kakigano, H.; Miura, Y.; Ise, T.; Uchida, R. DC micro-grid for super high quality distribution—System configuration and control of distributed generations and energy storage devices. In Proceeding of the 2006 37th IEEE Power Electronics Specialists Conference, Jeju, Korea, 18–22 June 2006. [CrossRef]
66. Jin, Z.; Savaghebi, M.; Vasquez, J.C.; Meng, L.; Guerrero, J.M. Maritime DC microgrids—A combination of microgrid technologies and maritime onboard power system for future ships. In Proceeding of the 2016 IEEE 8th International Power Electronics and Motion Control Conference (IPEMC-ECCE Asia), Hefei, China, 22–26 May 2016; pp. 179–184. [CrossRef]
67. Prenc, R.; Cuculić, A.; Baumgartner, I. Advantages of using a DC power system on board ship. *J. Marit. Transp. Sci.* **2016**, *52*, 83–97. [CrossRef]
68. Fregosi, D.; Ravula, S.; Brhlik, D.; Saussele, J.; Frank, S.; Bonnema, E.; Scheib, J.; Wilson, E. A comparative study of DC and AC microgrids in commercial buildings across different climates and operating profiles. In Proceeding of the 2015 IEEE First International Conference on DC Microgrids (ICDCM), Atlanta, GA, USA, 7–10 June 2015; pp. 159–164. [CrossRef]
69. Liu, X.; Wang, P.; Loh, P.C. A hybrid AC/DC microgrid and its coordination control. *IEEE Trans. Smart Gird* **2011**, *2*, 278–286.
70. Guerrero, J.M.; Jin, Z.; Liu, W.; Othman, M.B.; Savaghebi, M.; Anvari-Moghaddam, A.; Meng, L.; Vasquez, J.C. Shipboard microgrids: Maritime islanded power systems technologies. In Proceeding of the PCIM Asia 2016; International Exhibition and Conference for Power Electronics, Intelligent Motion, Renewable Energy and Energy Management, Shanghai, China, 28–30 June 2016; pp. 28–30.
71. Al-Falahi, M.D.A.; Tarasiuk, T.; Jayasinghe, S.G.; Jin, Z.; Enshaei, H.; Guerrero, J.M. Ac ship microgrids: Control and power management optimization. *Energies* **2018**, *11*, 1458. [CrossRef]
72. Liu, W.; Tarasiuk, T.; Gorniak, M.; Savaghebi, M.; Vasquez, J.C.; Su, C.L.; Guerrero, J.M. Power quality assessment in shipboard microgrids under unbalanced and harmonic AC bus voltage. *IEEE Trans. Ind. Appl.* **2019**, *55*, 765–775. [CrossRef]

73. Feste, M.D.; Chiandone, M.; Bosich, D.; Sulligoi, G. Evolution of the Trieste Port: A real-time system for a coordinated cold ironing. In Proceeding of the 2019 IEEE International Conference on Environment and Electrical Engineering and 2019 IEEE Industrial and Commercial Power Systems Europe (EEEIC/I&CPS Europe), Genova, Italy, 11–14 June 2019. [CrossRef]
74. Jin, B.Z.; Sulligoi, G.; Cuzner, R.; Meng, L.; Vasquez, J.C.; Guerrero, J.M. Next-generation shipboard DC power system. *IEEE Electrif. Mag.* **2016**, *4*, 45–57. [CrossRef]
75. D'Agostino, F.; Kaza, D.; Martelli, M.; Schiapparelli, G.P.; Silvestro, F.; Soldano, C. Development of a multiphysics real-time simulator for model-based design of a DC shipboard microgrid. *Energies* **2020**, *13*, 3580. [CrossRef]
76. Jin, Z.; Meng, L.; Vasquez, J.C.; Guerrero, J.M.; Jin, Z.; Meng, L.; Vasquez, J.C.; Guerrero, J.M. Specialized Hierarchical Control Strategy for DC Distribution based Shipboard Microgrids A combination of emerging DC shipboard power systems and microgrid technologies. In Proceeding of the IECON 2017-43rd Annual Conference of the IEEE Industrial Electronics Society, Beijing, China, 29 October–1 November 2017; pp. 10–12.
77. Kwon, K.; Park, D. Load frequency-based power management for shipboard DC hybrid power systems. In Proceeding of the 2020 IEEE 29th International Symposium on Industrial Electronics (ISIE), Delft, The Netherlands, 17–19 June 2020; pp. 142–147.
78. German-Galkin, S.; Tarnapowicz, D. Energy optimization of the 'shore to ship' system—A universal power system for ships at berth in a port. *Sensors* **2020**, *20*, 3815. [CrossRef]
79. Reed, G.F.; Grainger, B.M.; Sparacino, A.R.; Mao, Z.H. Ship to grid: Medium-voltage dc concepts in theory and practice. *IEEE Power Energy Mag.* **2012**, *10*, 70–79. [CrossRef]
80. Guide for Direct Current (DC). *Power Distribution Systems for Marine and Offshore Applications Direct Current (DC) Power Distribution Systems for Marine and Offshore Applications*; American Bureau of Shipping: Houston, TX, USA, 2018.
81. Zhaoxia, X.; Tianli, Z.; Huaimin, L.; Guerrero, J.M.; Su, C.; Member, S.; Vásquez, J.C.; Member, S. Coordinated Control of a Hybrid-Electric-Ferry Shipboard Microgrid. *IEEE Trans. Transp. Electrif.* **2019**, *5*, 828–839. [CrossRef]
82. Mutarraf, M.U.; Terriche, Y.; Niazi, K.A.K.; Khan, F.; Vasquez, J.C.; Guerrero, J.M. Control of hybrid diesel/PV/battery/ultra-capacitor systems for future shipboard microgrids. *Energies* **2019**, *12*, 3460. [CrossRef]
83. Ahamad, N.B.; Othman, M.; Vasquez, J.C.; Guerrero, J.M.; Su, C.L. Optimal sizing and performance evaluation of a renewable energy based microgrid in future seaports. In Proceeding of the 2018 IEEE International Conference on Industrial Technology (ICIT), Lyon, France, 20–22 February 2018; pp. 1043–1048. [CrossRef]
84. Gennitsaris, S.G.; Kanellos, F.D. Emission-aware and cost-effective distributed demand response system for extensively electrified large ports. *IEEE Trans. Power Syst.* **2019**, *34*, 4341–4351. [CrossRef]
85. Hamburg Port Authority. *Hamburg Port Authority Hamburg Is Staying on Course—The Port Development Plan to 2025*; Hamburg Port Authority: Hamburg, Germany, 2012; pp. 1–98.
86. Iris, Ç.; Lam, J.S.L. A review of energy efficiency in ports: Operational strategies, technologies and energy management systems. *Renew. Sustain. Energy Rev.* **2019**, *112*, 170–182. [CrossRef]
87. Elwany, M.H.; Ali, I.; Abouelseoud, Y. A heuristics-based solution to the continuous berth allocation and crane assignment problem. *Alex. Eng. J.* **2013**, *52*, 671–677. [CrossRef]
88. Liu, A.; Liu, H.; Tsai, S.B.; Lu, H.; Zhang, X.; Wang, J. Using a hybrid model on joint scheduling of berths and quay cranes-from a sustainable perspective. *Sustainability* **2018**, *10*, 1959. [CrossRef]
89. Wang, Y.; Ding, W.; Dai, L.; Hu, H.; Jing, D. How would government subsidize the port on shore side electricity usage improvement? *J. Clean. Prod.* **2021**, *278*, 123893. [CrossRef]
90. Xie, P.; Guerrero, J.M.; Tan, S.; Bazmohammadi, N.; Vasquez, J.C.; Mehrzadi, M.; Al-Turki, Y. Optimization-based power and energy management system in shipboard microgrid: A review. *IEEE Syst. J.* **2021**, 1–13. [CrossRef]
91. Huang, Y.; Wang, H.; Khajepour, A.; He, H.; Ji, J. Model predictive control power management strategies for HEVs: A review. *J. Power Sources* **2017**, *341*, 91–106. [CrossRef]
92. Sun, Z.; Li, L. Potential capability estimation for real time electricity demand response of sustainable manufacturing systems using markov decision process. *J. Clean. Prod.* **2014**, *65*, 184–193. [CrossRef]
93. Subramani, G.; Ramachandaramurthy, V.K.; Padmanaban, S.; Mihet-Popa, L.; Blaabjerg, F.; Guerrero, J.M. Grid-tied photovoltaic and battery storage systems with Malaysian electricity tariff—A review on maximum demand shaving. *Energies* **2017**, *10*, 1884. [CrossRef]
94. Hirsch, A.; Parag, Y.; Guerrero, J. Microgrids: A review of technologies, key drivers, and outstanding issues. *Renew. Sustain. Energy Rev.* **2018**, *90*, 402–411. [CrossRef]
95. Divshali, P.H.; Choi, B.J. Electrical market management considering power system constraints in smart distribution grids. *Energies* **2016**, *9*, 405. [CrossRef]
96. Huang, Y.; Lan, H.; Hong, Y.Y.; Wen, S.; Fang, S. Joint voyage scheduling and economic dispatch for all-electric ships with virtual energy storage systems. *Energy* **2020**, *190*, 116268. [CrossRef]
97. Othman, M.; Anvari-Moghaddam, A.; Ahamad, N.; Chun-Lien, S.; Guerrero, J.M. Scheduling of Power Generation in Hybrid Shipboard Microgrids with Energy Storage Systems. In Proceeding of the 2018 IEEE International Conference on Environment and Electrical Engineering and 2018 IEEE Industrial and Commercial Power Systems Europe (EEEIC/I&CPS Europe), Palermo, Italy, 12–15 June 2018. [CrossRef]

98. Rao, K.S.; Chauhan, P.J.; Panda, S.K.; Wilson, G.; Liu, X.; Gupta, A.K. Optimal scheduling of diesel generators in offshore support vessels to minimize fuel consumption. In Proceeding of the IECON 2015-41st Annual Conference of the IEEE Industrial Electronics Society, Yokohama, Japan, 9–12 November 2015; pp. 4726–4731. [CrossRef]
99. Pande, S.; Ghodekar, P.J.G. Computation of technical power loss of feeders and transformers in distribution system using load factor and load loss factor. *Int. J. Multidiscip. Sci. Eng.* **2012**, *3*, 22–25.
100. Chua, K.H.; Lim, Y.S.; Morris, S. Energy storage system for peak shaving. *Int. J. Energy Sect. Manag.* **2016**, *10*, 3–18. [CrossRef]
101. Uddin, M.; Romlie, M.F.; Abdullah, M.F. Performance assessment and economic analysis of a gas-fueled islanded microgrid—A malaysian case study. *Infrastructures* **2019**, *4*, 61. [CrossRef]
102. Robert, F.C.; Sisodia, G.S.; Gopalan, S. The critical role of anchor customers in rural microgrids: Impact of load factor on energy. In Proceeding of the 2017 International Conference on Computation of Power, Energy Information and Commuincation (ICCPEIC), Melmaruvathur, India, 22–23 March 2017; pp. 398–403. [CrossRef]
103. Davydova, A.; Chakirov, R.; Vagapov, Y.; Komenda, T.; Lupin, S. Coordinated in-home charging of plug-in electric vehicles from a household smart microgrid. In Proceeding of the 2013 Africon, Pointe aux Piments, Mauritius, 9–12 September 2013.
104. Ma, X.; Qu, H.; Pei, W.; Xiao, H. Optimal interactive operation of microgrid under demand response based on rolling optimization algorithm. *Energy Procedia* **2018**, *145*, 97–102. [CrossRef]
105. Chowdhury, A.H.; Alam, S.; Hossain, A. Home energy management for community microgrids using optimal power sharing algorithm. *Energies* **2021**, *14*, 1060.
106. Geerlings, H.; Heij, R.; Van Duin, R. Opportunities for Peak Shaving the Energy Demand of Ship-to-Shore Quay cranes at Container Terminals. *J. Shipp. Trade* **2018**, *3*, 1–20. [CrossRef]
107. Heij, R. Opportunities for Peak Shaving Electricity Consumption at Container Terminals. Applying New Rules of Operation to Achieve a More Balanced Electricity Consumption. Master's thesis, Delft University of Technology, Delft, The Netherlands, 2015.
108. Bunnonn, P.; Chalermyanont, K.; Limsakul, C. A computing model of artificial intelligent approaches to mid-term load forecasting: A state of the art survey for the researcher. *Int. J. Eng. Technol.* **2010**, *2*, 94–100. [CrossRef]
109. Hong, W.; Li, M.; Fan, G. *Short-Term Load Forecasting by Artificial Intelligent Technologies*; MDPI: Basel, Switzerland, 2019; ISBN 9783038975823.
110. Soliman Abdel-hady Soliman, A.M.A.-K. *8-Dynamic Electric Load Forecasting*; Electric Load Forecast: Seattle, WA, USA, 2010; pp. 291–352. ISBN 978012381543. [CrossRef]
111. Gross, G.; Galiana, F.D. Short-Term Load Forecasting. *Proc. IEEE* **1987**, *75*, 1558–1573. [CrossRef]
112. Samuel, I.A.; Emmanuel, A.; Odigwe, I.A.; Felly-Njoku, F.C. A comparative study of regression analysis and artificial neural network methods for medium-term load forecasting. *Indian J. Sci. Technol.* **2017**, *10*, 1–7. [CrossRef]
113. Alasali, F.; Haben, S.; Becerra, V.; Holderbaum, W. Optimal energy management and MPC strategies for electrified RTG cranes with energy storage systems. *Energies* **2017**, *10*, 1598. [CrossRef]
114. Alasali, F.; Haben, S.; Becerra, V.; Holderbaum, W. Day-ahead industrial load forecasting for electric RTG cranes. *J. Mod. Power Syst. Clean Energy* **2018**, *6*, 223–234. [CrossRef]
115. Colarossi, D.; Principi, P. *Feasibility Study of a Cold Ironing System and District Heating in Port Area*; Department of Industrial Engineering and Mathematic Sciences: Ancona, Italy, 2020; pp. 666–675. [CrossRef]
116. D'Agostino, F.; Schiapparelli, G.P.; Dallas, S.; Spathis, D.; Georgiou, V.; Prousalidis, J. On estimating the port power demands for cold ironing applications. In Proceeding of the 2021 IEEE Electric Ship Technologies Symposium (ESTS), Arlington, VA, USA, 3–6 August 2021; pp. 1–5. [CrossRef]
117. Peng, Y.; Li, X.; Wang, W.; Liu, K.; Bing, X.; Song, X. A method for determining the required power capacity of an on-shore power system considering uncertainties of arriving ships. *Sustainability* **2018**, *10*, 4524. [CrossRef]
118. José, E.; Gutierrez-Romero, J.; Esteve-Pérez, B.Z. Implementing onshore power supply from renewable energy sources for requirements of ships at berth. *Appl. Energy* **2019**, *255*, 113883. [CrossRef]
119. Hein, K.; Yan, X.; Wilson, G. Multi-objective optimal scheduling of a hybrid ferry with shore-to-ship power supply considering energy storage degradation. *Electronics* **2020**, *9*, 849. [CrossRef]
120. Mehrzadi, M.; Terriche, Y.; Su, C.L.; Xie, P.; Bazmohammadi, N.; Costa, M.N.; Liao, C.H.; Vasquez, J.C.; Guerrero, J.M. A deep learning method for short-term dynamic positioning load forecasting in maritime microgrids. *Appl. Sci.* **2020**, *10*, 4889. [CrossRef]
121. Hein, K.; Xu, Y.; Wilson, G.; Gupta, A.K. Coordinated optimal voyage planning and energy management of all-electric ship with hybrid energy storage system. *IEEE Trans. Power Syst.* **2021**, *36*, 2355–2365. [CrossRef]
122. Lai, C.S.; Locatelli, G.; Pimm, A.; Wu, X.; Lai, L.L. A review on long-term electrical power system modeling with energy storage. *J. Clean. Prod.* **2021**, *280*, 124298. [CrossRef]
123. Amirante, R.; Cassone, E.; Elia Distaso, P.T. Overview on recent developments in energy storage: Mechanical, electrochemical and hydrogen technologies. *Energy Convers. Manag.* **2017**, *132*, 372–387. [CrossRef]
124. Mutarraf, M.U.; Terriche, Y.; Niazi, K.A.K.; Vasquez, J.C.; Guerrero, J.M. Energy storage systems for shipboard microgrids—A review. *Energies* **2018**, *11*, 3492. [CrossRef]
125. Kim, K.; An, J.; Park, K.; Roh, G.; Chun, K. Analysis of a supercapacitor/battery hybrid power system for a bulk carrier. *Appl. Sci.* **2019**, *9*, 1547. [CrossRef]
126. Nayak, A.; Lee, S.; Sutherland, J.W. Storage trade-offs and optimal load scheduling for cooperative consumers in a microgrid with different load types. *IISE Trans.* **2019**, *51*, 397–405. [CrossRef]

127. Palensky, P.; Dietrich, D. Demand side management: Demand response, intelligent energy systems, and smart loads. *IEEE Trans. Ind. Inform.* **2011**, *7*, 381–388. [CrossRef]
128. Vahedipour-Dahraie, M.; Najafi, H.R.; Anvari-Moghaddam, A.; Guerrero, J.M. Study of the effect of time-based rate demand response programs on stochastic day-ahead energy and reserve scheduling in islanded residential microgrids. *Appl. Sci.* **2017**, *7*, 378. [CrossRef]
129. Commission, E. *Reducing Emissions from the Shipping Sector—European Commission*; European Commission: Brussels, Belgium, 2016; p. 1.
130. Sanz, J.F.; Matute, G.; Fernández, G.; Alonso, M.A.; Sanz, M. Analysis of european policies and incentives for microgrids. *Renew. Energy Power Qual. J.* **2014**, *1*, 874–879. [CrossRef]
131. Hansen, J.F.; Wendt, F. History and state of the art in commercial electric ship propulsion, integrated power systems, and future trends. *Proc. IEEE* **2015**, *103*, 2229–2242. [CrossRef]
132. Sahoo, S.; Yang, Y.; Blaabjerg, F. Resilient synchronization strategy for ac microgrids under cyber attacks. *IEEE Trans. Power Electron.* **2021**, *36*, 73–77. [CrossRef]
133. Rana, M.; Li, L.; Su, S.W. Cyber attack protection and control of microgrids. *IEEE/CAA J. Autom. Sin.* **2018**, *5*, 602–609. [CrossRef]
134. Colarossi, D.; Principi, P. Technical analysis and economic evaluation of a complex shore-to-ship power supply system. *Appl. Therm. Eng.* **2020**, *181*, 115988. [CrossRef]

energies

Article

Blockchain-Enabled Energy Demand Side Management Cap and Trade Model

Alain Aoun [1,*], Hussein Ibrahim [2], Mazen Ghandour [3] and Adrian Ilinca [1]

1 Department of Mathematics Computer Science and Engineering, Université du Québec à Rimouski (UQAR), Rimouski, QC G5L 3A1, Canada; adrian_ilinca@uqar.ca
2 Technology Institute of Industrial Maintenance (ITMI), 175 Rue de la Vérendrye, Cégep de Sept-Îles, Sept-Iles, QC G4R 5B7, Canada; hussein.ibrahim@itmi.ca
3 Faculty of Engineering, Lebanese University, Beirut 6573/14, Lebanon; ghandour@ul.edu.lb
* Correspondence: alain.aoun@uqar.ca; Tel.: +1-418-896-3996

Abstract: Global economic growth, demographic explosion, digitization, increased mobility, and greater demand for heating and cooling due to climate change in different world areas are the main drivers for the surge in energy demand. The increase in energy demand is the basis of economic challenges for power companies alongside several socio-economic problems in communities, such as energy poverty, defined as the insufficient coverage of energy needs, especially in the residential sector. Two main strategies are considered to meet this increased demand. The first strategy focuses on new sustainable and eco-friendly modes of power generation, such as renewable energy resources and distributed energy resources. The second strategy is demand-side oriented rather than the supply side. Demand-side management, demand response (DR), and energy efficiency (EE) programs fall under this category. On the other hand, the decentralization and digitization of the energy sector convoyed by the emersion of new technologies such as blockchain, Internet of Things (IoT), and Artificial Intelligence (AI), opened the door to new solutions for the energy demand dilemma. Among these technologies, blockchain has proved itself as a decentralized trading platform between untrusted peers without the involvement of a trusted third party. This newly introduced Peer-to-Peer (P2P) trading model can be used to create a new demand load control model. In this article, the concept of an energy cap and trade demand-side management (DSM) model is introduced and simulated. The introduced DSM model is based on the concept of capping consumers' monthly energy consumption and rewarding consumers who do not exceed this cap with energy tradeable credits that can be traded using blockchain-based Peer-to-Peer (P2P) energy trading. A model based on 200 households is used to simulate the proposed DSM model and prove that this model can be beneficial to both energy companies and consumers.

Keywords: energy; cap and trade; blockchain; demand-side management; energy policy; energy trading

Citation: Aoun, A.; Ibrahim, H.; Ghandour, M.; Ilinca, A. Blockchain-Enabled Energy Demand Side Management Cap and Trade Model. *Energies* **2021**, *14*, 8600. https://doi.org/10.3390/en14248600

Academic Editor: Ben McLellan

Received: 9 November 2021
Accepted: 7 December 2021
Published: 20 December 2021

1. Introduction

Electric utility companies worldwide deal with severe and recurring power crises caused by a combination of supply-side problems, including fuel supply challenges, maintenance requirements and unplanned outages, and a continuous increase in energy demand, driven by global demographic growth, electrification of the transportation sector, and greater need for heating and cooling. The dilemma of equilibrating the energy supply with energy demand can be targeted using two principal methodologies. The first way is to meet the demand by increasing the power supply, using fast-acting, low energy cost power generators. The second method to equilibrate the supply with the demand is to act on the demand side by managing the load to minimize the overall energy demand and clear the gap between the supply power and demand power [1]. This second methodology relies heavily on the participation of electricity consumers to reduce their electrical energy consumptions during peak times in exchange for an incentive compensation [2].

Nevertheless, when energy demand-side management (DSM) strategies are considered, the economic problem of free-riding emerges [3]. Free-riding usually refers to users who benefit from conservation subsidies without really contributing to decreasing energy consumption or paying for it. In demand-side management programs, free-riding occurs when consumers get incentives without really reducing their loads [4]. On the other hand, the rebound effect is another challenge when assessing the long-term goals and impacts of energy efficiency projects. In energy efficiency projects, the rebound effect, also known as the take-back effect, is the reduction in anticipated energy savings from the applied energy conservation measures or technology caused by a change in consumer's energy consumption behavior due to the applied measure itself [5]. This is a significant outcome of energy efficiency, which is commonly underestimated. Moreover, it characterizes the negative relationship between technology and consumption [6]. Thus, it is essential to develop a demand-side management program that can achieve the ultimate targets of reducing the overall energy demand and shaving the peak load while avoiding the free-riding and rebound effects. The model proposed in this article is a demand-side management model based on the cap-and-trade system, empowered by a peer-to-peer (P2P) blockchain-based credit trading platform that aims to limit the increase in energy demand.

On the other hand, new emerging technologies such as blockchain, Internet of Things (IoT), artificial intelligence (AI), machine learning, Big Data, etc. have the potential to disrupt the energy sector heavily and change it from a centralized, hierarchical supply chain to a decentralized, decarbonized and decentralized smart platform. Energy suppliers are in a continuous quest to reap greater productivity, improve safety and reduce energy costs. The digital transformation of the energy sector offers the means to achieve those goals. Furthermore, the digitalization of the energy sector is the cornerstone of the broad integration of distributed energy resources (DERs) in any electric grid, which aims to increase load flexibility and diversity in the power generation systems. This digital transformation also unlocks new potentials for users to manage their energy consumption and supply while offering them the possibility to become active stakeholders in electricity grid management. For electric utilities, digitalization permits better monitoring of the grid, faster detection of failures supplemented by an autonomous diagnosis and response to those faults, thus reducing power failures, grid downtime, and higher quality of service [7]. Additionally, the accelerated expansion of DERs, as a fundamental part of smart grids, can no longer be efficiently addressed using conventional centralized methods but instead requires a decentralized real-time control and supervision of all grid assets. Hence, new innovative management systems relying on decentralized networks are becoming a necessity. Blockchain technology could facilitate a fully decentralized energy system [8].

A blockchain is a decentralized network that relies on a distributed chronological ledger, hosted, updated, and validated by several peer nodes rather than a single centralized authority, acting as an immutable record of all transactions. The fact that blockchain substitutes the trusted third parties makes this technology a simple, fast, safe, and transparent means of transaction between peers. An example of the implementation of blockchain technology is the famous cryptocurrency Bitcoin. While conventional wire transfers require validation from a bank and take several days to be completed, Bitcoin transactions can be achieved in near real-time directly from peer to peer [9]. Since the appearance of Bitcoin in 2008 as a peer-to-peer (P2P) electronic cash system, blockchain, its underpinning technology, has gained broad interest from different businesses and industries. Blockchain technology promises a secure, near real-time, and low-cost method for conducting digital assets transactions [10]. It increases process automation while managing more significant volumes of data with limited human intervention at lower cost and risk. Meanwhile, energy companies are grappling with increased reporting, transparency, and security regulations, which incurs additional costs to the energy trading process and greater demand for personnel and resources. Blockchain technology can help target those challenges and have a significant positive impact on the energy sector [11]. Traditional transactional models rely on centralized, server-client-based architectures. Transactions between network nodes occur only through an intermediary third party required to establish trust, especially when unknown parties are involved. Nevertheless, the involvement of intermediaries induces additional commission fees in exchange for their services and increases the processing time required for transactions. Furthermore, since all transactions are managed and recorded using a central server, centralized networks suffer from a single point of failure. Alternatively, decentralized architectures, such as blockchain's P2P platform, offer a network of interconnected nodes that can interact directly, preserving the integrity of the grid even if several nodes are jeopardized or disconnected. Thus, P2P decentralized networks allow bridge or mitigate most of the problems associated with centralized networks [12].

Blockchain can transform the energy sector in harmony with the natural laws of growth. It provides an incremental, sequential, highly integrated approach to developing the energy sector's effectiveness and efficiency. This new technology can move the energy market progressively from a dependent market to an independent market to an interdependent market. The existing market is an entirely dependent market where consumers rely on utility companies and service providers. With the development of DER technologies, the market can evolve into an independent market where off-grid isolated micro-grids can survive. But with the integration of blockchain, the energy sector has the potential to metamorphose into an interdependent market ruled by the paradigm of we-we can do it, where people can cooperate and combine efforts, abilities, and resources to create something greater together.

In this article, a new demand-side management (DSM) program, based on the "Cap-and-Trade" concept merged with blockchain-enabled peer-to-peer (P2P) energy trading mechanism, is proposed. The second section of the article provides a review of the most recent researches on peer-to-peer energy trading, smart energy management, and innovative mechanisms used to mitigate challenges and limitations faced with traditional demand-side management models. In Section 3, blockchain technology and its components will be introduced. In Section 4, the proposed blockchain-enabled energy demand-side management Cap and Trade model is defined along with a proposed trading model that highlights its functionality. The proposed DSM model combines the features of blockchain P2P energy trading and the well-known Cap and Trade mechanism used for carbon emissions trading. Thus, to clearly explain the proposed DSM mechanism, at first the Cap-and-Trade mechanism is introduced, and secondly, the importance of adopting a P2P blockchain-based energy trading platform is highlighted. Finally, in Section 5, a simulation model using historical data from 200 households is implemented to test the effectiveness of the DSM mechanism.

2. Related Works

The language of innovation is expected in sustainable development policy contexts and load reduction, resulting from demand-side management mechanisms, including energy efficiency programs and demand response models. This can be perceived as an innovative equivalent of sustainable power generation development. Thus, in this framework, several works have already been developed to respond to challenges faced by existing demand-side management projects. The work conducted in [3] offers a two-step model to address the free-rider issue resulting from demand response programs. The first step focuses on predicting the customer's baseline load using a regression-based estimation model. The second step proposes an incentive paid to the consumer based on load reduction for a specific baseline rate. The proposed two-step method outperformed other approaches in terms of payment rule improvement.

Additionally, demand-side management is considered the central pillar of smart grids and distributed energy resources. Conversely, as presented in [13], the emergence of new technologies in smart grid settings has led to the advancement of the communication and control infrastructure, enabling a better exchange of information and data necessary to implement any demand-side management program properly. Similarly, article [14] proposes, in the frame of smart grids, a methodology to implement an active demand-side management model for households equipped with solar photovoltaic (PV) systems and battery energy storage systems. Nevertheless, the development of innovative demand-side management models is not just limited to the smart grid outline. Advanced technologies such as Artificial Intelligence (AI), machine learning (ML), blockchain, etc., have disrupted conventional demand-side management practices. As presented in [15], AI and ML have emerged as new enablers of demand response programs by tackling various challenges, limitations, and barriers. Among these barriers are the consumers' behavioral characteristics and preferences, a pricing model that responds to the consumers' expectations, the management of the demand load, and connected devices. They also contribute to setting an incentive reward program in a fair and economically efficient manner. In the same vein and as highlighted in [16], data analytics and ML can be employed to forecast energy demand, understand customer behavior, and tailor power generation solutions required in the future to respond to increasing demands.

However, recently blockchain has emerged as a new technology that can play an important role in smart grids and more specifically in advanced demand-side management mechanisms. As presented in [17], blockchain technology, especially when merged with advanced metering infrastructure (AMI), can deliver a transparent, secure, reliable, and timely energy flexibility to adapt consumers' energy load profiles to existing energy value chain stakeholders' capabilities. The work conducted in [18] suggests a blockchain-based DSM model using the Ethereum platform that matches energy demand and energy production at a smart grid level to validate this concept. The model improves feedback from DR enrolled consumers and aggregates and forecasts available DR loads while reducing the amount of energy flexibility needed for convergence. Similarly, article [19] uses a micro-grid with various residential load profiles to test a blockchain enhanced demand-side management mechanism that reduces peak-to-average ratio and smoothens the dips in the load profile caused by supply constraints.

Additionally, the proposed model optimizes the pay-off of both the energy provider and the consumer. Equally, Ref. [20] introduces a distributed demand-side management interconnecting, using a network of IoT smart meters, multiple households equipped with renewable energy sources in a single micro-grid. The proposed system minimizes the individual electricity cost for each household and the total cost of energy consumption for the entire micro-grid. The consumers aim to optimize their daily energy consumption in addition to their source of energy: self-generated energy from renewable energy sources, shared energy on the community microgrid, and energy provided by the utility. Each participant applies the best strategy that minimizes his energy consumption cost while maintaining his privacy of energy consumption.

On the other side, the application of blockchain technology in demand-side management is not restricted to the relationship between the energy service provider and the consumer but can cover machine-to-machine (M2M) interaction in the context of demand response. The work presented in [21] provides an example of M2M interaction where a power management system and a generator will cooperate to adjust the power generation trading over the blockchain. But then again, demand-side management is not only limited to demand response programs, and it also includes energy efficiency mechanisms. Even at this level, blockchain can play a major role in advancing energy conservation measures, as shown in [22].

Based on those above and even though only a little work has been conducted on applying advanced technologies, such as AI, ML, and blockchain in the field of DSM, the potential of these technologies in disrupting conventional DSM models is obvious and worthy of further investigation. Hence, the work presented here offers a new perspective for implementing the renowned emissions trading scheme known as Cap-and-Trade, as a demand-side management mechanism in the energy sector. It uses blockchain technology as a trading platform to limit the continuous increase in electricity demand faced by most utilities worldwide, as shown in Figure 1.

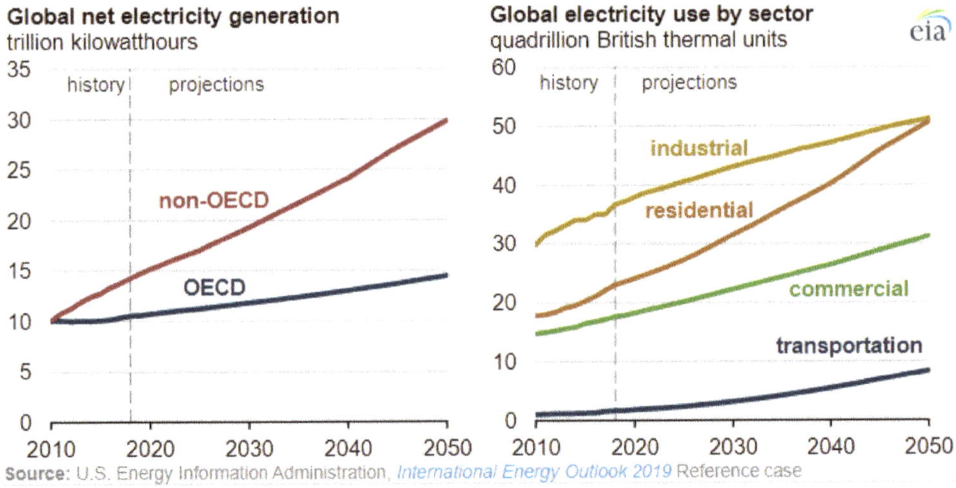

Figure 1. Global electricity generation and use.

3. Blockchain and Smart Contracts

3.1. Blockchain Technology

Satoshi Nakamoto first proposed the concept of blockchain in 2008 [23]. Blockchain technology is an additional layer on top of the internet. It is defined as the "internet of assets" as opposed to the world wide web that is recognized as the "internet of information" [24]. Blockchain offers the great potential of digitizing assets such as records, deeds, bonds, copyrights, currencies, art, real estate, carbon credits and energy while enabling a P2P trading system that doesn't rely on third parties. This process is known as tokenization. By converting physical assets to digital tokens, blockchain unlocks new values for real-world assets and enables trading them in real-time.

The value of the global market for blockchain technology is growing. Governments, utilities to academia, and civil organizations are accommodating a digital era in which blockchain is best known by cryptocurrencies like Bitcoin [25]. The blockchain has indeed based its reputation on the Bitcoin revolution, but the blockchain is not only about transferring token ownership. Blockchain technology can have significant social and economic impacts on established business practices. It offers an alternative means of transacting, sharing value, storing data, and doing business by eliminating the need for centralized entities. It is a decentralized trusted network, enabling anyone to digitize and save or transact data, assets, contracts, or value in a secure manner. Global revenues of blockchain technology are forecasted to grow in the coming years to more than $23 billion (U.S.) by 2023. The largest shares will come from the financial and energy sectors [26].

Traditional transactional models are based on a centralized structure. Transactions between network nodes occur through an intermediary third party. Intermediaries are most often required to establish trust between unknown involved parties. On the other side, one of the most notable features of blockchain is its decentralized structure (Figure 2). With blockchain, the trust between peers is empowered by mathematical algorithms and cryptography, and transactions are conducted from peer-to-peer, which makes blockchain most suitable for applications that meet the following criteria:

- Decentralized problems
- Peer-to-peer transactions
- Beyond boundaries of trust among unknown peers
- Require validation, verification, and recording of a time-stamped immutable ledger
- Autonomous operations guided by a rule structure and policies

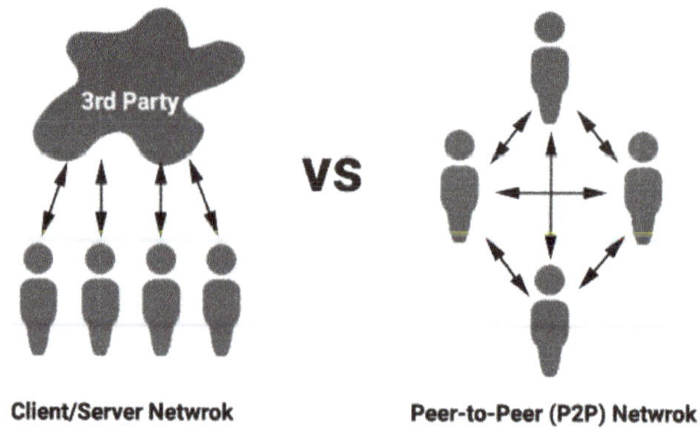

Figure 2. P2P network vs Client/Server network.

3.2. *What Is a Smart Contract?*

The concept of smart contracts refers to autonomous, self-enforcing programming codes that run on a blockchain network to simplify, govern and enforce agreements and transactions between untrusted peers without the need for a trusted intermediary [27].

The main problem with traditional client-server coded contracts is that when any of the parties involved has root access to the server, they can change the rules and conditions of the contracts. This is why in traditional contract models, a trusted third party or government is required. Intermediaries usually charge fees and can be considered a source of risk that potentially jeopardizes the confidentiality of any contract. This is a problem that blockchain and smart contracts can solve by offering a platform that enables parties to reach a consensus on the required set of rules or business policies and to jointly control the information. Hence, what blockchain and smart contracts offer is an evolution of the internet with immutable shared rules.

Bitcoin was the first blockchain to support basic smart contracts, providing a network capable of transferring value between participants. Its programming language enabled limited transaction features to be created (e.g., payment channels, escrows, multi-sign accounts, and time locks). Ethereum introduced the new feature of smart contracts. Its programming language enabled developers to build their decentralized applications (Dapps). At present, it is the most widely used smart contracts platform.

Smart contracts aim to make legal agreements self-executable by using computer code. Unlike conventional contracts drafted by lawyers, signed by stakeholders, and enforced by law, a smart contract establishes a relationship with cryptographic coding. Unlike a client-server-coded contract, smart contracts are immutable and unstoppable once deployed on the blockchain. Smart contracts offer the following benefits:

- Smart contracts are cryptographically secured, immutable, and enforced.
- Smart contracts are fast and inexpensive.
- They offer a multi-sig feature, which means that transactions are executed only when all approvals or signatures are provided.
- They are capable of managing agreements between users without human interruption.
- They may serve as part of other contracts (similar to how a software library works). Smart contracts can run independently and can automatically interact with other smart contracts.
- They store information, such as records, prices, energy consumption, etc., generated by the smart contract itself, fed by another smart contract or an outside oracle.

A smart contract is made up of two main parts: smart contract code and smart legal contracts. The smart contract code is the code that is stored, verified, and executed on the blockchain. A smart legal contract is a digital representation of a legally binding agreement using a smart contract code. A smart contract is created and signed by parties using their digital signatures. The terms of the agreement and the obligations of each participant are established and limited by program code instructions and functions. Once the agreement terms are satisfied, transactions are automatically executed as defined in the smart contract code.

Smart contracts are capable of redefining business relationships. Smart contracts offer viable solutions with lower transaction costs and risks [28]. Smart contracts' autonomous and self-enforcing nature makes business operations faster, safer, and less prone to human errors. Additionally, the fact that smart contract-managed processes require less human intervention allows reducing the overhead. Finally, since all transactions are recorded on a distributed ledger, it would be very difficult to tamper with the relevant data.

4. Blockchain-Enabled Energy Cap and Trade DSM Mechanism

4.1. Concept

The concept of sustainable consumption was first used on a global scale in Agenda 21, the action plan for sustainable development adopted by 179 heads of state at the Rio Earth Summit in 1992. For the first time, overconsumption in industrialized countries was identified as a direct driver of unsustainable development, and the idea of sustainable degrowth was introduced. Basically, the wished-for solution involved short-term and long-term market means to shift consumption patterns in an eco-efficient manner [29]. One way to achieve sustainable consumption is to use economic tools, such as subsidies, taxes, and charges, to impact the price of goods or services and directly alter consumers' choices [30]. Such market-based mechanisms minimize the impact of adverse market externalities and play an essential role in influencing purchasing patterns [31].

Nevertheless, economic tools can only impact consumers' preferences if the offered financial incentive is strong enough to overcome the threshold of the decision-making process. In other words, taxes should be adequately scaled to sway consumer-purchasing decisions. As a result, economic tools can be seen as mechanisms to render eco-friendly products financially appealing for consumers while making ecologically detrimental ones more expensive with the aim of discouraging their consumption. However, subsidies and taxes are not the only forms of economic tools. Trading or purchase schemes can also be used as financial strategies. The energy cap and trade scheme presented in this article falls under this category.

Often, prices of certain goods or services are increased to reflect their environmental cost. Primarily, these additional costs are determined by the price elasticity of the products in question. In other words, it depends on the percentage of consumers willing to decrease their consumption when prices increase at a certain rate. The price elasticity depends on the product or service itself and the household's income group. The cost of environmental policy initiatives is thus determined by users' willingness to alter their consumption and sacrifice some of their well-being in exchange for their environmental contribution. As a result, these increasing costs, which may not necessarily be evenly dispersed, have significant consequences for households [32,33]. Therefore, any economic tool needs to achieve a balanced distribution of the repercussions of the increased cost. One example of such behavior, is the Fuel Poverty Strategy, a UK effort that targets environmental policy distributional challenges [34]. This plan aims at all households whose energy use surpasses 10% of their income to meet their heating demands, with a special focus on disadvantaged groups.

People's habits in consuming energy should be altered to fight against energy demand increase. Habits are defined as the intersection of knowledge (what to do), skill (how to do), and desire (want to do) [35]. Any plan that aims to change people's habits or their energy consumption patterns should combine all three. Thus, considering these criteria, the proposed energy cap and trade scheme penalizes heavy electricity consumers and compensates low energy consumers. The work conducted in [36,37] reveals that as one's income rises, so does one's energy use. This growth, however, is not uniform. Energy consumption growth remains flat as income grows at the lowest end of the income spectrum. Energy consumption begins to rise only after a particular low-income threshold is reached. Therefore, indirectly the suggested energy cap and trade targets consumers with high income without impacting households majorly with relatively low income, which respects the equitability concept in the distribution of the increased costs.

The Cap-and-Trade mechanism is well known in the greenhouse gases emissions reduction and carbon credit markets. A cap-and-trade scheme allows governments to issue a limited number of carbon dioxide emission permits each year, essentially putting a "cap" on overall pollution. Companies that are part of the program can acquire and sell permits amongst themselves [38]. Because the overall supply of permits is limited, some businesses may determine that reducing their emissions is more cost-effective than purchasing permits. For example, investing in more efficient technology, reducing energy waste, or switching to renewable energy are all options that are less expensive than buying permits. The cap is then reduced each year, resulting in fewer and fewer permits available each year. As permits become more expensive, many businesses opt to minimize their emissions rather than purchase allowances. Cap-and-Trade is considered more flexible than flat regulations since they simultaneously combine sustainability goals, covered by the cap, and economic efficiency, covered by the trading feature, and offer the potential to achieve both [39].

In a Cap-and-Trade scheme, allocation is the process of distributing allowances to participating or involved entities. The allocation process should be governed by rules to ensure the fair distribution of allowances. There are three basic ways of allocation; allocating based on historical data (grandfathering), allocating based on benchmarking, and auction [40]. Grandfathering offers participating entities allowances based on their historical data from a pre-defined base year or period. Grandfathering increases the scheme's feasibility by avoiding significant upfront costs. However, grandfathering as a mechanism of allocation tends to reward historically high consumers and necessitates additional allowances for future entrants. Comparatively, using a benchmarking allocation method, allowances are distributed based on performance metrics. Benchmarking rewards efficient and innovative solutions and makes it easier for newcomers to fit in [40]. The third allocation method is selling allowances, usually by auction. The auction method is advantageous in reflecting the real requirement of installations for allowances and offers participating entities an equal chance to procure allowances. Furthermore, it generates revenue for the regulator, which may then be used to fund additional initiatives. Therefore, the choice of the allocation method has a significant impact on the cost-efficiency of any Cap-and-Trade mechanism.

Therefore, analogously to the mechanism applied to fight against the increase in greenhouse gas (GHG) emissions, we propose a cap-and-trade model to limit the increase in energy demand while combining both the goals of sustainability and cost-effectiveness. The proposed model is based on a benchmarking allowance system. Each category of users or each sector, residential, commercial, healthcare, industrial, etc., is allocated a cap for the monthly energy consumption based on a benchmark study of the relevant market. Suppose the monthly energy consumption of the involved entity is below the monthly cap value. In that case, the entity will pay the usual energy rate, but the entity will have to pay an incremented rate for each energy unit consumed above the cap value. Nevertheless, efficient energy consumers might have a total monthly energy consumption way below the energy cap. Thus, the difference between their anticipated monthly energy consumption and the cap value can be tokenized and sold to heavy energy consumers at any rate between the usual rate and the incremented rate (Figure 3). Thus, the proposed model generates two incentives for consumers. The first incentive is for consumers to lower their energy consumption not to exceed the set cap value. The second incentive is to improve their energy efficiency to maximize the number of energy tokens traded and thus lower their monthly energy bill. The suggested selling algorithm is defined in Algorithm 1.

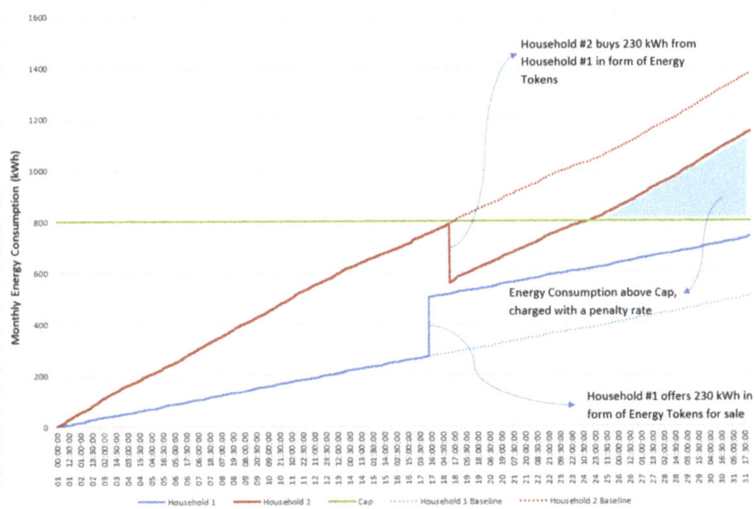

Figure 3. Energy Cap and Trade Energy Consumption Model.

However, the application of such a model can be very complex for utility companies and energy providers. Applying a cap-and-trade model for GHG emissions at the national level or large-scale companies might be feasible. Still, its implementation on a small retail scale with thousands and millions of consumers will undoubtedly prove to be a time-consuming, labor-intensive, challenging, and sophisticated process. For this reason, we propose using blockchain as a distributed ledger to govern the tokenization of traded energy. It handles payments, keeps an immutable record of all transactions, and manages the relationship between stakeholders: selsumers (consumers who are selling excess energy tokens), pursumers (consumers buying energy tokens), and the utility company.

The adoption of a blockchain P2P architecture for our energy cap-and-trade model offers several advantages. The anonymity of peers and security guaranteed by blockchain are among the most important features. Additionally, blockchain's greatest feature derives from the transparency of its distributed ledger shared by all participating nodes. It provides an unprecedented layer of accountability to any financial or business model, forcing all involved stakeholders to be accountable towards other involved parties. Moreover, the fact that each exchange of tokens is recorded on a blockchain creates an auditable trail of all transactions, which can improve security, prevent fraud, and verify the legitimacy of the traded asset. Finally, blockchain will help make the process more efficient and less costly by eliminating intermediaries or third parties.

Hence, the proposed blockchain-enabled energy Cap-and-Trade DSM mechanism is based on creating a monthly cap for consumers' energy consumption, allowing them to tokenize any unused energy, not exceeding the cap limit, and trading them with other interested buyers using a blockchain P2P energy trading platform.

Algorithm 1 Energy Cap and Trade

Initiate algorithm at time t
If t = end of billing period **then**
$\mathrm{R}(m,n)=0$
$t=0$
$i-1$
Goto 1
Else Goto 9
end if
If User = Seller **then**
Step 1: Seller Registration
Check if Seller Smart Meter is registered
Seller set γ_S^i and E_S^i
Offer is stored in matrix:$\mathrm{R}(1,i)=\gamma_S^i$ and $\mathrm{R}(2,i)=E_S^i$
Arrange matrix R from lowest to highest γ_s^i
$i=i+1$
End Seller Registration
Else
Step 2: Buyer Registration
Check if Buyer Smart Meter is registered
Seller set γ_B and E_B
Initialize $j=1$
If $j=i$ **then**
$t=t+1$
Goto 2
Else
If $\mathrm{R}(1,j)\leq\gamma_B$ **then**
If $\mathrm{R}(2,j)\geq E_B$ **then**
Buy from Seller the quantity E_B for γ_S^j
$\mathrm{R}(2,j)=\mathrm{R}(2,j)-E_B$
$t=t+1$
Goto 2
Else
Buy from Seller the quantity E_S^j
$\mathrm{R}(2,j)=0$
$E_B=E_B-E_S^j$
$j=j+1$
Goto 19
end if
Else
$j=j+1$
Goto 19
end if
End Seller Registration
end if
$t=t+1$
Goto 2

4.2. System Architecture and Functionality

To better understand how the proposed blockchain-enabled energy Cap-and-Trade DSM mechanism works, the architecture of the system, including all soft and hard components functionalities are detailed in this section.

The main target is to enable electric end-users to exchange electric energy tokens securely, in near real-time, and transparently. To achieve this, a Dapp will allow energy consumers and prosumers to participate easily and create their own decentralized and deregulated open P2P energy market.

A public decentralized virtual machine will operate as an autonomous agent that connects and matches selsumers (consumers willing to sell energy tokens) to pursumers (consumers willing to buy energy tokens) and conducts the financial settlement between them without the need for an aggregator, broker, or any type of intermediary. The decentralized virtual machine is nothing more than several smart contracts that take over the two major functions of today's energy retailers: billing and trading while keeping a transparent record of all executed transactions. These smart contracts are then paired with a user-friendly front-end application programming interface (API) that allows selsumers and pursumers to interact directly with the virtual machine. Thus, the virtual machine collects data, processes it, charges the pursumer for the bought electricity based on an agreed-upon rate, and pays the selsumers for the sold energy tokens. It manages the financial settlement between the two parties' digital wallets in an autonomous, secure, and completely transparent manner that doesn't require the intervention of any trusted third party. The proposed architecture is shown in Figure 4.

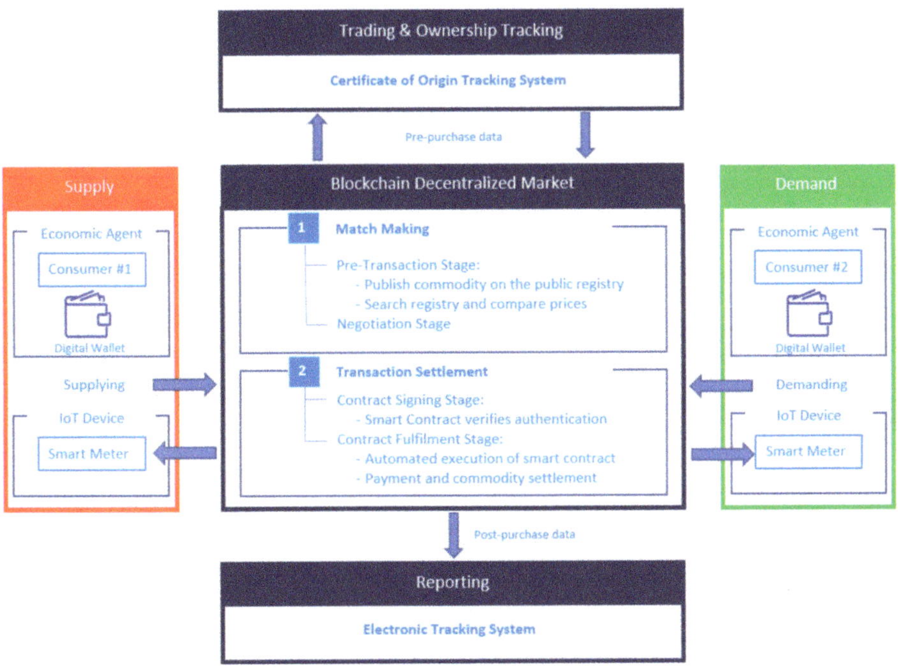

Figure 4. Trading model architecture.

The functionality of the model and the different steps of the process can be summarized as follows:

- Data collection is provided by IoT devices connected to Smart Meters
- Smart contracts manage the processing and storage of data on the blockchain network in an autonomous way
- Transactions between selsumers and pursumers are conducted in near real-time using smart contracts and via the Dapp itself.
- Data is recorded and encrypted using cryptography, guaranteeing its immutability and authenticity.
- Smart Contracts automate and ensure that all system functionalities, i.e., billing, trading, and reporting, are carried out correctly without the risk of human error.

The Solidity code, included in the Appendix A, represents the smart contract trading.sol that allows to create buy/sell orders and log them. In addition, this smart contract calls for another smart contract called Registeredusers.sol that governs the registration of smart meters for consumers who are willing to participate in the program and allocate an address for each smart meter.

5. Case Study

To test the functionality and effectiveness of the proposed DSM mechanism, we propose using a simulation model based on historical data from two hundred households and applying the proposed DSM's logic, outlined in Algorithm 1, to the adopted use case. However, to achieve that, the proposed selected use case model is defined in the first part of this section. In the second part, the simulation model and algorithm are transformed into a mathematical model that allows one to test the DSM mechanism under several scenarios. The ECT DSM mechanism is tested under different scenarios to evaluate its dynamicity and how the outcome will change when the system variables are modified. Finally, the simulation results, under different scenarios, are presented and analyzed.

5.1. Selected Model

Two hundred (200) households, randomly taken from a sample of housing units in the Residential Energy Consumption Survey in the United States' Midwest region, are considered to test the proposed energy demand-side management model. The chosen sample is part of the work conducted in [41]. The curves shown in Figure 5 represent the total monthly energy consumption of the 200 households registered over four months of the year (January, May, July, and October).

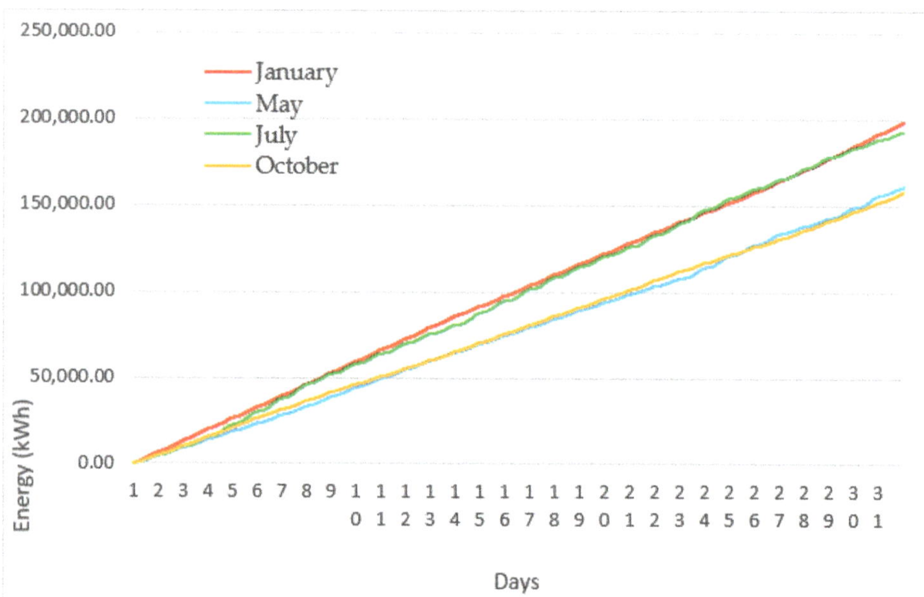

Figure 5. Daily cumulative total energy consumption of the 200 households.

The proposed Cap-and-Trade demand-side management model is verified for the above-defined energy consumption capping applications under different scenarios. In addition, the blockchain-enabled Cap-and-Trade DSM model is tested over four months of the year (January, May, July, and October), each representing a season to account for the energy seasonal variations. Table 1 shows the distribution of the households' monthly energy consumption for the selected months of the year.

Table 1. Distribution of households' monthly energy consumption.

Monthly Energy Consumption (kWh)	Number of Households			
	January	May	July	October
0–600	29	31	17	43
601–800	61	80	53	71
801–1000	54	59	58	60
1001–1200	11	21	32	14
1201–1400	9	6	22	6
1401–1600	15	1	9	3
1601–1800	9	0	5	1
1801–2000	3	2	2	1
2001–2200	2	0	2	1
>2200	7	0	0	0

The considered cases are validated according to three different strategies: aggressive, intermediate, and soft. The three strategies differ in the set cap value, cap charge rate, and average energy tokens selling price. The defined values for each simulation strategy are given in Table 2.

Table 2. Defined values for simulation strategies.

Parameter	Aggressive Strategy	Moderate Strategy	Soft Strategy
Cap Value as percentile of the Population's Monthly Energy Consumption	15 percentiles	30 percentiles	45 percentiles
Cap Charge	$0.50	$0.20	$0.12
Average Energy Token Price	$0.20	$0.15	$0.10

The outcomes of the simulation models are evaluated based on three criteria:

- The variation of the consumer's monthly electricity bill between the baseline period and under the cap-and-trade DSM program
- The variation of the total consumed monthly electric energy of the 200 households between the baseline period and under the cap-and-trade DSM program
- The variation of the total collected monthly electric bills by the utility between the baseline period and under the cap-and-trade DSM program

An average price for the traded energy tokens was considered to simplify the simulation. Optimization of the transactions between the selsumers and the pursumers is applied, which means that they sold energy tokens are equally distributed among all pursumers. Additionally, it is assumed that all available excess energy tokens are sold. Such an assumption is justified because, in the case of a large grid, there will always be a pursumer willing to buy the offered energy token at a lower price than the electricity tariff per kWh. In other words, if the supply in energy tokens is higher than the demand, all pursumers are served, and in case the demand is higher than the supply, all selsumers are served. This hypothesis defines the optimal case of the considered model.

Furthermore, it is considered that a certain percentage of the population will reduce their energy consumption in response to the applied cap and the surcharge for energy consumed above the monthly cap value. This assumption is defined by two the factors α and β, respectively the percentage of households reducing their energy consumption and the percentage by which the monthly energy consumption of these households is reduced. For simulation purposes, α is set at 20% and β at 25%.

5.2. Energy Cap and Trade Formulation

A mathematical model serves to test the above-defined algorithm. Accordingly, as previously described, after the application of the Energy Cap and Trade (ECT) demand-side management (DSM) program, it is expected that a certain percentage $\propto$ of the households will automatically respond to the program and reduce their energy consumption by a certain ratio defined as β_i. Moreover, it is assumed that the households that will respond to the applied ECT DSM program are part of the households whose monthly electrical energy consumption E_i exceeds the set cap value K. Households with $E_i < K$ will not have any interest in decreasing their monthly electricity consumption. For simulation purposes and to keep track of the number of households reducing their monthly electrical energy consumption, a counter n was created. The counter n will be increased by 1 at each iteration if the subject household i have a benchmarked monthly electrical energy consumption above the set cap and if the maximum number of expected households to reduce their energy consumption, defined by the percentage $\propto$, is not reached. The counter n will continue to increase until it is equal to the round down of $\propto \times N$ which represents the total number of households to reduce their monthly electrical energy consumption in response to the ECT DSM program.

$$n_i = q_i \times \left[\frac{[\alpha N] - n_{i-1}}{[\alpha N]} \right] + n_{i-1} + q_i \tag{1}$$

Hence, to identify the households that have decreased their monthly electrical energy consumption E_i, in response to the ECT DSM program, the index p_i is defined. This index equals 1 if the subject household i has reduced its monthly energy consumption and 0 if not. The index p_i is calculated using Equation (2):

$$p_i = \left(1 + \left[\frac{[\alpha N] - n_i}{[\alpha N]} \right] \right) \times q_i \tag{2}$$

$i \in N$ set of all households
$t \in T$ set of time intervals

$$\begin{cases} q_i = 1\ if\ E_i \leq K \\ q_i = 0\ Else \end{cases}$$

$$\begin{cases} p_i = 1\ if\ \textit{Household i reduced his electrical energy consumption} \\ p_i = 0\ Else \end{cases}$$

N: Population (in our case, the 200 households)
E_i: Household monthly electrical energy consumption in kWh
K: Defined electrical energy cap value in kWh
$\propto$: Percentage of households to respond to the new rate structure and reduce their monthly electrical energy consumption

The simulation aims to analyze the energy demand and economic impacts of the proposed ECT DSM program on the consumer and the utility company or energy provider. Two factors assess the effects on the electricity distribution company. The first one is the total energy consumed by the 200 studied households, and the second factor is the sum of the collected electricity bills from the 200 households. As for the consumers, the main criteria used for the evaluation of the model is the new monthly electric bill U'_i.

The new total energy demand E'_T at the utility-scale is given by Equation (3):

$$E\prime_T = \sum_{i \in N} (1 - p_i).E_i + \sum_{i \in N} p_i.\beta_i.E_i \tag{3}$$

The reduction in energy demand at the utility-scale is given by Equation (4):

$$R_E = \frac{E_T - E\prime_T}{E_T} \tag{4}$$

E_T: Baseline total monthly electrical energy consumption of the population in kWh
$E\prime_T$: Total monthly electrical energy consumption of the population in kWh after implementation of the Cap-and-Trade scheme
β_i: Percentage of monthly electrical energy reduction for each household
R_E: Total monthly electrical energy consumption reduction

The baseline monthly electric bill for household *i*, is given by Equation (5):

$$U_i = \gamma_U \times E_i^C \tag{5}$$

The new monthly electric bill for household *i*, is given by Equation (6):

$$U\prime_i = \gamma_U \times \left[q_i.K + (1 - q_i).E_i^C + E_i^S\right] - \gamma_S.E_i^S + \gamma_B.E_i^B + \gamma_C.a_i.\left(E_i^C - K - E_i^B\right) \tag{6}$$

U_i: Baseline total monthly electricity utility bill for household *i*
$U\prime_i$: Total monthly electricity bill after implementation of the Cap-and-Trade scheme for household *i*
γ_U: Utility electricity rate ($/kWh)
γ_S: Rate for sold electrical energy tokens ($/kWh)
γ_B: Rate for bought electrical energy tokens ($/kWh)
E_i^C: Total monthly electrical energy consumed by household *i* in kWh
E_i^S: Total monthly electrical energy sold by household *i* in kWh
E_i^B: Total monthly electrical energy bought by household *i* in kWh

$$\begin{cases} a_i = 1 \; if \; E_i^C - K > E_i^B \\ a_i = 0 \; Else \end{cases}$$

The reduction in the monthly electric bill for household *i*, is given by Equation (7):

$$R_i^B = \frac{U_i - U\prime_i}{U_i} \tag{7}$$

R_i^B: Total monthly reduction or increase on electricity bill of household *i*.

5.3. Simulation Results

As previously detailed, the simulation is conducted over four different months of the year to account for the seasonal effect on the energy demand. The obtained results are analyzed considering two different perspectives. The first is the utility company's perspective, and the second one is the consumer's perspective. From a utility perspective, the main target is to reduce the monthly electric energy consumption without heavily impacting the monthly turnover. The obtained results show that even when incorporating the selsumers as energy sellers and competitors to the utility company in terms of selling electricity at a lower rate to penalized pursumers, the monthly turnover of the utility is either comparable to the baseline (period without implementation of any DSM program), for the soft and moderate scenario, or exceeding the baseline for the aggressive scenario (refer to Figure 6).

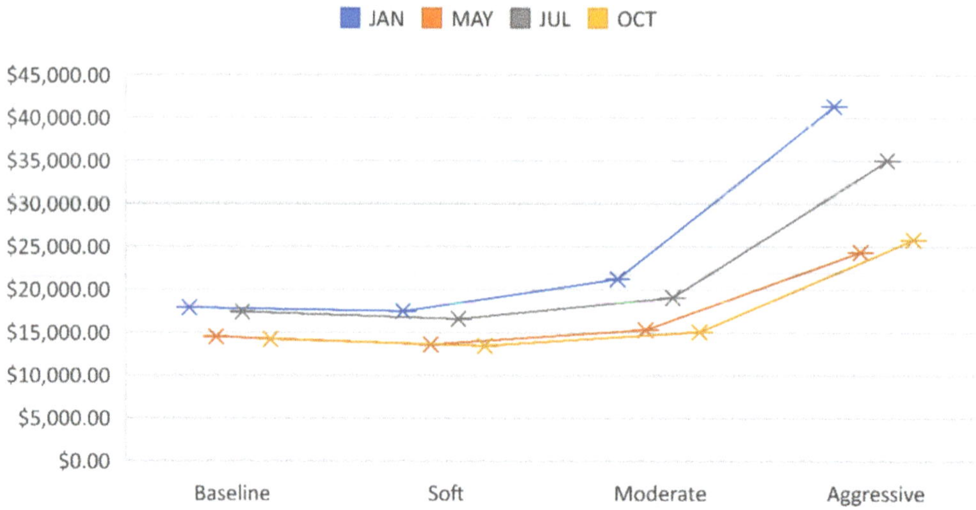

Figure 6. Utility monthly turnover in USD.

From the consumer's perspective, the simulation results show that the seasonal effect lightly affects the cap value (refer to Figure 7). Still, the same cannot be said for the traded energy tokens. Figure 8 shows a remarkable increase in the number of traded kWh during January and July. The main reason for this development in the number of traded kWh during January and July is the increase in heating and cooling loads during those two months. The high demand for heating and cooling energy is directly reflected in the monthly electric bills, which means that under a Cap-and-Trade scheme, heavy consumers will exceed the monthly cap value and consequently be penalized for the excess energy consumed. Thus, in the proposed ECT DSM program, those heavy consumers will try to buy energy tokens to minimize the additional charge applied to their monthly bills, which verifies the increase in the number of traded energy tokens.

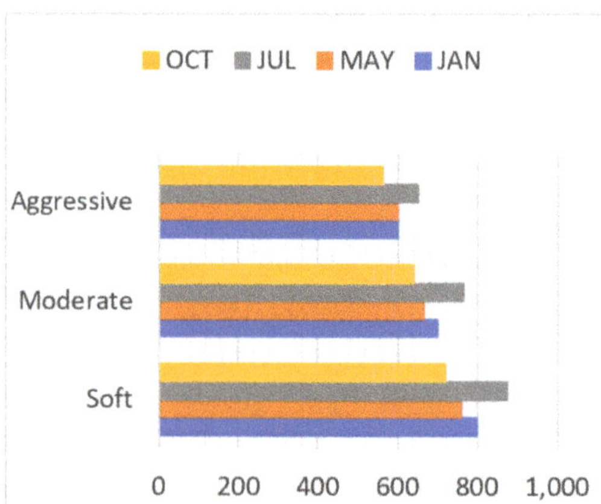

Figure 7. Monthly Cap Values in kWh.

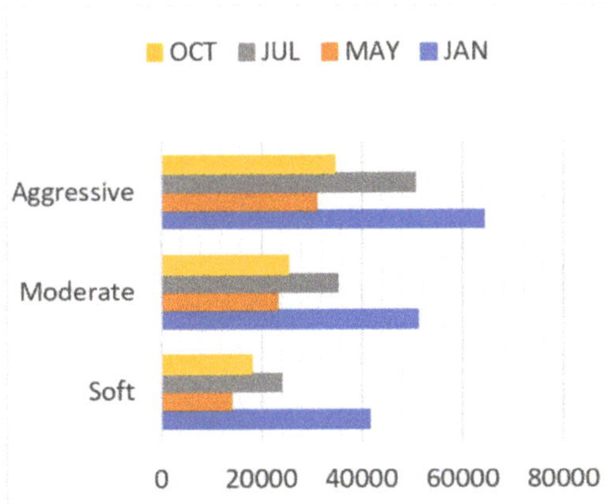

Figure 8. Total traded monthly Energy Credits in kWh.

Besides that, it is entirely reasonable to see that the number of selsumers is high in a soft scenario, where the cap value is relatively high, allowing consumers to earn a considerable number of tradeable energy credits. On the other hand, identically, the number of pursumers is low, whereas these numbers are reversed under the conditions of an aggressive scenario, as shown in Figures 9 and 10. Additionally, the proposed ECT DSM model minimizes the impact of free riders while rewarding efficient consumers and penalizing heavy consumers. Figures 11 and 12 show respectively the average monthly reductions on the selsumers' electricity bills and the average monthly increase on the pursumers' electricity bills. Under the conditions of the soft scenario, selsumers can achieve a reduction of approximately 9% on their monthly bills whereas the increase on pursumers' monthly bills is between 3% and 7%. With an intermediate scenario, the selsumers' reduction is between 16% and 18.6%, whereas the pursumers' increase is between 23% and 32.7%. Subsequently, the aggressive scenario allows selsumers to achieve higher reductions on their monthly electric bills and imposes higher penalties on heavy consumers where the increase on the monthly electric bill can reach 140% during January. On the other side, the applied ECT DSM, based on the previously detailed assumptions, can achieve a 6.04% to 8.79% reduction in the total monthly electrical energy consumption (refer to Figure 13).

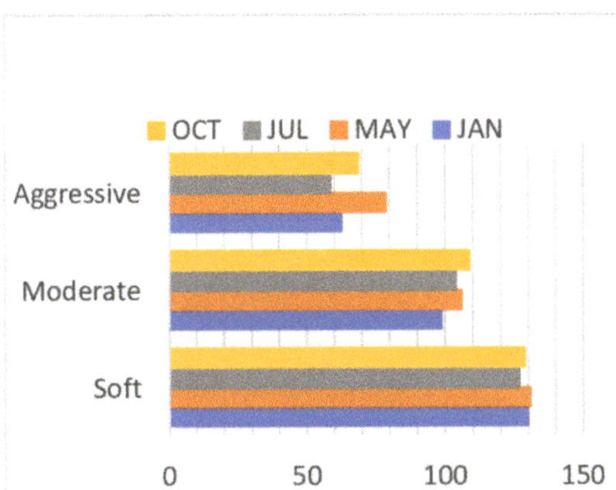

Figure 9. Number of Selsumers per month.

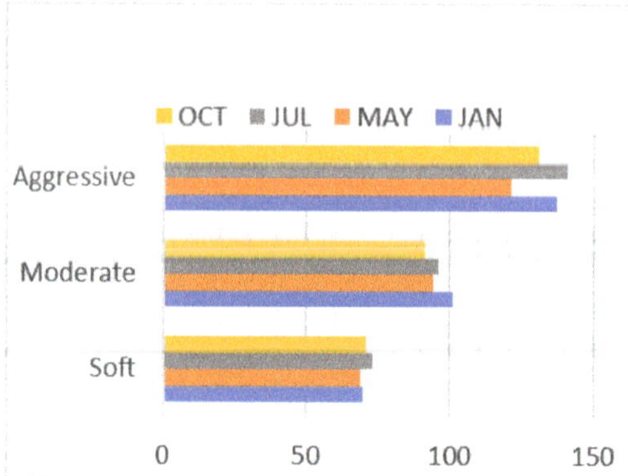

Figure 10. Number of Pursumers per month.

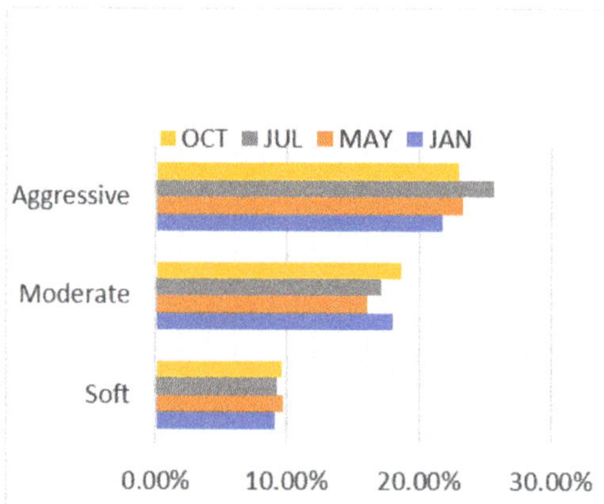

Figure 11. Average monthly reduction on selsumers' bills.

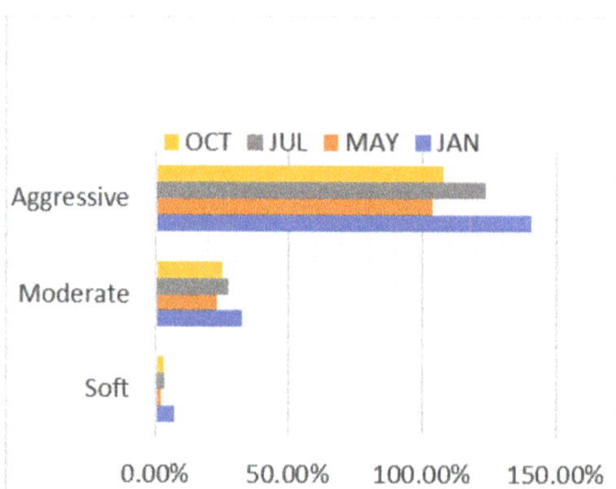

Figure 12. Average monthly increase on pursumers' bills.

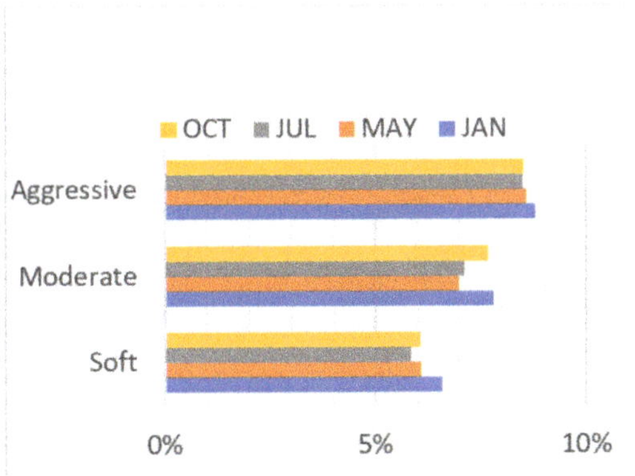

Figure 13. Total monthly energy reduction.

Lastly, Figure 14 compares the average monthly bills of prosumers under the different scenarios that the utility can adopt to minimize the yearly increase in electricity demand. The baseline scenario is the business-as-usual scenario where no demand-side management program is applied. The second scenario is the "Baseline + Cap" scenario. This scenario simulates the application of a cap value for the monthly energy consumption and penalizes consumers that exceed this cap without offering them any trading option. In that case, even though it is assumed that 20% of households will automatically reduce their monthly electricity consumption by 25% in response to the application of the cap value, the average monthly electricity bills have nearly doubled. This proves that such a model can have a severe negative impact on consumers and can be considered an unembellished DSM model. The other three scenarios represent the implementation of the ECT DSM program with respectively aggressive, intermediate, and soft conditions. As shown in Figure 14, the aggressive scenario can have similar adverse outcomes as the "Baseline + Cap" scenario. In contrast, the intermediate and soft scenarios can maintain an average monthly bill comparable to the baseline model to decrease the overall monthly energy demand, which can be considered a win-win situation for both the utility company and the consumers.

5.4. Sensitivity Analysis

As defined in Section 5.1, several parameters are considered in the proposed blockchain-enabled C&T energy trading model design. Thus, it is essential to determine and analyze the set of independent variables or inputs that affect the outcomes of our model. The parameters or variables identified as having an impact on the outcome of our model are defined as follows:

- Cap value
- Energy rates
- Percentage of people reducing their energy consumption
- The energy consumption reduction rate

Indeed, the set cap value is the most critical variable that can significantly impact the performance of the presented model. Hence in this context, Figure 15 represents the sensitivity analysis or the "What if" analysis that illustrates the impact of the set cap value on the model's outcomes.

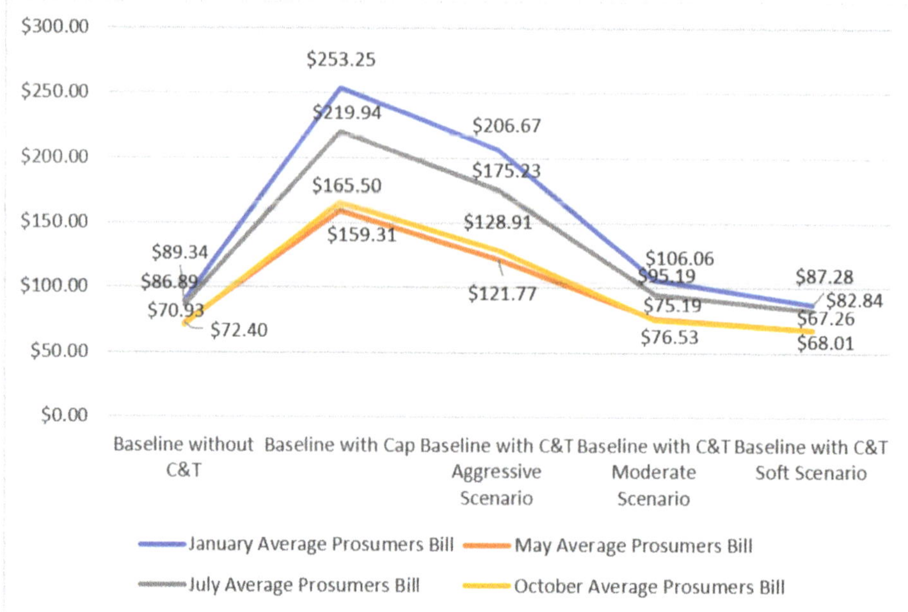

Figure 14. Average prosumer monthly bill in USD.

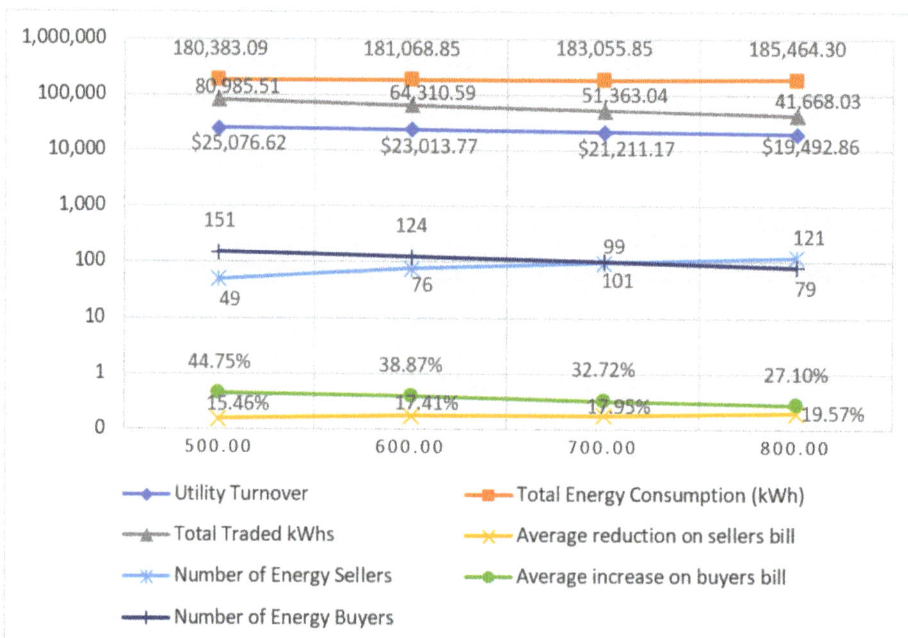

Figure 15. Energy Cap's impact on the model's outcomes.

5.5. Other Application: Peak Load Shaving Use Case

The presented and simulated model, in Section 5, is not the only application for the proposed blockchain decentralized application. The developed blockchain-enabled C&T

energy trading model can also be used for peak load shaving. The suggested peak load shaving model can operate as follows:

- Based on historical data of at least two complete cycles for each household, a cap value will be defined for the peak power during the peak time for every billing cycle
- When the power during the peak period for each household is lower than the set cap, the consumer earns energy tokens equivalent to the amount of energy reduction below the set cap value (as shown in Figure 16)
- The earned tokens can be exchanged with the utility company or the independent system operator (ISO) to pay the monthly energy bill or can be sold to other consumers
- Pursumers can benefit from the bought tokens to pay for the extra charge applied to their energy consumption during on-peak periods
- The complete energy trading process will be similar to the one detailed in Section 4.2

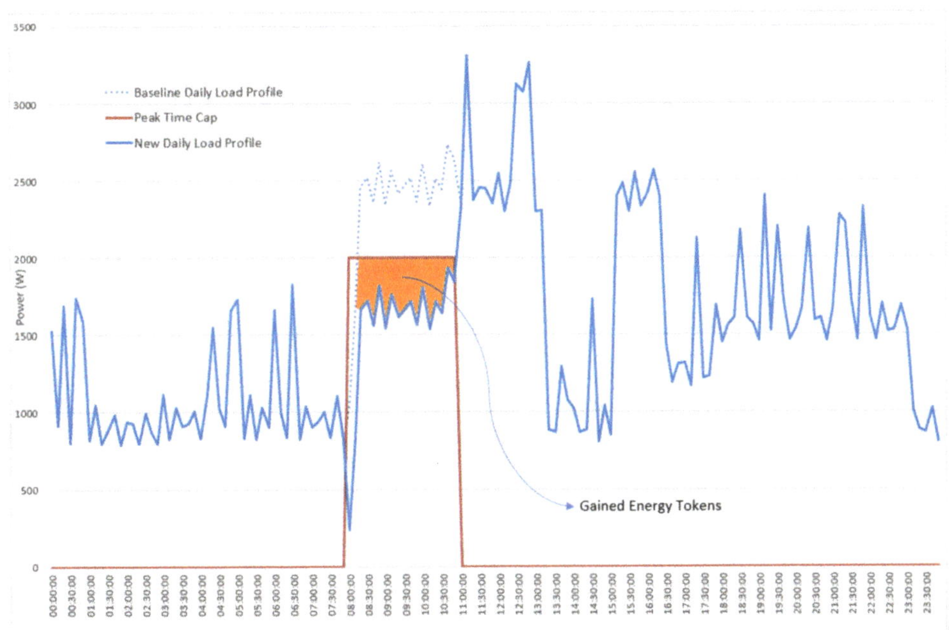

Figure 16. Energy Cap and Trade Peak Load Shaving Model.

Such a model can be used as a demand response program to incentivize consumers to lower their power peak demand during peak-time periods in exchange for tradable energy tokens and thus to contribute to the reduction of the daily peak load and minimizing the need for alternative means of power generation or a large spinning reserve.

6. Conclusions

This article presents a new DSM model that can help minimize the yearly increase in electricity demand, inspired from the concept of Cap and Trade previously applied to fight against the rise in carbon emissions, and benefiting from the emergence of blockchain technology and the tokenization of energy assets. The suggested ECT DSM concept minimizes the free-rider and rebound effects usually faced with conventional DSM programs. It is fundamentally based on integrating consumers as the main stakeholders in the energy supply and procurement process. It rewards efficient consumers and penalizes heavy consumers. Furthermore, it creates an open market for energy trading based on P2P energy trading. The prices are not controlled by a central entity, i.e., the utility company, but instead governed by the general rules of a deregulated supply and demand. The central concept is that the average energy consumer no longer acts as a passive user, located at the end of the energy supply chain, but rather as an integral part of a future smart digital grid, where more and more distributed energy resources (DERs) are being integrated. Such a consumer-driven energy system can benefit all stakeholders, as proven by the presented ECT DSM model.

The proposed DSM program was tested using 200 households as a test bank, and the model was simulated over four months to account for any seasonal effect. Moreover, three different scenarios, soft, intermediate, and aggressive, were taken into consideration to validate the most appropriate criteria to be applied to get the best-equilibrated outcomes for both the utility and the consumer. The obtained results proved the effectiveness of the model, especially for the soft and intermediate scenarios where the utility company can achieve a reduction between 6.04% and 7.8% on the total monthly energy consumption, without any considerable change in its total monthly turnover and without highly affecting the average monthly bills of the consumers. Furthermore, the sensitivity analysis showed the importance of properly selecting the cap value to achieve an optimal result for both the utility company and the consumers. This value has to be modified every few years to account for new market needs and control the energy demand.

Nevertheless, cash incentives alone are not sufficient to create a good DSM program. Still, they should be complemented by real-time feedback from energy meters and an updated billing system that considers an aggregate bill per consumer rather than individual bills based on single points of delivery (PoD). Such a model can help to protect the system against free riders. For example, a person or a family owning and occupying more than one household will not benefit from a double quota of energy credits and can be billed fairly for their aggregated monthly energy consumption as any other family who owns just one household.

The presented blockchain-enabled energy Cap-and-Trade model can be applied to several applications such as peak load shaving. Moreover, it can offer a new perspective for the utility's relationship with its customers, specifically for DSM projects. However, this model might be confronted by several challenges, such as the need for new legislation and regulations. Additionally, it is crucial to create awareness amongst people to embrace this change and appropriately interact with the system and the new technological innovation to extract the maximum benefits from the program. Otherwise, such a model might backfire and lead to negative results.

The work conducted in this article can serve as a test-bed to evaluate the proposed DSM mechanism for new applications, commercial or industrial, and various microgrids, to optimize the model's outcome especially by combining the impact of other distributed energy resources. Additionally, it will be beneficial to test the ECT DSM model under different rate structures such as Real-Time Pricing (RTP) and Time of Use (ToU), as well as applying the rules of open market demand and supply on the actual price of the traded energy tokens.

Author Contributions: Conceptualization, A.A.; methodology, A.A.; software, A.A.; validation, A.I., M.G. and H.I.; formal analysis, A.A.; investigation, A.A.; resources, A.A.; data curation, A.A.; writing—original draft preparation, A.A. and A.I.; writing—review and editing, A.A., M.G. and A.I.; visualization, A.A.; supervision, H.I., M.G. and A.I.; project administration, H.I., M.G. and A.I. All authors have read and agreed to the published version of the manuscript.

Funding: This research received no external funding.

Institutional Review Board Statement: Not applicable.

Informed Consent Statement: Not applicable.

Data Availability Statement: Data available on request due to restrictions. The data presented in this study are available on request from the corresponding author. The data are not publicly available since they are part of an ongoing project, and full data will be publicly published once the project is finalized.

Conflicts of Interest: The authors declare no conflict of interest.

Appendix A

The following code is written using Solidity a programming language used for coding smart contracts on the Ethereum platform. The smart contract trading.sol allows to create buy/sell orders and log them on the blockchain's shared ledger. In addition, this smart contract calls for another smart contract called Registeredusers.sol that governs the registration of smart meters for consumers who are willing to participate in the program and allocate an address for each smart meter.

Energy Trading Smart Contract—Solidity Code

```
Pragma solidity >=0.4.22 <0.6.0;
import "./Registeredusers.sol";
// Contract Registeredusers.sol registers a wallet address for each smart meter to enable the registered user to trade energy on blockchain
// Contract trading.sol manages and records energy buying or selling transactions
contract trading is Registeredusers {
    struct SellOrder {
    address seller;
        uint32 price;
        uint64 energy;
    uint64 timestamp;
}
struct BuyOrder {
    address seller;
    uint32 price;
    uint64 energy;
    address meterAddress;
    uint64 timestamp;
}
// Add registry address
address public utilityreg = 0x89205A3A3b2A69De6Dbf7f01ED13B2108B2c43e7;

SellOrder[] public sellOrders;
BuyOrder[] public buyOrders;

//stores the amount of energy supplied by the seller
BuyOrder[] public sellerEnergy;
mapping(address => uint) public sellIndex;
mapping(address => uint) public buyIndex;
event sellEvent(address indexed seller, uint32 indexed price, uint64 energy);
event buyEvent(address indexed seller, uint32 price, uint64 energy, address meterAddress);

function sellEnergy(uint32 aprice, uint64 aenergy, uint64 atimestamp) onlyRegisteredUsers public
{
// record the sell order
uint idx = sellIndex[msg.sender];
sellOrders.push(SellOrder({
    seller: msg.sender,
    price: aprice,
    energy: aenergy,
    timestamp: atimestamp
    }));
emit sellEvent(sellOrders[idx].seller, sellOrders[idx].price, sellOrders[idx].energy);
}
```

```
function buyEnergy(address aseller, uint32 aprice, uint64 aenergy, address mAddress, uint64
atimestamp) onlyRegisteredUsers public {
// find offer by seller (aseller)
uint idx = sellIndex[aseller];
require(0x0 != idx);
// check if any matching offer is available
if ((sellOrders.length > idx) && (sellOrders[idx].seller == aseller)) {
// check if price is matching
    require(sellOrders[idx].price == aprice);
    buyIndex[msg.sender] = buyOrders.length;
// record the buyer's choice
    buyOrders.push(BuyOrder({
        seller: aseller,
        price: aprice,
        energy: aenergy,
        meterAddress: mAddress,
        timestamp: atimestamp
        }));
emit buyEvent(aseller, aprice, aenergy, mAddress);
//checks if the consumer bought from the seller and stores it
// The array sellerEnergy in trading.sol stores the energy transaction
require(buyOrders[idx].seller == utilityreg);
sellerEnergy.push(BuyOrder({
    seller: aseller,
    price: aprice,
    energy: aenergy,
    meterAddress: mAddress,
    timestamp: atimestamp
    }));
} else {
revert();
}
}
}
```

References

1. Aoun, A.; Ibrahim, H.; Ghandour, M.; Ilinca, A. Supply Side Management vs. Demand Side Management of a Residential Microgrid Equipped with an Electric Vehicle in a Dual Tariff Scheme. *Energies* **2019**, *12*, 4351. [CrossRef]
2. U.S. Department of Energy (DOE). *Benefits of Demand Response in Electricity Markets and Recommendations for Achieving Them: A Report to the United States Congress Pursuant to Section 1252 of the Energy Policy Act of 2005*; U.S. DOE: Washington, DC, USA, 2006.
3. Lee, E.; Jang, D.; Kim, J. A Two-Step Methodology for Free Rider Mitigation with an Improved Settlement Algorithm: Regression in CBL Estimation and New Incentive Payment Rule in Residential Demand Response. *Energies* **2018**, *11*, 3417. [CrossRef]
4. Garbi, A.; Malamou, A.; Michas, N.; Pontikas, Z.; Doulamis, N.; Protopapadakis, E.; Mikkelsen, T.N.; Kanellakis, K.; Baradat, J.-L. BENEFFICE: Behaviour Change, Consumption Monitoring and Analytics with Complementary Currency Rewards. *Proceedings* **2019**, *20*, 12. [CrossRef]
5. Gillingham, K.; Rapson, D.; Wagner, G. The Rebound Effect and Energy Efficiency Policy. *Rev. Environ. Econ. Policy* **2016**, *10*, 68–88. [CrossRef]
6. Vivanco, D.F.; Kemp, R.; van der Voet, E. How to deal with the rebound effect? A policy-oriented approach. *Energy Policy* **2016**, *94*, 114–125. [CrossRef]
7. Abe, R.; Taoka, H.; McQuilkin, D. Digital Grid: Communicative Electrical Grids of the Future. In *IEEE Transactions on Smart Grid*; IEEE: Piscataway, NJ, USA, 2011; pp. 399–410.
8. Mika, B.; Goudz, A. Blockchain-technology in the energy industry: Blockchain as a driver of the energy revolution? With focus on the situation in Germany. *Energy Syst.* **2020**, *12*, 285–355. [CrossRef]
9. Thakkar, A. How Blockchain and Peer-to-Peer Energy Markets Could Make Distributed Energy Resources More Attractive. Bachelor's Thesis, Duke University, Durham, NC, USA.
10. Deloitte. Blockchain Application in Energy Trading. Available online: https://www2.deloitte.com/content/dam/Deloitte/global/Documents/Energy-and-Resources/gx-Blockchain-applications-in-energy-trading.pdf (accessed on 20 September 2021).

11. World Energy Council. "The Developing Role of Blockchain", White Paper in collaboration with Pricewaterhouse Coopers PwC. Available online: https://www.worldenergy.org/assets/downloads/Full-White-paper_the-developing-role-of-blockchain.pdf (accessed on 6 December 2021).
12. Babich, V.; Hilary, G. OM Forum—Distributed Ledgers and Operations: What Operations Management Researchers Should Know About Blockchain Technology. *Manuf. Serv. Oper. Manag.* **2020**, *22*, 223–240. [CrossRef]
13. Gaur, G.; Mehta, N.; Khanna, R.; Kaur, S. Demand side management in a smart grid environment. In Proceedings of the 2017 IEEE International Conference on Smart Grid and Smart Cities (ICSGSC), Singapore, 23–26 July 2017; pp. 227–231.
14. Di Santo, K.G.; Di Santo, S.; Monaro, R.M.; Saidel, M.A. Active demand side management for households in smart grids using optimization and artificial intelligence. *Measurement* **2018**, *115*, 152–161. [CrossRef]
15. Antonopoulos, I.; Robu, V.; Couraud, B.; Kirli, D.; Norbu, S.; Kiprakis, A.; Flynn, D.; Elizondo-Gonzalez, S.; Wattam, S. Artificial intelligence and machine learning approaches to energy demand-side response: A systematic review. *Renew. Sustain. Energy Rev.* **2020**, *130*, 109899. [CrossRef]
16. Vahidi, B.; Dadkhah, A. New Demand Response Platform with Machine Learning and Data Analytics. In *Demand Response Application in Smart Grids*; Springer: Singapore, 2020; pp. 113–137.
17. Hajizadeh, A.; Hakimi, S.M. Blockchain in decentralized demand-side control of microgrids. In *Blockchain-Based Smart Grids*; Academic Press: Cambridge, MA, USA, 2020; pp. 145–167.
18. Alladi, T.; Chamola, V.; Rodrigues, J.J.P.C.; Kozlov, S.A. Blockchain in Smart Grids: A Review on Different Use Cases. *Sensors* **2019**, *19*, 4862. [CrossRef]
19. Noor, S.; Yang, W.; Guo, M.; van Dam, K.H.; Wang, X. Energy Demand Side Management within micro-grid networks enhanced by blockchain. *Appl. Energy* **2018**, *228*, 1385–1398. [CrossRef]
20. Afzal, M.; Huang, Q.; Amin, W.; Umer, K.; Raza, A.; Naeem, M. Blockchain Enabled Distributed Demand Side Management in Community Energy System with Smart Homes. *IEEE Access* **2020**, *8*, 37428–37439. [CrossRef]
21. Wu, X.; Duan, B.; Yan, Y.; Zhong, Y. M2M Blockchain: The Case of Demand Side Management of Smart Grid. In Proceedings of the 2017 IEEE 23rd International Conference on Parallel and Distributed Systems (ICPADS), Shenzhen, China, 15–17 December 2017; pp. 810–813.
22. Khatoon, A.; Verma, P.; Southernwood, J.; Massey, B.; Corcoran, P. Blockchain in Energy Efficiency: Potential Applications and Benefits. *Energies* **2019**, *12*, 3317. [CrossRef]
23. Nakamoto, S. Bitcoin: A Peer-to-Peer Electronic Cash System. 2008. Available online: https://bitcoin.org/bitcoin.pdf (accessed on 20 September 2021).
24. Mougayar, W. *The Business Blockchain: Promise, Practice, and Application of the Next Internet Technology*; John Wiley and Sons: Hoboken, NJ, USA, 2016.
25. German-Mexican Energy Partnership and Florence School of Regulation. Blockchain Meets Energy—Digital Solu-tions for a Decentralized and Decarbonized Sector. June 2019. Available online: https://fsr.eui.eu/wp-content/uploads/Blockchain_meets_Energy_-_ENG.pdf (accessed on 6 December 2021).
26. Maher, A.; Moorsel, A. Blockchain based smart contracts: A systematic mapping study. *arXiv* **2017**, arXiv:1710.06372.
27. Detecon Consulting. The Advantages of 'Blockchain Smart Contracts'. 12 December 2012. Available online: https://www.detecon.com/en/knowledge/advantages-blockchain-smart-contracts (accessed on 20 September 2021).
28. Seyfang, G. *The New Economics of Sustainable Consumption*; Palgrave Macmillan: London, UK, 2009.
29. Scholl, G.; Rubik, F.; Kalimo, H.; Biedenkopf, K.; Söebech, Ó. Policies to promote sustainable consumption: Innovative approaches in Europe. *Nat. Resour. Forum* **2010**, *34*, 39–50. [CrossRef]
30. OECD. Promoting Sustainable Consumption: Good Practices in OECD Countries. 2008. Available online: https://www.oecd.org/greengrowth/40317373.pdf (accessed on 20 September 2021).
31. OECD. The Distributional Effects of Environmental Policy. 2006. Available online: https://www.oecd.org/env/tools-evaluation/36830749.pdf (accessed on 20 September 2021).
32. CEDD. Préservation de L'environnement, Équité Et Accès Aux Services Essentiels. Conseil Économique Pour Le Développement Durable. 2011. Available online: https://www.ecologique-solidaire.gouv.fr/sites/default/files/CEDD%20-%20Pr%C3%A9servation%20de%20l%E2%80%99environnement%2C%20%C3%A9quit%C3%A9%20et%20acc%C3%A8s%20aux%20services%20essentiels.pdf (accessed on 20 September 2021).
33. UK Department of Energy and Climate. Available online: www.decc.gov.uk/en/content/cms/funding/fuel_poverty/strategy/strategy.aspx (accessed on 20 September 2021).
34. Khandker, S.R.; Barnes, D.F.; Samad, H.A. *Welfare Impacts of Rural Electrification: A Case Study from Bangladesh*; World Bank Policy Research Working Paper Series; The World Bank: Washington, DC, USA, 2009.
35. Lankton, N.K.; Wilson, E.V.; Mao, E. Antecedents and determinants of information technology habit. *Inf. Manag.* **2010**, *47*, 300–307. [CrossRef]
36. Khandker, S.R.; Barnes, D.F.; Samad, H.A. *Energy Poverty in Rural and Urban India: Are the Energy Poor Also Income Poor? World Bank Policy Research Working Paper No. 5463*; The World Bank: Washington, DC, USA, 2010. [CrossRef]
37. Residential Energy Consumption Survey (RECS); US Department of Energy, Energy Information Administration. 2009. Available online: http://www.eia.gov/consumption/residential/ (accessed on 20 September 2021).
38. Goulder, L.H.; Schein, A.R. Carbon taxes versus cap and trade: A critical review. *Clim. Chang. Econ.* **2013**, *4*, 1350010. [CrossRef]

39. Wang, M.; Zhou, P. Impact of Permit Allocation on Cap-and-trade System Performance under Market Power. *Energy J.* **2020**, *41*. [CrossRef]
40. Groenenberg, H.; Blok, K. Benchmark-based emission allocation in a cap-and-trade system. *Clim. Policy* **2002**, *2*, 105–109. [CrossRef]
41. Muratori, M. *Impact of Uncoordinated Plug-In Electric Vehicle Charging on Residential Power Demand—Supplementary Data*; National Renewable Energy Laboratory-Data (NREL-DATA): Golden, CO, USA, 2017.

energies

MDPI

Article

Developing Novel Technologies and Services for Intelligent Low Voltage Electricity Grids: Cost–Benefit Analysis and Policy Implications

Alemu Moges Belay [1,*], Sanket Puranik [1], Ramon Gallart-Fernández [2], Heidi Tuiskula [1], Joaquim Melendez [3], Ilias Lamprinos [4], Francisco Díaz-González [5] and Miha Smolnikar [6,7]

1 Smart Innovation Norway, Håkon Melbergs vei 16, 1783 Halden, Norway; sanket.puranik@smartinnovationnorway.com (S.P.); heidi.tuiskula@smartinnovationnorway.com (H.T.)
2 Estabanell Energia, C. del Rec, 28, 08401 Granollers, Spain; rgallart@estabanell.cat
3 Institut d'Informàtica i Aplicacions, University of Girona, Campus Montilivi, 17003 Girona, Spain; quimmel@eia.udg.edu
4 Intracom S.A. Telecom Solutions, 19.7 km Markopoulou Ave., GR-19002 Peania Athens, Greece; labil@intracom-telecom.com
5 Centre d'Innovació Tecnològica en Convertidors Estàtics i Accionaments (CITCEA-UPC), Department of Electrical Engineering, Universitat Politecnica de Catalunya ETS d'Enginyeria Industrial de Barcelona, C. Avinguda Diagonal, 647, Pl. 2, 08028 Barcelona, Spain; francisco.diaz-gonzalez@upc.edu
6 ComSensus d.o.o., Brezje pri Dobu 8a, SI-1233 Dob, Slovenia; miha.smolnikar@comsensus.eu
7 Jožef Stefan Institute, Jamova cesta 39, SI-1000 Ljubljana, Slovenia
* Correspondence: alemu.belay@smartinnovationnorway.com

Citation: Belay, A.M.; Puranik, S.; Gallart-Fernández, R.; Tuiskula, H.; Melendez, J.; Lamprinos, I.; Díaz-González, F.; Smolnikar, M. Developing Novel Technologies and Services for Intelligent Low Voltage Electricity Grids: Cost–Benefit Analysis and Policy Implications. *Energies* **2022**, *15*, 94. https://doi.org/10.3390/en15010094

Academic Editor: Miguel Jiménez Carrizosa

Received: 1 November 2021
Accepted: 17 December 2021
Published: 23 December 2021

Abstract: The paper presents a set of prototype smart grid technologies and services and validates the economic viability of the proposed solution using cost–benefit analysis (CBA). The study considered the EU-funded project called RESOLVD and implemented the technologies and services in a real-life pilot. The paper focuses on the analysis of technological solutions which enhance the operational efficiency and the hosting capacity of low-voltage electricity distribution grids. The solutions provided better integration of a hybrid battery storage system, with the grid interfacing power electronics, smart gateways for the interconnection of assets at the grid edge, and sensors enhancing infrastructure observability and control. The result from the CBA indicates the economic viability of the project, high scalability, and replicability. The economic benefits were realized with the breakeven value of eight secondary substations (SS) and 16 feeders. The scenario test on the DSO's willingness to pay for the software as a service (SaaS) revealed that the payback period can further be reduced by almost half with a higher internal rate of return (IRR) and net present value (NPV). Both the CBA and scenario tests showed RESOLVD solution can become more economically viable when deployed in largescale. Moreover, the CBA results provide evidence to the energy policy by allowing DSOs to consider both CAPEX and OPEX for better investment decisions. Further, the paper proposes an alternative business approach that shifts from grid reinforcement to service provision. The paper also discusses the research implications on energy policy and business.

Keywords: cost–benefit analysis; smart grid; hosting capacity; efficiency; technology; economic; the EU; willingness to pay; policy

1. Introduction

The green transition of energy through low-carbon pathways moves the electricity sector towards higher penetration of distributed energy resources (DERs). According to the EU's amended Energy Efficiency Directive (EU) 2018/2002 and 2030 climate and energy framework of 2014, the EU has set a target for renewable energy of at least 32.5% and 32% share by 2030. This implies an increase in renewable resources (RES) connected to the distribution grid and will require the distribution system operator (DSO) to expand, upgrade, and modernize the infrastructure [1]. Under high DERs, DSOs face multifaceted challenges in managing LV grid networks, as they lack high-resolution grid observability and advanced control and planning systems. Challenges include frequent voltage variations, congestions, reverse power flows, and reactive power balancing [2–5]. The authors of [4] discussed how technical solutions, by providing new information about the state of medium-voltage (MV) and low-voltage (LV) networks, can improve the quality of network management.

Along with better management of the grid, DSOs need to increase their hosting capacity cost-efficiently in the part of the grid that expects more challenges as a result of the growth of intermittent renewable energy sources and mobility's electrification to provide the distributed asset flexibility. This can be achieved by activating flexibility and coupling demand and generation [6,7]. This implies directly acting on energy assets (special focus is put on storage) and the participation of consumers/prosumers through demand response [8]. Storages are claimed to be the way to mitigate some of the distribution grid challenges [5] but are expensive [9]. To overcome all challenges, cost-effective advanced technologies need to be developed, tested, and the investment should be economically justified using a detailed cost–benefit analysis (CBA).

Nevertheless, there is no standard methodology to assess the economic viability of the technology developed, especially in the highly regulated business of the DSO. However, there are some techno-economic analysis methods proposed in the electricity domain. However, only a small amount of the research is linked to grid-scale electrical energy projects. The authors of [10] conducted a social cost–benefit analysis and developed a framework that helps the regulatory regime to assess and decide on investing grid-scale electrical energy storage (EES). The authors of [11] applied CBA to analyze the deployment of advanced metering infrastructure (AMI) by focusing on loans to check the worst-case feasibility of the project. Although CBA has been applied in smart grid projects, there is limited research focusing on technical projects.

This paper considers the case of the H2020 RESOLVD project, and describes the technologies developed to solve LV grid operation challenges. These technologies developed in the project fall under the category of (a) use of hybrid storage systems, (b) observability of the grid, (c) integration of IT systems, and (d) decision support for control of the grid. The paper presents the outcome of a detailed CBA that was performed to explore if the investment in the proposed solution is economically viable. Further, the paper presents a scalability analysis and scenario test for willingness to pay by DSO. The paper does not only validate and justify the economic soundness of intelligent LV grid management, but also lays a foundation to give insight about the future flexibility services by paving the way to reach high DER penetration through increased hosting capacity and efficiency. This shall further help DSOs to consider OPEX-related investments as a key business consideration in addition to the traditional grid upgrade (CAPEX). This enables and has implications on efficiently managing the grid and to take the advantage of the combination of the two, i.e., total expenditure (TOTEX). Furthermore, the results substantiate and shed light on how to improve the energy policy and contribute to the green transition through high DER penetration.

1.1. Objective and Research Questions

The objective of the paper is to demonstrate and validate the proof of concept of RESOLVD solutions and provide evidence of its economic viability in managing an intelligent LV grid operation in the context of distributed networks. This is achieved by measuring and quantifying defined KPIs for all technologies developed and implemented on the selected pilot. The research objective is achieved by answering the following four thematic research questions (RQ):

- RQ1. How can the DSO manage intelligently the LV grid and improve the grid performance in terms of efficiency and hosting capacity?

- RQ2. What concrete and quantifiable benefits (based on defined KPIs) can be achieved from the proposed RESOLVD solution?
- RQ3. Are the RESOLVD hardware and software solutions economically viable and can their adoption be economically justified when compared to the business-as-usual (BAU) scenario, i.e., without RESOLVD? What will be the impact of scaling up towards large-scale deployment of RESOLVD solutions and what is its sensitivity with respect to reduction in major costs, such as software and battery?
- RQ4. What implications does RESOLVD have to the energy policy, technology, and businesses?

To answer these questions the research framework shown in Figure 1 is followed.

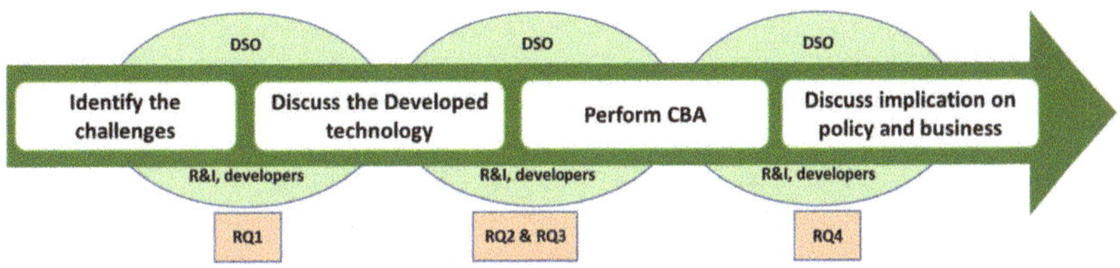

Figure 1. Research framework to answer research questions of the paper.

1.2. Paper Organization

Section 2 is a literature review on aspects of intelligent distribution grids that are of interest for RESOLVD, i.e., grid observability, integration of energy storage systems, decision support systems for the DSOs, and integration of IT systems at the DSO level. Additionally, Section 2 provides a brief review on cost–benefit analysis (CBA) for relevant, low technology readiness level (TRL) solutions. Section 3 presents the RESOLVD concept and solution as implemented at a demonstration site. Section 4 describes the methodology used for performing CBA. Section 5 provides results and a discussion of the research questions of the paper. Section 6 presents the implication of RESOLVD technologies on business and policy. Finally, Section 7 concludes the paper.

2. Review on Intelligent Distribution Grids

2.1. Observability of the Grid

The traditional unidirectional power flow in power networks, from large generation assets to end consumers in the grid, is being progressively replaced by bidirectional ones because of the increasing penetration of distributed energy resources and active demand management of consumers. The exploitation of various types of energy storage could improve to reduce the controllable loads to support the active system management, requirement for real-time knowledge on the distribution network state [12], and monitoring of dynamic and transient events affecting the quality of supply [13]. The prerequisite towards achieving comprehensive observability of the power network is the deployment of advanced measurement technology and the provision of appropriate connectivity solutions to support data aggregation and processing. Providing the means of monitoring, analysis, prediction, control, and automation in the segment between the secondary substation and consumer's/producer's connection point, which is a great part of increased system dynamics that exists, allows the operator to provide resilient services, better utilization of assets, and potential delay of investment in infrastructure reinforcement, fostering the innovative solution to invest in real smart assets.

A power network is considered observable if it is possible to calculate its quasi-steady state using the available set of measurements and knowledge on current configuration [14]. The approach can either be model-driven, where graph theory, line models, and instantaneous measurements are used to estimate the state, or data-driven, where a numerical model is built using historical measurement data and instantaneous measurements are then used to derive the state from the model. The smart meter data provided by the Advanced Metering Infrastructure (AMI) are increasingly used to support various applications, such as load profiling or forecasting. Moreover, power quality meters (PMQs) are typically deployed to support the site or assets monitoring, and (micro) phasor measurement units (PMUs) are emerging to support the wide-area situation awareness. Although there are new approaches, technologies, and proposed observability solutions that could be deployed in the energy domain, the results from such technologies should be further validated and justified. In this connection, some of the proposed solutions include an automatic target detector in infrared imagery [15], a multi-remote-sensor approach, such as a deep-learning-based burned area mapping using the synergy of the satellite data and images of Sentinels 1 and 2 [16]. Moreover, there is a new method that could be potentially adopted, i.e., the use of multi-camera multi-player tracking with deep player identification [17]; however, the cost-effectiveness in a distributed energy grid context should be assessed.

2.2. Integration of Energy Storage Systems

The applications of energy storage systems (ESS) in networks are diverse and have been widely discussed in the literature [18,19]. ESS is required to provide power and energy services to upgrade the electrical grid and achieve a neutral role for DSOs and highlight its developed infrastructures. From a techno-economic perspective, there is no optimal battery type for all types of applications. Additionally, the stacking of multiple services simultaneously can result in a higher return on investment [20,21]. Thus, the development of hybrid energy storage solutions is suggested based on the combination and synergistic exploitation of various storage technologies (each providing different services). The realization of hybrid storage systems mainly depends on the ability to electrically interconnect and manage a heterogeneous grouping of storage technologies. This is triggered by advanced power electronics [22,23] and associated power-sharing algorithms [24–26] to manage energy in such hybrid systems to provide a variety of energy services (e.g., voltage compensation, storage, phase balancing, etc.). Aspects to consider for such power-sharing algorithms are usually the main technical performance of each storage technology embedded into the hybrid solution (e.g., energy and power ratings, cyclability, time response), and degradation mechanisms. Based on such aspects and criteria based on the power time profile, the setpoint for the hybrid solution can be evaluated—or stratified—to identify fast time-varying and low time-varying components to associate to, for instance, a lithium-ion battery and a lead-acid battery, respectively, pursuing an optimal performance. All in all, this comprises a challenging decision process involving technical aspects and providing business opportunities.

2.3. Decision Support Systems for the DSOs

The notion of a decision support system (DSS) embraces a variety of applications oriented towards short-term (i.e., grid operation), medium-term (i.e., maintenance), and long-term (i.e., grid planning) functions that usually also include visualization and simulation capabilities.

In this paper, we focus on the short-term time frame and we present DSSs that allow the operation of a smart grid in a more efficient way by scheduling grid configuration and controllable assets at the day-ahead time frame.

Such DSSs usually combine power flow solvers with algorithms that aim to maximize a specific utility function. Existing studies focus on determining the optimal static configuration of a network to minimize power losses [27–31]. The authors of [30] showed that voltage control and power losses were improved with a simultaneous reconfiguration and an optimal location of the distributed generation assets (DG). The same formulation was used in [29], but a different solver algorithm with better performance was used. Usually, the optimization problem involves the co-ordination of the operation of ESS with grid reconfiguration (i.e., switchgear operation), resulting in a multi-objective function. To decouple grid configuration from charge/discharge of ESS, a two-stage optimization approach has been applied [32]. In this approach, firstly, the grid configuration is set (on an hourly basis) to avoid possible critical events (congestions and over/sub voltages), and once this problem is solved, the ESS is scheduled to minimize energy exchange and losses at the substation level. The approach in [33] also considers the existence of interruptible loads that transfer the problem to the demand side and incorporate the dispatchable load costs.

2.4. Integration of IT Systems at the DSO Level

As distribution systems gradually evolve into smart distribution systems, the DSOs' operations centers that control them evolve in parallel, incorporating new functions. The separate IT systems operating in these control centers become more streamlined, communicating seamlessly to provide an integrated monitoring and management system. More advanced applications and analytical software provide more sophisticated analyses and automated operations. The control systems of operation centers not only help to make the grid smarter, but also help to improve support for decision-makers responsible for operations, maintenance, and planning [34]. Among the IT systems present at a DSO's operations center, the distribution management system (DMS) is considered as one of the most critical components for the efficient operation of the distribution grid. As a fully functional unit, a DMS may include various applications, such as fault detection, isolation, and service restoration, voltage variance optimization, power flow management, etc.

According to [35], a growing number of DSOs have DMS running in their operations center. However, participants also indicated that challenges remain in developing a highly reliable and functional DMS in different respects (e.g., system integration, communication). This is just one side of the coin. In many cases, the DSO IT ecosystem is composed of many different legacy systems and subsystems, rather than a monolithic DMS. Examples of separate (nonintegrated) IT systems include customer information system; geographic information system; trouble calls center; workforce management; switch order management; AMI; SCADA, etc. On top of the above, there are several non-interconnected advanced applications that are useful for the DSOs, both at the planning and operation phase (outage planning, grid resources optimization, power quality analysis, distribution grid state estimation, etc.). These and other applications are often in need of real-time information, such as the status of switches, voltage regulator taps, capacitor banks, fault location, the status of distributed resources, and customer demand/load. This signals the need for an information exchange bus. In addition, ultimately, there is a need for integration of the decision support functions (analytics plane) with the actions' execution functions (control plane).

2.5. Cost–Benefit Analysis for Relevant Low TRL Solutions

The development and adoption of smart grid technologies need economic validation. Some researchers have developed cost–benefit analysis (CBA) frameworks for smart grid applications. For example, the authors of [36] applied CBA to assess system management and planning tools for the electrical grid. The authors of [37] proposed, in their theoretical paper, a multi-agent systems (MAS) approach applied to the operation of power grids. The authors of [10] applied CBA on grid-scale electrical energy storage (EES) and validated the results using simulations. Similarly, authors of [38] used CBA on a virtual power plant including solar PV, flow battery, heat pump, and demand management and attempted to validate it by simulation.

In general, most of the CBAs for assessing the investment in smart grid technologies are still too conceptual and theoretical, and include many assumptions. Moreover, although there are various reports on such CBAs, the research on complete CBA and investment analysis, including scalability and willingness to pay, is not covered adequately in the literature. Unlike previous CBAs of smart grid technologies, this paper focuses on investments in the LV part of the grid.

3. RESOLVD Concept and Demonstration Pilot

3.1. The Project Concepts

The RESOLVD project aimed to improve the efficiency and the hosting capacity of distribution networks in a context of highly distributed renewable generation by introducing flexibility and control in the low-voltage (LV) grid without the participation of consumers. Specifically, the project tackles the following challenges:

- Improvement of low-voltage grid observability.
- Critical event forecasting. This includes day-ahead forecasting of congestions and over/under voltages due to daily variations in demand profiles with the presence of PV generation.
- Early fault detection by exploiting the grid observability and redundancy of measurements.
- Island management; this includes the detection of uncontrolled islanding situations and the management of ESS and frequency control in controlled LV islands.
- Power quality improvement and loss reduction; this includes compensation of harmonics reactive power, and phase balancing with power electronics (PED).
- Reduction of technical losses in the network by reducing imported energy and maximizing the use of renewable energy locally produced.
- Cyber security; the design of subsystems and further integration has been submitted to rigorous analysis and tests to assure the cybersecurity of the whole solution.

To tackle these challenges, the project developed and deployed hardware and software solutions to smartly manage LV distribution grid operations and improve the grid performance. The project developed an advanced power electronic device to control and manage hybrid storage (lithium and lead-acid batteries) at the customer premises and two substations with four feeders.

Advanced power electronics with integrated storage management capabilities provide both switching and energy balancing capacities to manage energy locally. Enhanced observability of low voltage grids has been achieved by introducing cost-effective high-resolution instruments (PMUs and PQM) and exploiting smart meter data to perform energy forecasting (generation and demand aggregated at bus level). Thus, enhanced observability and improved energy control allow the introduction of day-ahead operation planning in the low voltage by scheduling three elements:

- Switchgears: to reconfigure the grid to avoid critical events (congestions and/or over/sub voltages)
- Batteries: schedule the charge/discharge of batteries according to efficient operation goals (reduce energy exchange at the substation level, peak reduction, maximal use of RES locally generated)
- Power electronics: its activation allows reduction of losses and improves quality by compensating harmonics and reactive, balancing phases, and compensating voltage variation by injecting reactive power.

Optimal scheduling and activation of such assets are performed by decision support modules that provide services for day-ahead energy forecasting (generation and demand) based on smart meter data, detection of abnormal operation as faults and critical events (congestions and over/sub-voltage variations), and grid operation scheduling under different optimization criteria (critical event avoidance, loss and peak reduction, or maximal consumption of RES locally generated). Thus, RESOLVD builds up an enhanced low-voltage grid monitoring with (local) wide area monitoring (WAMs) capabilities to detect faults, forecast critical events (congestion voltage variations), and correct power quality issues, resulting in a better quality of supply that is more secure and efficient. RESOLVD is conceived as a set of subsystems fully interoperable with the legacy systems (such as SCADA, DMS, MDMS, and GIS) that provides enhanced low-voltage observability and day-ahead scheduling capabilities to optimally manage the grid.

The interoperation of these technologies in a secure and safe way has been implemented by considering current communication standards and putting the CIM model on the basis of the development. Integration of the overall system has been performed with an enterprise service bus (ESB) that supports all the data transactions and co-ordinates the invocation of modules according to specific decision support tasks. Next, Figure 2 shows in green the RESOLVD components and, in light blue, existing legacy systems [39].

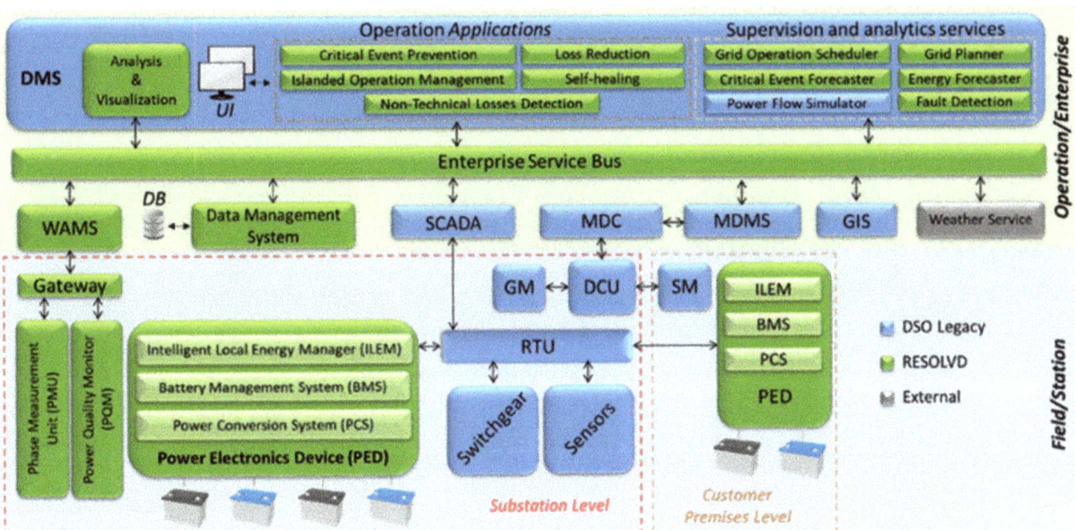

Figure 2. RESOLVD high-level architecture.

3.2. The Demonstration Pilot

The RESOLVD technology has been tested in a pilot integrated into a semi-urban area of the distribution grid operated by Estabanell Energía (EyPESA) in the province of Barcelona (Catalonia, Northeast Spain). The pilot (see Figure 3 for the schematic diagram) covers two secondary substations (SS) feeding 25 consumers and five PV plants. It is composed of two low-voltage (LV) three-phase lines, deriving from two different secondary substations (SS-A and SS-B). Blue arrows represent the supply points, while the yellow circles stand for the PV installations. The LV allows reconfiguration by operating three switchgears, also permitting operation in a ring configuration. Residential consumers are normally single-phase fed (phase + neutral or two phases).

In Table 1, the values of the rated power of the transformers and the total contracted and generation power installed are summarized. It should be noted that the MV/LV transformers are oversized with respect to the load of the area. This is because, some years ago, the two substations had to also supply two high-load factories that have now closed.

Table 1. Rated power and contracted power values of SS-528 and SS-0030.

Secondary Substation (SS)	SS (A) 528	SS (B) 0030
Transformer rated power	250 kVA	630 kVA
Total contracted power	138.6 kW	127.8 kW
Contracted power of line in the pilot	58.3 kW	56.9 kW
PV power installed in the pilot	12.5 kW	9.9 kW

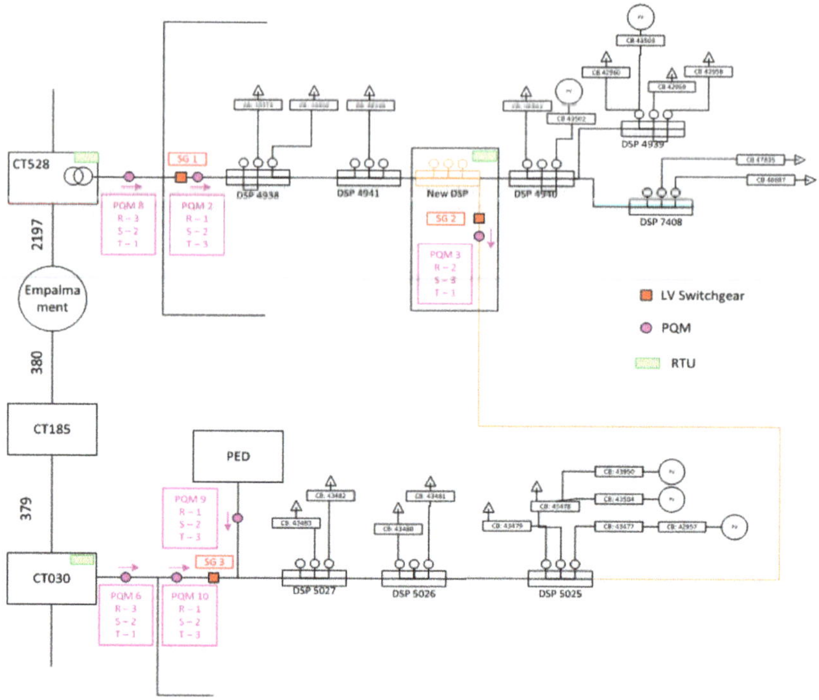

Figure 3. Schematic of the pilot area of RESOLVD.

Figure 4 shows a simplified schematic representation focusing on power electronic devices' (PED) connection with the pilot area. It should be noted that the power electronics and the batteries are installed at the beginning of one line fed by SS-B. This scheme represents the possible grid configurations that can be implemented during the test phase.

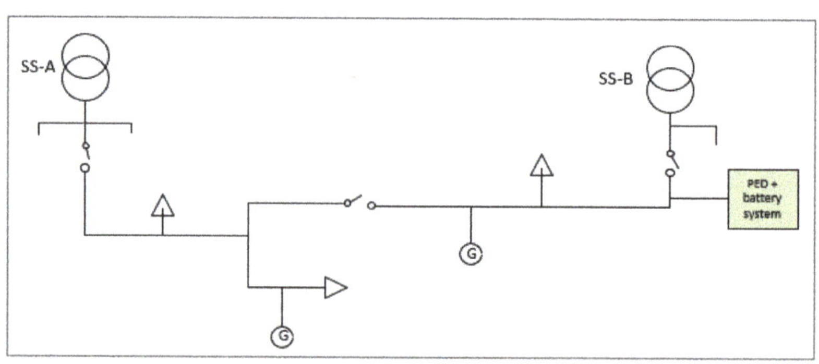

Figure 4. Simple pilot schematic showing the location of PED connection.

Smart meters and power quality monitors installed at the beginning of the two feeders and in the point of common coupling of the PED were installed to collect data used during the validation phase. The PED and the batteries were installed in the SS-B (Figure 5). This location has been selected for space reasons, as it consists of a two-floor building, with an empty second floor. Both the PED and batteries through the battery management system (BMS) were integrated with the SCADA through specific remote terminal unit modules.

Figure 5. A picture of SS-B where the PED and the batteries were installed.

4. Cost–Benefit (CBA) Methodology

4.1. CBA Process Steps for Smart Grid Projects

CBA is an analytical tool to evaluate investments and to support policy decisions for society's welfare [40]. In general, CBA concepts investigate the opportunity cost in a long-term perspective, which depends on a set factor, such as the sector, discount rate, forecast future costs, and other economic indicators quantified in monetary terms. The EU BRIDGE initiative and the European Commission's Joint Research Centre (JRC) recommended the CBA framework to be followed for smart grid projects [41]. In RESOLVD, similar process steps were followed by characterizing the project, quantifying the benefits, and, finally, comparing the benefit to cost to see if the benefit outweighs the investment cost (see linked RESOLVD activities in Figure 6 highlighted in blue color).

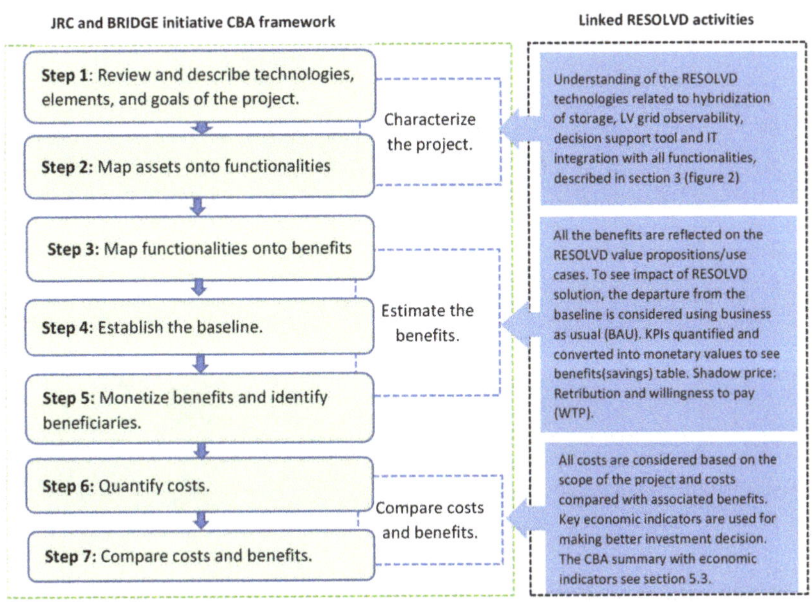

Figure 6. CBA process steps used in RESOLVD are based on the JRC and BRIDGE initiative.

Indicators used to perform CBA are defined in Section 4.2. Using the indicators two scenario tests were performed. The first scenario test focused on assessing the scale-up potential of the RESOLVD solution from an economic point of view. Here, the solution was scaled up from a small pilot site to selected numbers of sub-stations (SS) of the DSO. The scenario test assessed the willingness to pay (WTP) for solutions under the software as a service (SaaS) model. The WTP measures the maximum amount DSO would be willing to pay to gain the desired outcome or, alternatively, the maximum amount that DSO would be willing to pay to avoid undesirable outcomes. In the second scenario test, it was assumed that software developed under the project is provided as a service rather than as a product to the DSO. This results in different cost structures of having an annual service fee for software instead of one lump-sum cost.

4.2. Economic Indicators for CBA

In this paper, the net present value (NPV), the internal rate of return (IRR), and payback period were considered for performing the CBA analysis. NPV is calculated using formula (3), while IRR is equal to discount rate when NPV is zero. The payback period is defined as the time taken by net benefits to overcome initial investments. The benefit–cost ratio is calculated based on the sum of all present values of quantified benefits divided by the present values of all costs. All cashflows are aggregated considering the interest rate at a time '*t*'.

$$\text{Benefit–cost ratio} = \frac{|Present\ value[Quantified\ Benefits]|}{|Present\ value[Costs]|} \tag{1}$$

$$\text{Benefit–cost ratio} = \frac{\sum_{t=0}^{N} \frac{|CF_t[Benefits]|}{(1+i_t)^t}}{\sum_{t=0}^{N} \frac{|CF_t[Costs]|}{(1+i_t)^t}} \tag{2}$$

The CBA or Benefit-cost ratio can be simplified further and presented in Equation (3) including the initial investment.

$$NPV = \sum_{t=0}^{T} \left\{ \frac{B_t - C_t}{(1+r)^t} \right\} - I_0 \tag{3}$$

where NPV = net present value; B_t = annual benefits at time t; C_t = annual costs at time t; T = lifetime of the project; I_0 = initial investment cost; r = discount rate; and CF_t = cashflow at time '*t*'.

Based upon consultation with DSO (also the pilot owner of the project) project lifetime was considered as 10 years, and a discount rate of 4% is favored considering current market conditions. In the EU, the projects are usually discounted using the social discount rate of 3–5% depending on the cohesion and member state countries (JRC 2012). Thus, the assumption of a 4% discount rate is within the normal range in the EU.

Quantification of costs and benefits:

The major costs and benefits are listed in Appendix A and are summarized as follows:

- Costs: all major costs for the hardware and software development, installation, personnel, services cost, such as cyber security.
- Shadow price: for calculating willingness to pay for SaaS, a shadow price is considered based upon other existing software used by the DSO.
- Benefits from technical indicators: in the project, a total of 12 indicators, nine technical KPIs, including the three control indicators (CI), have been defined, which are converted into monetary values. The process of converting these indicators to monetary value is qualitatively described in Table 2. Details on the project indicators can be found under public deliverable D5.1 of the RESOLVD project [42].
- Benefits from remuneration: these are considered based on the DSOs savings from the remuneration formula and described in the next sub-section.

Table 2. Qualitative description of converting project indicators into benefits with linked remuneration variables.

Project Indicators	Basic Assumptions Considered to Calculate the Benefit	Linked Remuneration Variables
KPI1: Power loss reduction due to waveform quality improvement	The annual average electricity increases in the pilot feeders (17 kW) multiplied by the % loss and compared with the change from the business-as-usual scenario (BAU).	P_n^i
KPI2: Improvement of the energy profile in the secondary substations	In SS where PEDs are managing renewable sources. The PV energy does not flow from all distribution infrastructures from TSO link.	data [1]
CI1: Efficiency rate of the PED and the energy storage system	Consumption of Vdc bus, batteries technologies and PED auxiliary internal consumptions.	$DESP_{n,15\rightarrow n-2}$, RI_n^i
KPI3: Increase in DER hosting capacity in LV network	PED, batteries, and auxiliary system internal consumptions.	$DESP_{n,15\rightarrow n-2}$
KPI4: Reduction in DSO investment	In terms of traditional infrastructure investment and assuming an average distance of 100 m multiplied by new LV feeder cable plus the cost of assets upgrade.	$DESP_{n,15\rightarrow n-2}$
CI2: DSO operation expenditures with respect to the BAU solutions	Regarding the remuneration for the operation and maintenance, the rule in force until 2020 (RD 1048/2013 + Order IET/2660/2015) differentiates between O&M of electrical assets (which is remunerated by unit value) and O&M of nonelectrical assets (which is remunerated by invoice). The O&M amount of electrical assets ranges from 3% to 5% of the unit investment reference value.	RI_n^i
KPI5: Percentage of improvement in line voltage profiles with power injection and consumption	The real and reactive power not supplied from the primary substation and provided from PED in the maximum consumption time and assuming that the active power was charged in the storage system when there was a surplus in the valley period.	-
KPI6: Rate of prevented critical events in the LV grid due to forecasting and remote control of grid actuators	Assuming average time response as a traditional way, the customer from the fault to fuse restoration is two hours. Moreover, the usual time that a user calls DSO to communicate the fault is 10 min on average. Average energy in the pilot feeders multiplied by the events that feeder fuses are blown per year assumed by DSO.	-
KPI7: Quality of online event detection in LV grid	Fast detection of short circuits and severe faults that lead to protection (fuses). The time difference between rapid detection (the time of occurrence) compared to a conventional method in which fuse actions are protections where they are reported. I have considered it to involve a reduction in response time x kWh considering the number of end-users affected due to fuse blow.	Q_n^i
KPI8: Quality and time needed for awareness and localization of grid fault MV grid	To localize an MV fault and isolate part of the grid on average from the historic SCADA registers in 5 h, consider wage of operator per hour. As result of this KPI, it is possible locate the fault and save in average 3.5 h. Moreover, it avoids the assets stress.	Q_n^i
KPI9: Quality of LV grid operation in island mode	It was just tested during the functional test. The functional testcreated an electrical island for 85 min.	Q_n^i, $REVU_n^i$ & $COMGES_n^i$
CI3: Waveform quality in LV grid	Only affect in the part of the upstream grid from PED until the primary substation is connected the pilot feeder.	P_n^i

Benefit from remuneration.

Some of the benefits relate to different KPIs and the benefits are calculated based on the assumption of the remuneration formula. In this connection, the Spanish Comisión para lospara los Merecados y la Compenténcia (CNMC) establishes for calendar years, by resolution and after hearing the interested parties, the remuneration recognized to each DSO for the distribution activity. Before November 15 of each year, the CNMC will submit to a public hearing the total remuneration proposal to be received the following year by each of the DSOs. The remuneration of the distribution activity recognized to the DSO

(i) and in the year (n) for the performance of its activity, year $n-2$ will be determined by formulation (4).

$$R_n^i = \underbrace{RI_n^i + DESP_{n,15\to n-2}^{\ i} + TER_{n,15\to n-2}^{\ i}}_{\text{Financial business}} + \underbrace{COMGES_n^i + REVU_n^i}_{\text{O\&M}} + \underbrace{ROTD_n^i}_{\substack{\text{Commercial}\\ \text{cycle}}} + \underbrace{P_n^i + Q_n^i}_{\substack{\text{Efficiency}\\ \text{incentives}}} \tag{4}$$

where:

R_n^i: is the return on investment to be received by company (i) in year (n).

RI_n^i: is the investment remuneration to be received by DSO (i) in year (n) for the facilities of distribution company i with commissioning prior to 1 January 2015.

$DESP_{n,15\to n-2}$: is the investment remuneration to be received in the year (n) for the maneuver's offices and distribution energy control center that have been carried out by the distribution company i in the years 2015 to $n-2$, both inclusive.

$TER_{n,15\to n-2}$: is the return on investment to be received in year n for the properties and lands associated with new electrical installations that have been carried out by the DSO (i) in the years 2015 to $n-2$, both inclusive.

$COMGES_n^i$: is the manageable component of the remuneration to be received in year (n), which will be calculated for each DSO (i).

$REVU_n^i$: is the remuneration for the extension of the useful life for year n that a DSO (i) will receive for all those distribution facilities, which, having exceeded their regulatory useful life, remain in service in year $n-2$, if availability is proven effective of each of said facilities.

$ROTD_n^i$: is the term of remuneration for other regulated tasks that the DSO (i) has to receive for the year n for the development of these tasks in year $n-2$.

P_n^i: is the term of incentive or penalty for the reduction of losses passed on to the DSO (i) in year (n), associated with the level of losses in its network between years $n-4$ and $n-2$.

Q_n^i: is the term of incentive or penalty to the quality of the service based on the DSO (i) year (n) associated with the supply quality indicators obtained by DSO i between years $n-4$ and $n-2$.

According to the Spanish remuneration equation, the KPIs of the RESOLVD project impact according to the following Table 2.

5. Results and Discussion

The results and discussion section are organized to respond point-by-point with the respective research questions presented in Section 1.1. This section discusses the four RESOLVD technologies with their novelty and main findings from the project to address the DSO challenges: (a) advanced power electronic device (PED) and storage, (b) observability, (c) decision support system (forecasting and flexibility control), and IT integration in DSO level.

Key results of the research from the RESOLVD project are the development of advanced technologies and integration for smart grid management, demonstration of these technologies in the real-life pilot, and testing the economic viability of the proposed solution supported by the DSO for the future investment in LV grid in a distributed network. The detailed findings of the research are as follows.

5.1. Developed Technology, Novelty, and Main Findings (Linked to RQ1)

Power electronic device (for multiple storage integration): As a result of the whole RESOLVD solution, and due to the PED, the DSO can improve the performance of the grid in terms of energy efficiency, since active and reactive power flows within the grid can be optimized. Energy efficiency was improved as: (i) since the PED is connected close to the consumption points in the network and partially fed them, distribution losses are minimized; (ii) the PED has been proved effective in managing voltage levels at the LV side of the transformer, and this also minimizes grid currents and transformer losses. In addition to energy efficiency, the hosting capacity of renewables of the network has been improved. Following the control setpoints from the DSO, the PED minimized transformer loading and mitigated current harmonic content in the network due to, for instance, PV inverters and nonlinear loads connected nearby. All in all, this permits the DSO to defer in time network reinforcements, while also hosting more DER. Further, the PED has been effective in acting as a voltage source in case of a mains failure. This is an attractive service for a DSO in a rural area, as in the RESOLVD project. Figure 7 below shows the integration of different batteries and the context of utilization.

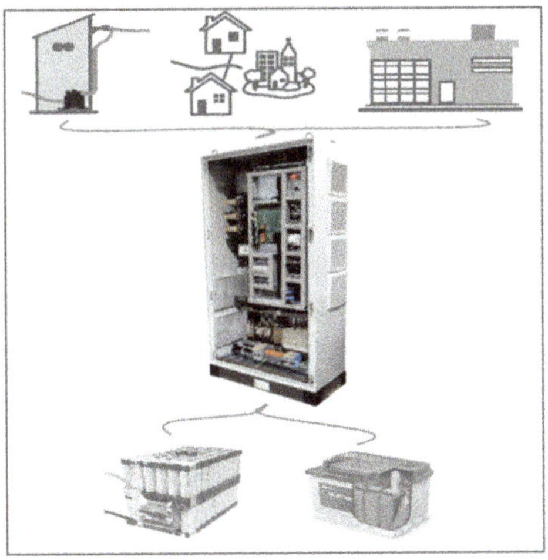

Figure 7. PED as developed in the project shows possibilities to integrate multiple batteries (at the bottom), as well as to use them in multiple contexts (on top).

The two battery types embedded into the PED are operated synergistically by a power-sharing algorithm explained in [43]. This algorithm optimizes how to split the total power the PED is to exchange with the network among the two battery types, considering their efficiency, degradation mechanisms, and state of charge. Degradation is associated to two phenomena: power stress and depth of discharge of cycles; and lithium-ion batteries are more resilient against degradation than lead-acid batteries. Thus, the power requested to the lead-acid batteries is limited, and also the depth of discharge (the minimum admissible state of charge is limited at 30%), while high-peak power needs are associated with lithium-ion batteries. This way, lead-acid batteries smoothly provide part of the total power needs the PED should exchange with the network, thus reducing the stress for the lithium-ion batteries that, in turn, are subjected to more stringent power requirements.

The front-end power electronics of the PED and the power inverters interfacing with the network are equipped with advanced algorithms to improve network waveform quality. This enables management of the reactive power, load balancing of the three phases of the network, and cancelling current harmonics. Relevant test data conducted on the performance of the PED while balancing loads of a three-phase network are conducted and presented in [44].

Sensors for observability: In designing the RESOLVD observability solution to support the implementation of use cases and testing of the KPIs, the emphasis has, on one hand, been given to the realization of appropriate metrology instruments and, on the other hand, the capability of efficient data communications and processing. The PQM instrument has been designed to observe the power quality parameters, and thereby the quality of supply or the impact of PED service in characteristic locations of the network. To support also direct interfacing to grid assets under control, it embeds a gateway functionality supporting the legacy control protocols. The embedded computing functionality allows for the implementation of moderate control schemes and schedules to allow for an autonomous operation. The design of the PMU instrument and edge computing solution target efficient implementation of advanced situation awareness applications in the distribution grid segment. The particular focus has been given to fault detection and localization to support fast service restoration [45]. Given the vast amount of measurement data produced by the PMU and no dedicated connectivity solution on disposal at deployment locations, the application consists of several interacting distributed services, each responsible for a specific sub-function. The data are preprocessed on edge for the purposes of event detection and model building, event classification is realized based on raw data snapshots aggregated from all affected locations, and localization is performed using the information on grid configuration, line models, and the pre- and post-event conditions. The connectivity towards the control center application is, for both instruments, realized via 4G cellular network. The three key components of observability developed in the project are shown in Figure 8.

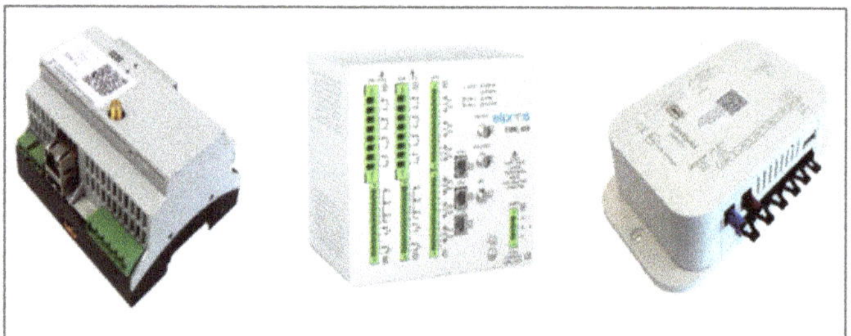

(a) (b) (c)

Figure 8. Observability-related sensors developed in the project: (**a**) PQM, (**b**) PMU, and (**c**) gateway.

Decision support toolkit (forecasting and flexibility control): Grid operation scheduling avoids exporting energy (exported 0) and, at the same time, storage is managed to reduce peaks of demand. Thus, the main contribution to loss reduction is achieved by flattening the curve. This is quantified, on average, by reducing the peak around 18% and the variability of the curve by 43%. A generalized reduction in peaks at the low-voltage grid supposes a substantial increase in efficiency of overall power system (transport losses depends on the square of current) and the consequent reduction in emissions (according to the energy mix). However, this has a cost. In the experimental pilot, the consumption of the PED and battery is around 6.5% of total energy managed (more than 4% is consumed by the PED) in the pilot. This supposes increasing the total demanded energy by around 4%. This elevated percentage is mainly because the PED is somehow oversized, and it is capable of managing larger peaks of power and amounts of energy than the pilot demand in the current state. This increase could be compensated by an increase in local production, resulting in an increase in hosting capacity. Due to the increase in energy required to run the PED and battery, it is recommended to limit the operation of the GOS only when the excess of PV is enough to compensate for this consumption. During the test, only dates 5, 8, 10, and 12 February 2021 produce an excess of generation (between 5.8 and 9.6 kWh), and this excess does not compensate the losses in the PED and the batteries (between 18.1 and 18.6 kWh). The technology developed in RESOLVD enabled the DSO to achieve the following key findings and improvements: hourly energy (aggregation of SM at bus level) forecasting error (MAPE) of 20%, peak reduction by 18%, and 0% exported energy (at SS level). Besides the key findings, the technology enables provision of the additional features revealed, such as 97.5% accuracy on fault detection (high impedance faults), 3.8% false alarm, and missed detection of 0%.

Novelty in Integrating IT systems: RESOLVD implemented a solution for integration of the novel applications of the project, not only in between them, but also with legacy systems of the DSO, by implementing an innovative and standards-based enterprise services bus (ESB) solution. Being compliant with the IEC 61968 standard, this solution comprises an integration middleware with a focus on the smart grids domain.

5.2. *Quantified RESOLVD Benefits (Linked to RQ2)*

All savings from each KPI and control indicator are aggregated to realize the direct benefits of the project. When available, control indicators based on existing technical standards have been used to validate the fulfillment of technical conditions (e.g., EN-50160 is the power quality standard that specifies the acceptable ranges for the parameters that define the supplied electric waveform). The benefits derived from the analyzed KPIs are quantified and summarized in Table 3 below. The results in this table show the benefits, in EUR, calculated from the changes in business-as-usual scenarios (with and without RESOLVD). Most of the KPIs are quantified but few KPIs are converted into two benefits based on the assumption of measuring two KPIs. For example, KPI-7 and KPI-8 are related to the devices and measure the accuracy of the software solution. In this case, for the CBA, hypotheses have been taken based on the benefit to locate an electrical fault and simplify the onsite work operation into a reduction in the asset stress. As some of the consortium members are business entities, some of the values are anonymized and described in some ranges instead of specific monetary values (Table 3).

The quantified benefits after the RESOLVD project implementation were calculated based on the KPIs defined and compared with the business-as-usual scenarios (BAU). All monetary values are registered in Table 3. Since some of the savings and benefits are business sensitive, the results are put in ranges for anonymity after agreement with partners. For example, in connection to the hosting capacity (HC) of the grid and in line with the project objective, the KPI-03 is defined to show the increase in DER hosting capacity in an LV network (%kVA with respect to the current maximum limit) and calculated as below.

$$\%Change\ gained = \frac{HC\ with\ RESOLVD - HC\ without\ RESOLVD}{HC\ without\ RESOLVD} \times 100$$

$$\frac{189\ \text{kVA} - 138\ \text{kVA}}{138\ \text{kVA}} \times 100 = 36.95\%$$

$$\text{Cable capacity: } 330\ \text{A}$$

$$330\ \text{A} \times 400\ \text{V} \times 1732 = 228{,}624\ \text{kVA}$$

$$\text{Assuming the power factor cosFi} = 1 \rightarrow 228{,}624\ \text{kW}$$

$$229\ \text{kW} \times 36.95\% = 84.5\ \text{kW of capacity}$$

$$84.5\ \text{kW} \times 24\ \text{h} \times 365\ \text{days} = 740.220\ \text{kWh year}$$

$$740.220\ \text{kWh year} \times 0.0350333\ € = 25{,}932\ €/\text{year i.e.,} >25{,}000\ \text{Euros/year.}$$

Similarly, the savings from the rest of the KPIs are calculated and the results are summarized in Table 3 below. For some of the KPIs, the average day-ahead price of the Spanish market is used (see Appendix B).

5.3. *Cost–Benefit Analysis (Linked to RQ3)*

5.3.1. Economic Viability of RESOLVD Technology

The CBA result clearly showed the economic viability of RESOLVD solution and the high scalability and replication potential of the project (see Table 4). The major savings come from increased hosting capacity, reduction of DSO investment through increased grid infrastructure flexibility, and the way to upgrade traditional assets by smart devices with inherent ability for real-time monitoring and control. This is relatively in line with one of the core project objectives of RESOLVD, and, therefore, savings from increased hosting capacity (see Figure 9) are considered to comprise a scale-up scenario with respect to the increased pilot size. This is carried out according to the DSO investment policy, which is up to 10% of the yearly investment on the grid infrastructure/upgrade and by ensuring the required additional costs are considered in the analysis. Because of high investment cost at the early stage of the project, the first CBA with the current pilot size of two substations (SS) showed the benefit/cost ratio less than 1. However, as the pilot size increases (scale-up) with all associated costs, such as a total cost for PMU, PQM, batteries, and PED, included, the benefit/cost ratio becomes greater than 1, with increased internal rate of return (IRR), NPV, and shorter payback periods (see Table 4 summary of CBA). The breakeven threshold is reached when RESOLVD solutions deploy on more than eight secondary substations (SS).

Table 3. Summary of quantified benefits from RESOLVD solution.

Category	Indicator	Business as Usual (BAU) without RESOLVD	After (with) RESOLVD	Saving in EUR/Year
Efficiency	KPI-01: Power loss reduction due to waveform quality improvement (W/Ω)	Unknown	166.7 W/Ω	<1000
	KPI-02: Improvement of the energy profile in the secondary substations Sub-indicators: (i) Losses T&D (%) (ii) Locally generated energy use (%) (iii) Maximum peak (%)	(i) Losses T&D is 7.3% at each SS, which is at SS-00528: 9.49 kW, SS-0030: = 8.3 kW.; (ii) Locally generated energy use is: 2.8% and 1.27%;	(i) SS-00528: Not affected because PED is not interacted as a normal exploitation. SS-0030 reduced T&D losses = 0.5% (ii) and (iii) Locally generated energy use and maximum peak (%) has no registered changes	1000–5000
Planning	CI-01: Efficiency rate of the PED and the energy storage system Note: The benefit of this indicator to DSO is in higher self-consumption rate (kW).	Without PED (no PED implemented)	Change registered: −4.2 kW	<1000
	KPI-03: Increase in DER hosting capacity in LV network (%kVA with respect to the current maximum limit)	138 kVA	189 kVA Change registered: 36.95%	>25,000
	KPI-04: Reduction in DSO investment (%)	Retribution	Retribution change registered: 22%	>25,000
	CI-02: DSO operation expenditures with respect to the BAU solutions (EUR per year)	The O&M amount of electrical assets ranges from 3% to 5% of the unit investment reference value.	Considering an investment of EUR 57.000 as a usual new building of network assets at LV level.	1000–5000
Quality of service	KPI-05: Percentage of improvement in line voltage profiles with power injection and consumption (%)	NA	0.03	<1000
	KPI-06: Rate of prevented critical events in the LV grid due to forecasting and remote control of grid actuators Sub indicators: (i) Precision of forecasting (%) (ii) Time to respond to critical event (minutes)	(i) No previous data available; (ii) It was taking 120 min	−15–40% (*) (**) −2 min Change registered: 118 min N.B. (*) (**) are kept anonymous.	<1000
	KPI-07: Quality of online event detection in LV grid Sub-indicators: (i) Accuracy (%) (ii) Precision (%) (iii) Miss Detections (%) (iv) False Alarms (%) (v) Detecting time (s) (vi) Informing delay (s)	(i) No previous data (ii) No previous data (iii) No previous data (iv) No previous data (v) No previous data (vi) 120 min	(i) 97.5% (ii) 93.5% (iii) 0% (iv) 3.8% (v) 1 s (vi) 35–400 s	<1000
	KPI-08: Quality and time needed for awareness and localization of grid fault MV grid (seconds)	NA because RESOLVD is focused on LV level. This KPI, is focused in MV level. Moreover, the KPI measures quality of effectivity of PMU	2.04 s Impact MV (indirect benefit to the DSO due to PMU)	1000–5000 (indirect saving from MV)
	KPI-09: Quality of LV grid operation in island mode (hours)	0.33 h (avg. from previous year)	1.417 h Changed registers: 1.087 h	1000–5000
	CI-03: Waveform quality in LV grid (%)	NA (no data before RESOLVD project)	0.12%	<1000

N.B, (*) and (**) are the assumptions made based on: (*) Forecasting performance has large variability depending on the node and time. Large errors are due to 24h delay in having SM data available and low consumption in most consumption points. (**) Critical events were simulated in the pilot by changing line parameters, demonstrating the availability to detect and mitigate. Given the number of tests is not representative enough, the paper considers 2 incidents to calculate KPI6.

Table 4. Summary of CBA with scale-up of the pilot size.

Indicators	2SS-4 Feeders (Current Pilot)	8SS-16 Feeders	10SS-20 Feeders	12SS-24 Feeders	24SS-48 Feeders
NPV	−459 k	−57 k	97 k	226 k	1054 k
IRR	−11%	2%	6%	9%	23%
Payback period	>10 years	9th year	8th year	5th year	4th year
Benefit–cost ratio	0.5 < 1	1.1 > 1	1.3 > 1	1.4 > 1	2.4 > 1
Recommendation	Not worth investing in the current pilot size.	Risky to continue with the investment.	Relatively lower IRR; however, it is worth continuing with the investment.	It is worth continuing with the investment.	It is worth continuing with the investment.

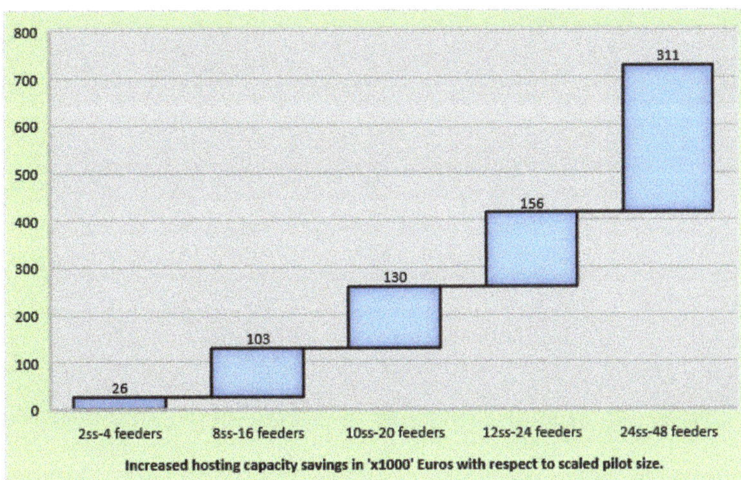

Figure 9. Increased hosting capacity with scale-up of RESOLVD pilot.

Considering the increase in the hosting capacity due to RESOLVD solution (Figure 10), the research summarized the economic indicators from the CBA in Table 4 below.

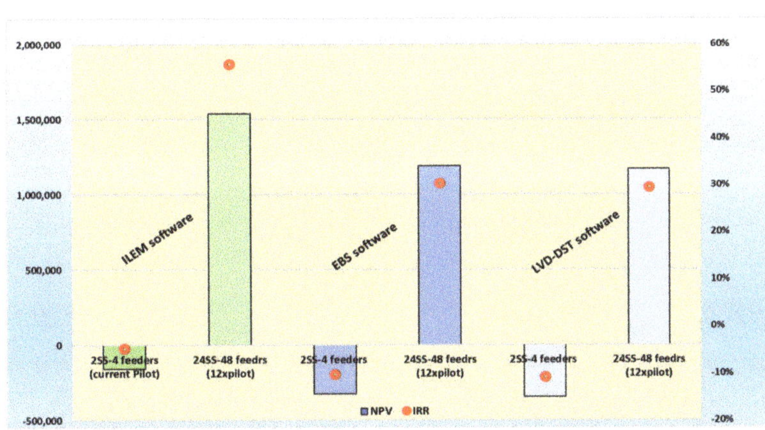

Figure 10. NPV (in EUR, left side of the figure) and IRR (% in the right side) based on WTP scenario for the three-software platform when the pilot is scaled up 12 times the pilot size.

5.3.2. Scenario Test Based on the DSO Willingness to Pay (WTP) of the Software Service

From the DSO long- and medium-term energy policy perspective, to have cost-effective, reliable smart grid solutions and have security in energy supply, the WTP scenario test on major costs is needed. In this regard, one of the major costs in RESOLVD solution is software development and associated IT platform solutions. For this test, the Intelligent Local Energy Manager (ILEM) software cost was considered, which enables control and management of the PED with or without storages thanks to the link with the energy management systems (EMS) and SCADA or advanced distribution management systems (ADMS). The main reason for this test is that the regulation does not allow the DSOs to own a battery and can get the ILEM service through licensing. However, the basis for defining willingness to pay is not straightforward and, therefore, it needs to benchmark with other licensing costs. In this connection, the DSO considered similar software service costs, such as the annual licensing fee for smart meter management platforms, which is calculated based on the number of the smart meters (see Table 5).

Table 5. A basis for DSO's willingness to pay for the software as a service fee.

Number of Supply point	<10,000	10,000–50,000	50,000–100,000	100,000–500,000	500,000–1,000,000	>1,000,000
Cost per supply point	EUR 1	EUR 0.2	EUR 0.1	EUR 0.02	EUR 0.01	EUR 0.001

Based on this assumption, the CBA is recalculated to test the scenario at which the DSO has a willingness to pay for the software service cost benchmarked from a similar software platform. The CBA with respect to WTP showed a positive economic indicator if the RESOLVD solutions are deployed on more than 8SS, and the payback period reduced to 2–3 years. This will give flexibility for the DSO to get the software service by involving other technology providers or aggregators while all regulatory boundaries are respected, including the battery ownership. However, the willingness to pay scenario test for the three-software platform is carried out and summarized in Table 6.

Table 6. Summary of CBA for willingness to pay scenario of the three RESOLVD software. 2SS-4 feeders represent current pilot size, and values are rounded to the nearest thousand (k).

Indicators	**ILEM**		**IT Integration Software**		**LVD-DST (Decision Support Toolkit)**	
	2SS-4 Feeders	**8SS-16 Feeders**	**2SS-4 Feeders**	**8SS-16 Feeders**	**2SS-4 Feeders**	**8SS-16 Feeders**
NPV	−156 k	305 k	−323 k	78 k	−343 k	58 k
IRR	−5%	17%	−11%	6%	−8%	6%
Payback period	>10 years	4th year	>10 years	6th year	>10 years	7th year
Benefit–cost ratio	0.8 < 1	1.8 > 1	0.6 < 1	1.3 > 1	0.6 < 1	1.3 > 1
Recommendation	Not worth investing with the current pilot size.	It is worth continuing with the investment.	Not worth investing with the current pilot size.	Longer payback period but worth continuing with the investment.	Not worth investing with the current pilot size.	Longer payback period but worth continuing with the investment.

Extending the WTP scenario test with 24 SS-48 feeders, the CBA showed the payback period further reduced by close to half, with increased NPV and IRR for each software developed compared to the assumption the complete solution developed at the DSO premise (see Figure 10).

5.4. Replication and Market Potential of RESOLVD Solution

The replication and market potential of the RESOLVD solution can be looked at in two perspectives. The first one is based on the number of secondary substations available in the EU and EEA. In this regard, there is an estimated number of 4.5 million secondary substations (SS). However, as the case pilot depicted, there is the possibility that RESOLVD solution may not apply to all these substations. For example, referring to the DSO where the

RESOLVD pilot was implemented, only 35% SS from the total available 907 SS could apply the solution due to different factors (e.g., physical available surfaces, the structure and topology of the distribution networks). Based on similar assumption of 35%, RESOLVD can be replicated to a large number of SS in the EU and receive a proportional benefit from increased hosting capacity, DSO investment reduction, and other associated direct and indirect benefits. From a market potential point of view, in the EU, there are about 190 large DSOs (≥100 K customers) and more than 2400 small and medium DSOs which have less than 100 K customers (See Table 7). Only within partners' countries there are 27 large DSOs and 640 medium and small DSOs. Considering all these DSOs together with other stakeholders (aggregators, business entities/sister companies, third parties, etc.), RESOLVD could contribute to the EU targets of renewables by increasing the hosting capacity to reach high DERs and create an impact to the society at large.

Table 7. Number of DSOs in partner countries and EU and EEA for RESOLVD replication and market.

Partners	Population (MN)	DSO (<100 K Customers)	DSO (>100 K Customers)	TWh
Spain	47	349	5	278
Norway	5.4	150	7	118
Austria	8.8	138	13	61
Greece	11.1	2	1	45
Slovenia	2	1	1	13
Total	74.3	640	27	515
EU and EEA	520MN (inc. UK)	2400	190	2700

6. RESOLVD Implications on Business and Policy (Linked to RQ4)

6.1. Business Implication

From WTP scenario analysis, it is found that the SaaS approach is more cost-attractive than the traditional approach of owning software. In consultation with DSO, five additional benefits of the SaaS approach have been identified, making it a more attractive option from a business perspective. These additional benefits are: (i) DSOs can obtain a predictable payment, which can be consistently budgeted in each term, (ii) SaaS removes the burden of maintaining on-site servers and software, freeing up IT staff to work on more strategic initiatives and ensuring greater solution uptime, (iii) DSOs do not have to worry about unexpected expenses associated with unplanned maintenance and/or breakdown, (iv) DSOs have access to updated software all the time, rather than buying new updates, and (v) cloud-based software can be updated in the background without any disruption to the existing deployment. Considering this attractiveness, technology developers have become interested in exploring the option of providing a complete RESOLVD solution as a service. This has been identified as an area of further research.

PED (along with storage) has shown the potential to provide multiple flexibility services to the distribution grids. It can also provide flexible services to other sectors, such as buildings, industries, and EV charging stations. Thus, there could be multiple beneficiaries of the PED, but the cost needs to justify investment for other beneficiaries. One way to make PED more cost attractive would be to stack multiple values together. From a policy perspective, DSO needs to procure these services through market mechanisms and should act as a neutral market facilitator of such services. However, market mechanisms to procure flexible services are not yet clearly defined at a local level in most of the member states. Absence of such local market limits DSOs' ability to procure services and limits benefits derived from the value stacking capabilities of PED. Thus, for market adoption of technology such as PED, it is crucial that regulations on local energy markets are defined sooner.

6.2. Policy and Regulatory Implications

The adoption of solutions such as those developed by the RESOLVD project is directly correlated with the specific targets set by the EU and the policy linked with the measures aiming to increase the efficiency and hosting capacity of the distribution grids. The CBA outcomes presented in this paper give interesting insights, specifically to the DSOs, as to the impact that the RESOLVD solutions can have in achieving those targets/plans. The positive CBA shows that research results from the project are cost-effective in improving the hosting capacity of the existing distribution grid infrastructure, thereby deferring grid capacity investments.

From a techno-economic point of view, DSOs constantly look for ways to lower the total cost of infrastructure investments (capital expenditure, i.e., CAPEX). As demand for new infrastructure grows, DSOs look for ways to improve their management and mitigate investment risk. The CAPEX of grid expansion can be deferred by implementing technologies that improve operational efficiency of the grid, i.e., having operation expenditure (OPEX). However, DSO being a regulated business, the remuneration for investment in most of the member states in the EU is capital expenditure (CAPEX)-based. This limits DSOs to invest in innovations that are OPEX-based. An optimal mix of CAPEX and OPEX depends upon local conditions and, thus, CEER, in its recommendation to the EU commission, proposes allowing DSO to decide optimal mix. In this regard, a total expenditure (TOTEX) approach is considered promising to support innovations, as well as reduce overall costs associated with DSO business. This business approach is in line with the proposed change in business model by [46,47].

According to the new Spanish law 3/2013 (under Royal Decree-Law 1/2019), DSO remuneration may incorporate incentives, which may have positive or negative signs, to the reduction of costs of the electricity system derived from its operation or other objectives (see Section 4.2 for remuneration formula). RESOLVD results of CBA provide evidence that OPEX-based investments can be cost-effective in both increasing hosting capacity and deferring gid upgrade.

7. Conclusions

The EU energy policy principles to overcome the challenges in the field of energy include speeding up the slow progress in energy efficiency and alleviating the challenges emanating from the increasing share of renewables in a cost-effective way. This research provides concrete evidence for energy policymakers, DSOs, and technology developers by validating a complete proof of concept of RESOLVD solution in the real-life pilot to improve energy efficiency and hosting capacity with reduced investment cost.

The IT and smart LV grid solutions developed, implemented, and tested in the project RESOLVD justified the economic viability of a complete solution. The research impact was substantiated by doing a detailed cost–benefit analysis (CBA) and the results revealed that the solution offered by RESOLVD is economically viable with high scale-up and replication potential. The major savings were found from increased renewables hosting capacity and reduction in DSO investment (see Tables 4 and 6). Both direct and indirect benefits quantified based on carefully defined KPIs, control indicators (CI), and shadow price (retribution and willingness to pay) showed positive results. These results substantiate the relevance of the proposed solution and give evidence for policymakers and other DSOs so that they can enable a speed up and facilitate the transition from the traditional grid upgrade towards intelligent grid operation and management.

Another interesting finding from the DSO, technology providers, and future smart grid business standpoint of view is the scenario test of software as a service (SaaS). The willingness to pay (WTP) scenario test based on SaaS showed that the three-software developed in the project (ILEM, ESB, and LVD-DST) reduced the payback period significantly (about 3 years on average), with higher NPV, IRR, and benefit–cost ratio. This has strong implications to the DSOs and technology developers to plan for the large-scale deployment of the solution considering the whole value chain, including manufacturing of the necessary hardware and software components for the complete solution. This paves a way forward and provides an in-depth insight to exploit the benefits gained from grid operational expenditure (OPEX) and transforms the traditional CAPEX to TOTEX. From the DSO perspective, intelligent local energy managers under the SaaS plan can make the solution economically attractive in the short term not only from the NPV values, but also from the need of high interactions with the end-user and proximity of the ILEM.

To summarize, the findings in this research strengthen the EU policy in the context of improved energy efficiency, increasing hosting capacity, and promoting low-carbon and clean energy technologies to drive the energy transition through the RESOLVD cost-effective solution, which, in turn, improves competitiveness. Further, it facilitates and speeds up the process of ensuring energy security through solidarity and co-operation between EU countries, which is one the EU energy policies (e.g., in RESOLVD, seven partners from Spain, Norway, Austria, Slovenia, and Greece involved).

Author Contributions: Conceptualization, A.M.B. and S.P.; methodology, A.M.B.; software, J.M., I.L., M.S., and F.D.-G.; validation, R.G.-F.; formal analysis, A.M.B. and S.P.; investigation, A.M.B.; resources, H.T.; data curation, A.M.B. and S.P.; writing—original draft preparation, A.M.B. and S.P.; writing—review and editing, H.T., A.M.B., S.P., J.M. and I.L.; visualization, J.M., R.G.-F., A.M.B. and S.P.; supervision, H.T.; project administration, H.T. and S.P.; funding acquisition, J.M. and H.T. All authors have read and agreed to the published version of the manuscript.

Funding: The RESOLVD project has received funding from the European Union's Horizon 2020 research and innovation program under grant agreement No. 773715. The information and views set out in this study are those of the author(s) and do not necessarily reflect the official opinion of the European Union. Neither the European Union institutions and bodies nor any person acting on their behalf may be held responsible for the use which may be made of the information contained therein.

Institutional Review Board Statement: Not applicable.

Informed Consent Statement: Not applicable.

Data Availability Statement: Not applicable.

Conflicts of Interest: The authors declare no conflict of interest.

Abbreviations

AMI	Advanced Metering Infrastructure
BAU	Business As Usual
BMS	Battery Management System
CAPEX	Capital Expenditures
CBA	Cost–Benefit Analysis
CIM	Common Information Model
DER	Distributed Energy Resource
DG	Distribution Generation
DMS	Distribution Management System
DSO	Distribution System Operator
EES	Electrical Energy Storage
EF	Energy Forecaster
GIS	Geographic Information System
GW	Gateway
IED	Intelligent Electronic Device
ILEM	Intelligent Local Energy Manager
JRC	Joint Research Centre
KPI	Key Performance Indicator
LV	Low Voltage
MDMS	Metering Data Management System
MV	Medium Voltage
OPEX	Operating Expenditure

PCS	Power Conversion System
PED	Power Electronic Device
PMU	Phasor Measurement Unit
PQM	Power Quality Monitoring
RQ	Research Question
RES	Renewable Energy Source
SaaS	Software-as-a-Service
SCADA	Supervisory Control and Data Acquisition
SG	Smart Grids
SGT	Smart Grid Technologies
SGAM	Smart Grid Architecture Model
SM	Smart Meter
SS	Secondary Substation
TOTEX	Total Expenditure
TSO	Transmission System Operator
WAMS	Wide Area Monitoring System
WTP	Willingness-To-Pay

Appendix A

Table A1. Some list of cost and benefit attributes.

Investment Costs (Development and Service)	Costs in EUR
Intelligent Local Energy Manager (ILEM) development (PED + Personnel+ batteries)	xxxxxxx
Remote Terminal Unit (RTU)	xxxx
Low voltage switchgear (LVSG)	xxxx
Phasor Measurement Unit (PMU)	xxxx
Phasor quality meter (PQM)	xxxx
Operation cost(effort)	xxxx (DSO retribution)
PSS power system supply	xxxx
Gateway	xxxx
Router	xxxx
EBS software platform (development cost)	xxxxxxx
LVD-DST software platform/forecasting and scheduling	xxxxxxx
Cyber security service	xxxxxx
Other investment costs	
Replacement cost	xxxx
Residual cost	(xx)
Revenue/Savings	**Savings in EUR per Year**
KPI-1: Power loss reduction due to waveform quality improvement	xxx
KPI-2 Improvement of the energy profile in the secondary substations	xxxx
CI-1 Efficiency rate of the PED and the energy storage system	-xxx
KPI-3 Increase of DERs hosting capacity in LV network	xxxxx
KPI-4 Reduction of DSO investment	xxxxx
CI-2 DSO operation expenditures with respect to the BAU solutions	xxxx
KPI-5 Percentage of improvement in line voltage profiles with power injection and consumption	xx
KPI-6 Rate of prevented critical events in the LV grid due to forecasting and remote control of grid actuators	x
KPI-7Quality of online event detection in LV grid	x
KPI-8 Quality and time needed for awareness and localization of grid fault MV grid (per incident 1785)	xxxx
KPI-9 Quality of LV grid operation in island mode	xxxx
CI-3 Waveform quality in LV grid	xx

Appendix B

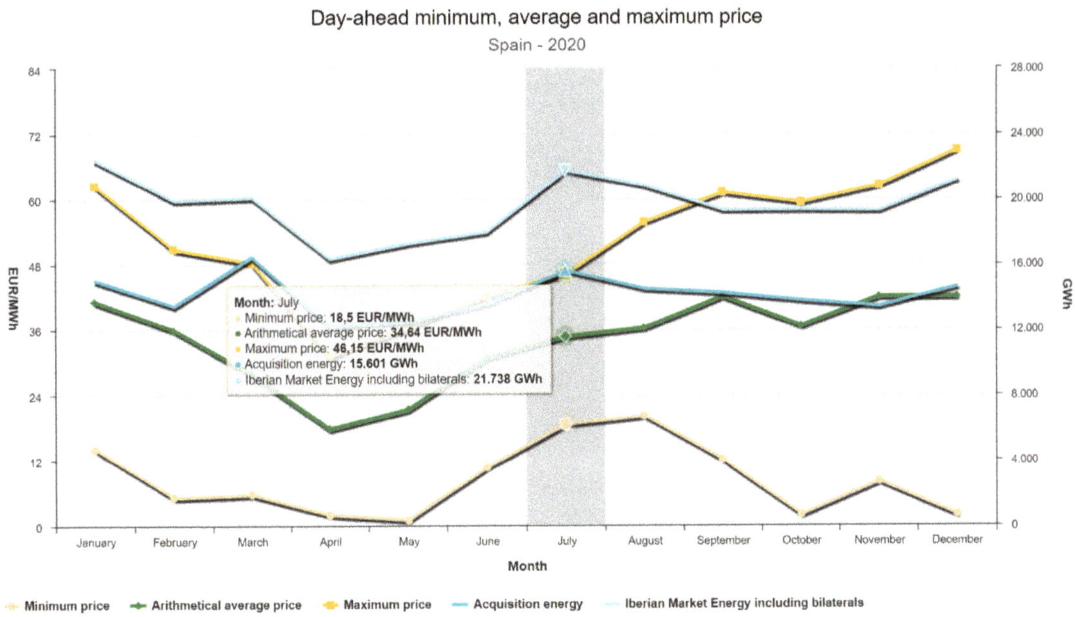

Figure A1. Average price used for some of the KPIS to quantify the benefits (source: OMIEData).

References

1. Cambini, C.; Meletiou, A.; Bompard, E.; Masera, M. Market and regulatory factors influencing smart-grid investment in Europe: Evidence from pilot projects and implications for reform. *Util. Policy* **2016**, *40*, 36–47. [CrossRef]
2. Obi, M.; Bass, R. Trends and challenges of grid-connected photovoltaic systems—A review. *Renew. Sust. Energy Rev.* **2016**, *58*, 1082–1094. [CrossRef]
3. Ayodele, T.R.; Jimoh, A.; Munda, J.L.; Agee, J.T. Challenges of grid integration of wind power on power system grid integrity: A review. *Int. J. Renew. Energy Res.* **2012**, *2*, 618–626.
4. Noske, S.; Falkowski, D.; Swat, K.; Boboli, T. UPGRID project: The management and control of LV network. *CIRED-Open Access Proc. J.* **2017**, *2017*, 1520–1522. [CrossRef]
5. Manzoni, A.; Castro, R. Microgeneration impact on lv distribution grids: A review of recent research on overvoltage mitigation techniques. *Int. J. Renew. Energy Res.* **2016**, *6*, 117–131.
6. Bødal, E.F.; Del Granado, P.C.; Farahmand, H.; Korpås, M.; Olivella, P.; Munné, I.; Lloret, P. Challenges in Distribution Grid with High Penetration of Renewables. Deliverable from H2020 Project INVADE. 2018. Available online: https://h2020invade.eu/wp-content/uploads/2017/05/D5.1-Challenges-in-distribution-grid-with-high-penetration-of-renewables.pdf (accessed on 30 March 2021).
7. Martinot, E. Grid Integration of Renewable Energy: Flexibility, Innovation, and Experience. *Annu. Rev. Environ. Resour.* **2016**, *41*, 223–251. [CrossRef]
8. Albadi, M.; El-Saadany, E. A summary of demand response in electricity markets. *Electr. Power Syst. Res.* **2008**, *78*, 1989–1996. [CrossRef]
9. Díaz-González, F.; Sumper, A.; Gomis-Bellmunt, O. *Energy Storage in Power Systems*; John Wiley & Sons: West Sussex, UK, 2016; p. 314.
10. Sidhu, A.S.; Pollitt, M.G.; Anaya, K.L. A social cost–benefit analysis of grid-scale electrical energy storage projects: A case study. *Appl. Energy* **2018**, *212*, 881–894. [CrossRef]
11. Padmini, V.; Omran, S.; Chatterjee, K.; Khaparde, S.A. Cost Benefit Analysis of Smart Grid: A Case Study from India. In Proceedings of the 2017 North American Power Symposium (NAPS), Morgantown, WV, USA, 17–19 September 2017; pp. 1–6.
12. Kuhar, U.; Kosec, G.; Svigelj, A. *Observability of Power-Distribution Systems: State-Estimation Techniques and Approaches*; Springer: Cham, Switzerland, 2020.
13. Saldaña-González, A.E.; Sumper, A.; Aragüés-Peñalba, M.; Smolnikar, M. Advanced Distribution Measurement Technologies and Data Applications for Smart Grids: A Review. *Energies* **2020**, *13*, 3730. [CrossRef]

14. Pokhrel, B.R.; Bak-Jensen, B.; Pillai, J.R. Integrated Approach for Network Observability and State Estimation in Active Distribution Grid. *Energies* **2019**, *12*, 2230. [CrossRef]
15. Zhang, R.; Xu, L.; Yu, Z.; Shi, Y.; Mu, C.; Xu, M. Deep-IRTarget: An Automatic Target Detector in Infrared Imagery using Dual-domain Feature Extraction and Allocation. *IEEE Trans. Multimed.* **2021**. [CrossRef]
16. Zhang, Q.; Ge, L.; Zhang, R.; Metternicht, G.I.; Du, Z.; Kuang, J.; Xu, M. Deep-learning-based burned area mapping using the synergy of Sentinel-1&2 data. *Remote Sens. Environ.* **2021**, *264*, 112575.
17. Zhang, R.; Wu, L.; Yang, Y.; Wu, W.; Chen, Y.; Xu, M. Multi-camera multi-player tracking with deep player identification in sports video. *Pattern Recognit.* **2020**, *102*, 107260. [CrossRef]
18. Díaz-González, F.; Sumper, A.; Gomis-Bellmunt, O.; Villafafila-Robles, R. A review of energy storage technologies for wind power applications. *Renew. Sustain. Energy Rev.* **2012**, *16*, 2154–2171. [CrossRef]
19. Bullich-Massagué, E.; Cifuentes-García, F.-J.; Glenny-Crende, I.; Cheah-Mañé, M.; Aragüés-Peñalba, M.; Díaz-González, F.; Gomis-Bellmunt, O. A review of energy storage technologies for large scale photovoltaic power plants. *Appl. Energy* **2020**, *274*, 115213. [CrossRef]
20. Brogan, P.V.; Best, R.; Morrow, J.; Duncan, R.; Kubik, M. Stacking battery energy storage revenues with enhanced service provision. *IET Smart Grid* **2020**, *3*, 520–529. [CrossRef]
21. Forrester, S.P.; Zaman, A.; Mathieu, J.L.; Johnson, J.X. Policy and market barriers to energy storage providing multiple services. *Electr. J.* **2017**, *30*, 50–56. [CrossRef]
22. Jing, W.; Lai, C.H.; Wong, W.S.H.; Wong, M.L.D. A comprehensive study of battery-supercapacitor hybrid energy storage system for standalone PV power system in rural electrification. *Appl. Energy* **2018**, *224*, 340–356. [CrossRef]
23. Ren, G.; Wang, H.; Chen, C.; Wang, J. An energy conservation and environmental improvement solution-ultra-capacitor / battery hybrid power source for vehicular applications. *Sustain. Energy Technol. Assess.* **2021**, *44*, 100998. [CrossRef]
24. Sellali, M.; Betka, A.; Djerdir, A. Power management improvement of hybrid energy storage system based on H ∞ control. *Math. Comput. Simul.* **2020**, *167*, 478–494. [CrossRef]
25. Jing, W.; Lai, C.H.; Wong, W.S.H.; Wong, M.L.D. Dynamic power allocation of battery-supercapacitor hybrid energy storage for standalone PV microgrid applications. *Sustain. Energy Technol. Assess.* **2017**, *22*, 55–64. [CrossRef]
26. Sinha, S.; Bajpai, P. Power management of hybrid energy storage system in a standalone DC microgrid. *J. Energy Storage* **2020**, *30*, 101523. [CrossRef]
27. Pegado, R.; Ñaupari, Z.; Molina, Y.; Castillo, C. Radial distribution network reconfiguration for power losses reduction based on improved selective BPSO. *Electric Power Syst. Res.* **2019**, *169*, 206–213. [CrossRef]
28. Li, P.; Xu, D.; Zhou, Z.; Lee, W.-J.; Zhao, B. Stochastic Optimal Operation of Microgrid Based on Chaotic Binary Particle Swarm Optimization. *IEEE Trans. Smart Grid* **2015**, *7*, 66–73. [CrossRef]
29. Rajaram, R.; Kumar, K.S.; Rajasekar, N. Power system reconfiguration in a radial distribution network for reducing losses and to improve voltage profile using modified plant growth simulation algorithm with Distributed Generation (DG). *Energy Rep.* **2015**, *1*, 116–122. [CrossRef]
30. Rao, R.S.; Ravindra, K.; Satish, K.; Narasimham, S.V.L. Power Loss Minimization in Distribution System Using Network Reconfiguration in the Presence of Distributed Generation. *IEEE Trans. Power Syst.* **2012**, *28*, 317–325. [CrossRef]
31. Lopez, E.; Opazo, H.; Garcia-Santander, L.; Bastard, P. Online Reconfiguration Considering Variability Demand: Applications to Real Networks. *IEEE Trans. Power Syst.* **2004**, *19*, 549–553. [CrossRef]
32. De Quevedo, P.M.; Contreras, J.; Rider, M.J.; Allahdadian, J. Contingency Assessment and Network Reconfiguration in Distribution Grids Including Wind Power and Energy Storage. *IEEE Trans. Sustain. Energy* **2015**, *6*, 1524–1533. [CrossRef]
33. Cong, P.; Tang, W.; Zhang, L.; Zhang, B.; Cai, Y. Day-Ahead Active Power Scheduling in Active Distribution Network Considering Renewable Energy Generation Forecast Errors. *Energies* **2017**, *10*, 1291. [CrossRef]
34. Taylor, T.; Ohrn, M. Network Management for Smart Grids—Innovative Operations Centers to Manage Future Distribution Networks. 2009. Available online: https://library.e.abb.com/public/461c2ae39130ceafc125762d0047f01f/45-49%203M901_ENG72dpi.pdf (accessed on 30 March 2021).
35. Wang, J.; Chen, C.; Lu, X. *Guidelines for Implementing Advanced Distribution Management Systems-Requirements for DMS Integration with DERMS and Microgrids*; Argonne National Laboratory: Argonne, IL, USA, 2015.
36. Alaqeel, T.A.; Suryanarayanan, S. A comprehensive cost–benefit analysis of the penetration of Smart Grid technologies in the Saudi Arabian electricity infrastructure. *Util. Policy* **2019**, *60*, 100933. [CrossRef]
37. Basso, G.; Gaud, N.; Gechter, F.; Hilaire, V.; Lauri, F. A framework for qualifying and evaluating smart grids approaches: Focus on multi-agent technologies. *Smart Grid Renew. Energy* **2013**, *4*, 33933. [CrossRef]
38. Behi, B.; Baniasadi, A.; Arefi, A.; Gorjy, A.; Jennings, P.; Pivrikas, A. Cost–benefit analysis of a virtual power plant including solar PV, flow battery, heat pump, and demand management: A western australian case study. *Energies* **2020**, *13*, 2614. [CrossRef]
39. Frigola, J.M.; Kokos, I.; Tuiskula, H.; Marksteiner, S.; Sumper, A.; Marksteiner, S.; Gallart, R.; Smolnikar, M.; Torrent, F.F. RESOLVD-Renewable Penetration Levered by Efficient Low Voltage Distribution Grids. Specifications and Use Case Analysis. In Proceedings of the 25th International Conference on Electricity Distribution, Madrid, Spain, 3–6 June 2019.
40. Sofia, D.; Gioiella, F.; Lotrecchiano, N.; Giuliano, A. Cost–benefit analysis to support decarbonization scenario for 2030: A case study in Italy. *Energy Policy* **2019**, *137*, 111137. [CrossRef]

41. Giordano, V.; Onyeji, I.; Fulli, G. *Guidelines for Conducting a Cost–Benefit Analysis of Smart Grid Projects*; Joint Research Centre: Brussels, Belgium, 2012.
42. Candido, L.; Gallart, R.; Coral, C.; Palma, V.; Girbau, F.; Ferrer, A.; Meléndez, J. Validation Set Up Report: Description of Scenarios, Tests and Validation Indicators; RESOLVD Deliverable D5.1. 2019. Available online: https://resolvd.eu/wp-content/uploads/2021/10/D5.1-FV.pdf (accessed on 31 October 2021).
43. Díaz-González, F.; Aragüés-Peñalba, M.; Girbau-Llistuella, F.; Llonch-Masachs, M.; Sumper, A. A Power Sharing Algorithm for a Hybrid Energy Storage System Based on Batteries. In Proceedings of the 2019 IEEE PES Innovative Smart Grid Technologies Europe (ISGT-Europe), Bucharest, Romania, 29 September–2 October 2019.
44. Llonch, M.; Girbau-Llistuella, F.; Díaz-González, F.; Sumper, A.; Aragüès-Peñalba, M. Deliverable 2.5: Functional Laboratory Tests on the Prototype, RESOLVD Project. Available online: https://resolvd.eu/documents/ (accessed on 31 October 2021).
45. Sodin, D.; Rudež, U.; Mihelin, M.; Smolnikar, M.; Čampa, A. Advanced Edge-Cloud Computing Framework for Automated PMU-Based Fault Localization in Distribution Networks. *Appl. Sci.* **2021**, *11*, 3100. [CrossRef]
46. Van den Oosterkamp, P.F.; de Joode, J.; van der Welle, A.J.; Lenstra, W.J.; Koutstaal, P.R.; van Hussen, K.; Haffner, R. *The Role of DSOs in a Smart Grid Environment*; Ecorys: Amsterdam, The Netherlands, 2014.
47. Azizivahed, A.; Arefi, A.; Ghavidel, S.; Shafie-khah, M.; Li, L.; Zhang, J.; Catalão, J.P. Energy management strategy in dynamic distribution network reconfiguration considering renewable energy resources and storage. *IEEE Trans. Sustain. Energy* **2019**, *11*, 662–673. [CrossRef]

energies

Article

Optimal Configuration and Sizing of Seaport Microgrids including Renewable Energy and Cold Ironing—The Port of Aalborg Case Study

Nur Najihah Abu Bakar [1,2,*], Josep M. Guerrero [1,*], Juan C. Vasquez [1], Najmeh Bazmohammadi [1], Muzaidi Othman [2], Brian Dalby Rasmussen [3] and Yusuf A. Al-Turki [4]

1 Center for Research on Microgrids (CROM), AAU Energy, Aalborg University, 9220 Aalborg, Denmark; juq@energy.aau.dk (J.C.V.); naj@energy.aau.dk (N.B.)
2 Faculty of Electrical Engineering Technology, University Malaysia Perlis (UniMAP), Kampus Pauh Putra, Arau 02600, Perlis, Malaysia; muzaidi@unimap.edu.my
3 Port Facility and Environment Management, Port of Aalborg, Langerak 19, 9220 Aalborg, Denmark; bdr@portofaalborg.com
4 Center of Research Excellence in Renewable Energy and Power Systems, Department of Electrical and Computer Engineering, Faculty of Engineering, K. A. CARE Energy Research and Innovation Center, King Abdulaziz University, Jeddah 21589, Saudi Arabia; yaturki@kau.edu.sa
* Correspondence: nurnbab@energy.aau.dk or nurnajihah@unimap.edu.my (N.N.A.B.); joz@energy.aau.dk (J.M.G.)

Citation: Bakar, N.N.A.; Guerrero, J.M.; C. Vasquez, J.; Bazmohammadi, N.; Othman, M.; Rasmussen, B.D.; Al-Turki, Y.A. Optimal Configuration and Sizing of Seaport Microgrids including Renewable Energy and Cold Ironing—The Port of Aalborg Case Study. *Energies* **2022**, *15*, 431 . https://doi.org/10.3390/en15020431

Academic Editor: Miguel Jiménez Carrizosa

Received: 19 October 2021
Accepted: 4 January 2022
Published: 7 January 2022

Abstract: Microgrids are among the promising green transition technologies that will provide enormous benefits to the seaports to manage major concerns over energy crises, environmental challenges, and economic issues. However, creating a good design for the seaport microgrid is a challenging task, considering different objectives, constraints, and uncertainties involved. To ensure the optimal operation of the system, determining the right microgrid configuration and component size at minimum cost is a vital decision at the design stage. This paper aims to design a hybrid system for a seaport microgrid with optimally sized components. The selected case study is the Port of Aalborg, Denmark. The proposed grid-connected structure consists of renewable energy sources (photovoltaic system and wind turbines), an energy storage system, and cold ironing facilities. The seaport architecture is then optimized by utilizing HOMER to meet the maximum load demand by considering important parameters such as solar global horizontal irradiance, temperature, and wind resources. Finally, the best configuration is analyzed in terms of economic feasibility, energy reliability, and environmental impacts.

Keywords: cold ironing; energy management system; optimal sizing; renewable energy sources; seaport microgrids; maritime; HOMER

1. Introduction

Ports worldwide have different sizes, operations, geological, geographical features, and a variety of energy sources that will affect their power demand as well as energy production. The main sources of energy supply come from the utility grid and diesel generators, constantly emitting greenhouse gas emissions. Moreover, the energy sector is encountering primary energy depletion, considering that the growing load demand is exceeding power generation. With a rising awareness of the jeopardy from resource depletion issues and environmental pollution, many ports around the world are taking action toward the zero-carbon footprint goal. The urge to use alternative clean energy resources makes microgrids one of the best solutions for future green seaports. In 2019, Denmark showed a remarkable result by generating half of its electricity from wind and solar power [1]. The advantage of the climate's constant breezes and bluster in this country makes the wind turbine practical for use and well established. The Danish maritime industry has set a target of 70% reduction in CO_2 emission by implementing wind energy and energy management systems in the ports [2].

Although microgrids are widely used in different land applications, there are limitations for their real implementation in the seaport sector. This situation is a good opportunity to introduce microgrids into the seaports, but at the same time it is challenging to access the references in terms of system requirements, optimal framework, stability, and maintenance needs. Especially during the initial planning phase, it is a vital task to find the optimal design of the seaport microgrid with a compatible configuration and the right sizing for each component.

In recent years, a lot of effort has been devoted to the optimal design of microgrids. Di Wu et al. [3] conduct a study on a two-stage stochastic mixed-integer programming technique for finding the optimal size of multiple distributed energy resources (DERs). The proposed model considers the optimal outcome in terms of economic benefits and system resilience. A comparative study in [4] evaluates the optimal framework for a standalone microgrid by using four different metaheuristic algorithms. The merit of the study is taking into account the uncertainty in RES power production and load consumption. Moreover, a sensitivity analysis for several battery prices and capacities is provided to find the best framework. The same technique is applied in [5] but with a three-level planning framework for the extreme event cases in a microgrid power system. The purpose of the study is to preserve system security for better power delivery. Furthermore, although numerous studies are available for the optimal sizing of batteries for microgrids, the battery sizing problem is still far from being mature. Different research strategies and schemes are applied in [6–10] to identify the optimal design for ESSs. Despite extensive research on design and optimal sizing of microgrid technologies in various applications, there is still a shortfall of this research in seaport microgrids. As the seaport itself is a very complex system consisting of various dynamic loads in different regions (shore and seaside), the maritime industry has traditionally been slow and reluctant in acknowledging this technology. However, seaport microgrids have been attracting a great deal of attention over recent years to improve maritime power systems' performance.

Therefore, this paper presents a simulation-based method using the HOMER optimizer to investigate the seaport microgrids configuration and sizing problem in terms of cost minimization, energy production, and environmental impact. Seaport microgrids in this paper refer to the integration of microgrid power systems into seaports with cold ironing facilities.

The evolvement of trading activities worldwide is increasing the reliance on maritime transportation. This situation resulted in the growing concern about the carbon footprint of ships when docked at ports [11]. When a ship is at berth, the auxiliary engine keeps on to supply onboard power demand [12]. It consumes a huge amount of heavy diesel oil depending on the berthing hours and ship's power requirement that emits hazardous gases and degrades air quality. Hence, ships' electrification alternative comes to the scene to reduce the air pollution during berthing by using cold ironing technology. Cold ironing is an emission-free technology that prevents vessels from continuously burning fossil fuels by supplying onboard electric power directly from the onshore grid [13]. The auxiliary engine is turned off (cold process), but the ship can continue to operate normally since the switchboard draws the power from the shore side (ironing). There are three types of cold ironing topologies: (1) centralized cold ironing, (2) distributed cold ironing, and (3) DC distribution cold ironing [14]. According to D. Colarossi and P. Principi [15], the size of the ship power plant varies with the type of the ship, with a typical distributed power system of 400 V. On the shore side, ships are generally connected with a medium voltage (MV) of 6.6 kV/11 kV. In order to step down or step up to the desired voltage level, transformers play a vital role in the system. Nevertheless, there are some obstacles in terms of frequency where most of the ports use 60 Hz, whereas Europe and Asia's ports use 50 Hz [13]. When the frequency on the shore side and the frequency on the ship do not match, a frequency converter is required. Figure 1 illustrates a typical cold ironing connection from the shore side to the shipside.

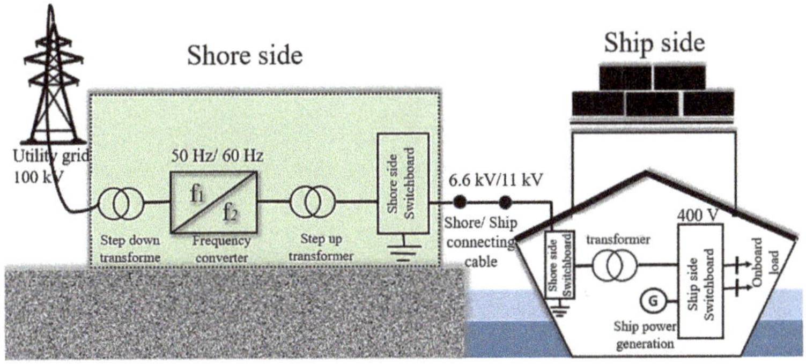

Figure 1. Typical cold ironing system from shore side to the seaside.

One of the challenges for port operation is the uncertainty in power demand, where the volume of traffic of berthing ships can suddenly change. Another scenario is the docking of large ships with heavy loads at the same time where the load demand might be larger than the available supply. Energy security to ensure the reliable port operation will be in low efficiency if cold ironing solely depends on the grid supply. These challenges are main motivations of this paper for investigating seaport microgrids with cold ironing facilities.

The emerging concept of the microgrid into the cold ironing is beneficial for the maritime industry. Collaboration between these two electrification technologies provides seaports with several advantages including resiliency, emission control, and economics. Cold ironing can cause power disturbances if the continuous high demand from the ships during berthing overtakes the supply capacity that cold ironing can provide. In the case of power breakdown, many of the port operations will be affected, leading to losses of billions of dollars. In this case, local distributed energy from the seaport microgrid is capable to offer the required energy in time of stress [16], thereby increasing energy security and resiliency. In addition, seaport microgrids can achieve high levels of port electrification by embracing cold ironing technology which the grid is unable to support. In terms of emission control, cold ironing itself eliminates a portion of emission by shutting down the conventional auxiliary engines while docking [17]. Seaport microgrids result in more pollution reduction by providing zero-carbon power from renewable energy sources. Moreover, seaport microgrids are economically efficient to reduce ports' operation costs, decrease peak-hour demand, and have the potential to sell back energy to the grid in case of having excess power. The contribution of this paper is three-fold:

- First, this paper attempts to integrate two of the most noteworthy maritime electrification technologies (cold ironing and microgrid) to enhance the sustainability of seaports energy systems. Although the microgrid concept is widely used in land-based applications, this technology remains scarce for the seaport sector.
- Second, this paper presents an optimal design for the seaport microgrid with the least cost providing a comparative study between three models.
- Third, in response to the three major concerns of the seaport sector, an optimal configuration for a seaport microgrid is provided with an analysis of economic feasibility, energy reliability, and environmental impacts. This analysis aims to investigate how integrating the microgrid concept into seaport applications may resolve the above-mentioned maritime issues.

The remainder of this paper is structured as follows. In Section 2, a review of the design optimization and sizing approaches for microgrids is provided. The methodology used for the design of the seaport microgrid, design parameters, and optimizer are introduced in

Section 3. Afterwards, the outcome from the proposed design is discussed and analyzed in Section 4. Finally, in Section 5, all significant findings of the paper are summarized.

2. Sizing

The optimal design and operation of microgrids have recently been the subject of extensive research. This is supported by an increasing trend of publications and research findings in this area. Ports are critical to the global economy, accounting for a big percentage of global trade and transportation. As a result, ports are preoccupied with providing labor for processing and handling goods, as well as other port-related services. Today's aggressive development in seaport trading necessitates an efficient power system capable of covering ports' electricity needs. Here, comes the concept of hybrid microgrid systems, in which the power generation is a mix of available clean energy sources with or without grid connection. The process to plan and develop this power system at seaports involves preliminary actions such as modeling, data collection, load and generation forecasting, initial simulation, evaluation, and performance assessment.

Sizing is a vital task to identify the optimal system configuration and the right capacity of the components to fulfill the load demand. Moreover, optimization is required to ensure that the system operates at high efficiency to maximize economic benefits while minimizing energy consumptions and environmental footprint. The study of microgrid systems in harbor areas by A. Roy et al. [18] emphasizes the importance of knowing the load demands and several evaluation criteria such as economic, pollution, and reliability, as well as geographical information to determine the sizing of the microgrid.

However, energy demands in seaports are highly dynamic and uncertain due to a variety of unpredictable factors such as the port's daily routine, activity handling, and environmental variables (weather conditions, temperature, and sea waves). Accordingly, there are sudden uncertain high loads that affect the stability of the system [19]. Considering this situation, ensuring the availability of power supply is significant to prevent disturbance in the maritime power system. P.Xie et al. [20] identify three common objectives and seven constraints of the seaport sector, which are summarized in Table 1. Based on the selected objectives and constraints, an optimization algorithm is developed and simulated to find the optimal design.

Table 1. Common objectives and constraints used in the port optimization problems.

Objectives	Constraints
▪ Fuel consumption minimization. ▪ Environmental footprint reduction. ▪ Economic investment minimization.	▪ Power and energy balance. ▪ Restraints for power quality. ▪ Restraints of power plants. ▪ Restraints of ESSs. ▪ Environmental constraints. ▪ Ship voyage constraints. ▪ Constraints for the auxiliary system.

For this purpose, both the computational resources and the required data related to the system are significant in the modeling phase. However, there are real-time data limitations, especially in the port sector. To overcome this challenge, computational simulation is a useful method to formulate and evaluate the microgrid performance before being implemented in real applications. The HOMER optimizer is one of the most widely used techniques in microgrid designs, which allows a flexible power system design with an integrated weather database for the RESs components. Table 2 presents an overview of the conducted studies on microgrid system designs using HOMER software in different sectors. Design objectives, system configuration, investigated sensitivity cases, site location, and the related sector are identified for each study.

Table 2. Summary of the Microgrid case study using HOMER optimizer. (Abbreviation: Photovoltaic (PV)).

References	*Sector*	Objectives	Configuration	Sensitivity Analysis	Case Study Location
[21] 2018	Residential Cottages	- Cost and emission minimization - Nonderivative optimization	A standalone MG including PV, diesel generator, ac load, lead-acid battery, li-ion battery, and power converter	- Fuel price - PV generation	Kea, Greece
[22] 2020	General	- Optimizing the size of the MG components	A grid-connected MG including wind turbine, PV, battery, load, and power converter	N/A	Bahir Dar City, Ethiophia
[23] 2016	City, general	- Optimal sizing - Optimal management of RESs and storage systems to fulfill the load demands and reduce the dependency on fossil fuels	A standalone MG including wind turbine, PV, microturbine, battery, and fuel cell	N/A	Nain, Iran
[24] 2016	ATM machine	- Feasibility analysis of solar- wind-diesel hybrid power system with maximum utilization of non-conventional generation systems while minimizing the total system cost	A standalone MG including PV, wind turbine, diesel generator, power converter, battery, and ac load	N/A	Vatar, Kolhapur
[25] 2011	Forest	- Analyzing real-time dynamic data	A standalone MG including PV, wind turbine, hydro, diesel generator, power converter, battery, ac load, and fuel cell	- Renewable resources - Hourly load data - PV array lifetime	Kondapalli, India
[26] 2020	Rural area (residential)	- Developing a microgrid, to explore the effect of certain problems such as power price, grid failure frequency, and grid mean repair time and studying its effects on cost (total operating cost, total capital cost, net present cost), electricity production, and unmet load	A grid connected MG including PV and battery	- Grid failure frequency - Grid mean repair time	South Africa
[27] 2020	Agricultural Load (residential and water pumping)	- Investigating the feasibility of the hybrid system	A standalone MG including battery, PV, diesel generator, ac load, water pumping load, and power converter	- Variations in PV cost - Diesel fuel price - Maximum annual capacity shortages (MACS)	Ein Albaida, Palestine
[28] 2017	Remote area	- To study two sizing methods for a standalone hybrid generation system using basic equations and Simulink Design Optimization (SDO) and HOMER optimizer	A standalone MG including Hydrokinetic, PV, diesel generator, battery, and ac load	N/A	Isla Santay (Guayaquil)
[29] 2018	Seaport	- Energy planning	A grid-connected MG including PV, wind, battery, ac load, and power converter	N/A	Copenhagen, Denmark
[30] 2021	Domestic load	- Minimizing the system cost	A standalone MG Biomass, PV, wind, battery, dual power converter, electrical load, and dumb load	N/A	Yanbu, Saudi Arabia

This paper develops a simulation-based method to determine the best configuration and overall sizing for a hybrid generation system in a seaport microgrid by utilizing the HOMER optimizer. The flow of the proposed methodology from the MG structure design, parameter selection, simulation, and sensitivity analysis are given in the following sections.

3. Optimization Framework

3.1. Structure of the Proposed Seaport Microgrid

In the proposed seaport microgrid design approach, RESs and ship demand during berthing are considered. The Port of Aalborg has been selected as a case study, which is located approximately at latitude 57°3.0′ N and longitude 10°3.2′ E, where energy sources mainly come from the utility grid, diesel generators, wind turbines (WTs), and PV systems. The Port of Aalborg handles a wide variety of goods and services such as container, cargo, railway, road, cruise, ships, and custom warehouse. However, in the modeling process, only the required power of ships during berthing is considered as it imposes a high energy demand on the port.

This paper aims to find the optimal seaport microgrid configuration and the optimal size of each component with the goal of cost minimization. Moreover, identifying the lowest net present cost (NPC) for the candidate architecture is a vital step in the seaport microgrid planning process. For instance, the payback period is very important for the investment by stakeholders as the energy consumption would be free during the rest of the project's lifetime. From the environmental perspective, RESs in microgrids provide the solution to the natural resources depletion issue and offer a green port landscape. Figure 2 illustrates the overview of the seaport microgrid configuration. The schematic design shows that the proposed system in this paper is connected with the main grid and consists of PV, WT, and a diesel generator, lithium-ion battery as a storage solution to complement RESs, a power converter between AC/DC busbars, and ships as electrical loads.

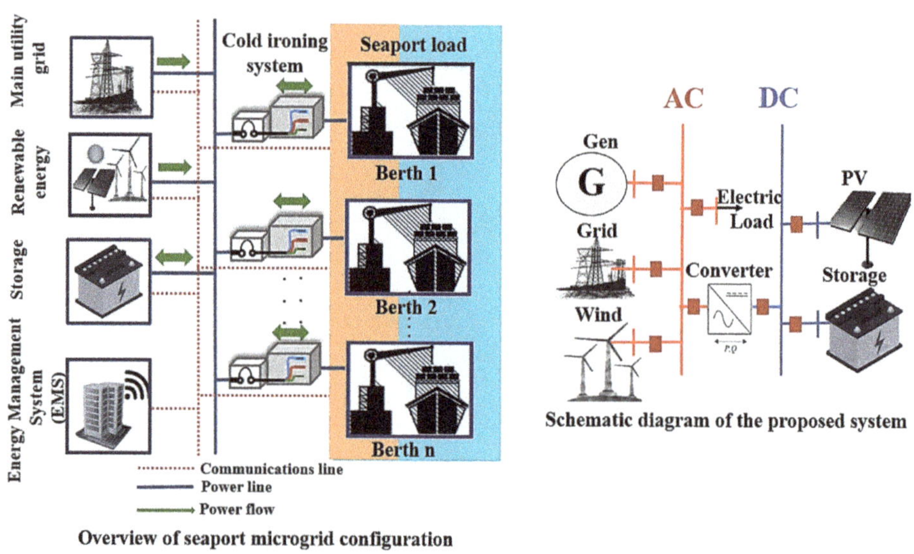

Figure 2. Overview and schematic diagram of the proposed system.

3.2. Load Profile

In the Port of Aalborg, ships are among the big energy consumers. Conventionally, when a ship is at berth, the auxiliary engines are turned on to support some basic functions and auxiliary loads in the ship that need electrical power. However, burning fuel by diesel generators harms the environment. Nowadays, with the green maritime goals of the shipping industry to reduce ship emission, cold ironing that also known as the onshore power supply (OPS) has come to the scene [31]. This technology allows ships to shut down their engines while berthed and plug into a power source at the shoreside [15]. In this way, ships have an adequate power supply to cover the onboard energy demand such as emergency equipment, cooling, heating, lighting, and refrigeration without the need to burn diesel fuels.

The load profile of ships during berthing varies according to three factors including time of berthing, the number of ships berthing per time, and the required power by a particular ship. There are different times of berthing for each ship and various numbers of ships berthing at the port from time to time. N. Ahamad et al. [32] summarize the average time of berthing and typical power requirement for different types of vessels, as listed in Table 3.

Table 3. Average berthing time and average power requirement for various types of vessels.

Type of Ships	Average Time Berthing (hrs)	Average Power
Chemical and other tankers	24–28	5 MW–6 MW
Bulk carrier	52	N/A
Container	21	1 MW–4 MW
General cargo	25	300 kW–6 MW
Ferries and RoRo	24	700 kW
Cruise	28	7 MW

The load profile used in the simulation in this study is shown in Figure 3. Peak hours are observed to be between 6 a.m. to 7 p.m. and the load slowly reduces afterward. According to this load profile, the average energy consumption is 23,977 kWh/day, which indicates that the average power per hour is 999.07 kW with a peak value of 2734.2 kW. Even though the power consumed by ships at the port depends on a few factors mentioned above during different periods and seasons of the year, it can be generally assumed that the peak load occurs during the afternoon as fewer ships are berthing between 8.00 p.m. and 5.00 a.m.

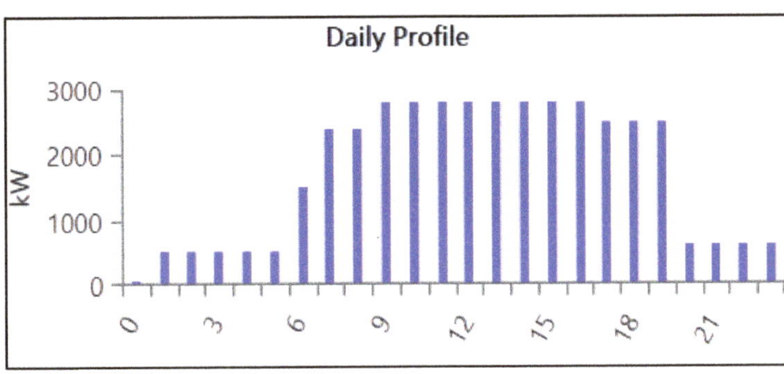

Figure 3. Hourly load profile in one day.

3.3. Meteorological Data

3.3.1. Solar Radiation and Temperature

All the meteorological data used in this simulation such as solar global horizontal irradiance (GHI) data, wind resources, and temperature information are from the NASA Prediction of Worldwide Energy Resources (POWER) database. The extracted data are related to the site location specified by the port coordination so that the approximation of the WT and PV output power produced in the simulation will be more accurate. The bar chart in Figure 4 shows that the highest daily radiation readings occur in May, June, and August. It is because Denmark is in the midst of its summer season at the time. Most of the days are sunny with a longer daylight period. Meanwhile, in other seasons, sun irradiation is low because it is cloudy and rainy most of the days. The irradiance ranges from 0.35 kWh/m^2/day to 6.06 kWh/m^2/day with an annual average of 3.02 kWh/m^2/day. The orange line indicates that the maximum and the minimum clearness indexes are 0.55 and 0.338, respectively.

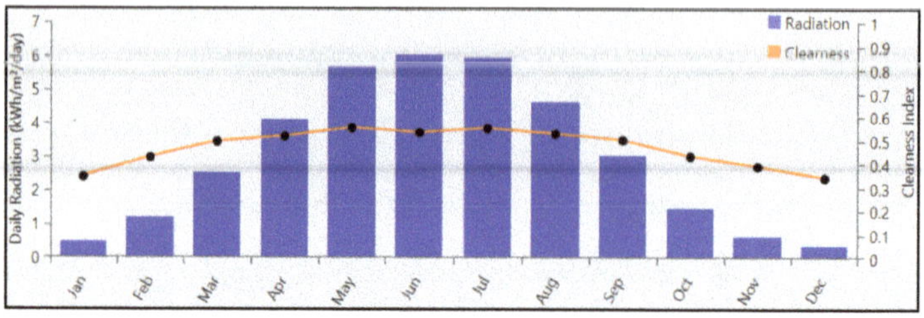

Figure 4. Solar GHI resources in one year.

The bar chart in Figure 5 displays the readings of the monthly average air temperature for this location. The range of temperature is between 0.94 °C and 16.66 °C with an annual average of 8.36 °C. The warmest months are June, July, August, and September, with an average temperature of 14.11 °C, 16.66 °C, 16.66 °C, and 13.44 °C, respectively. The temperature begins to fall sharply at the end of autumn and reaches its lowest point in February (0.94 °C).

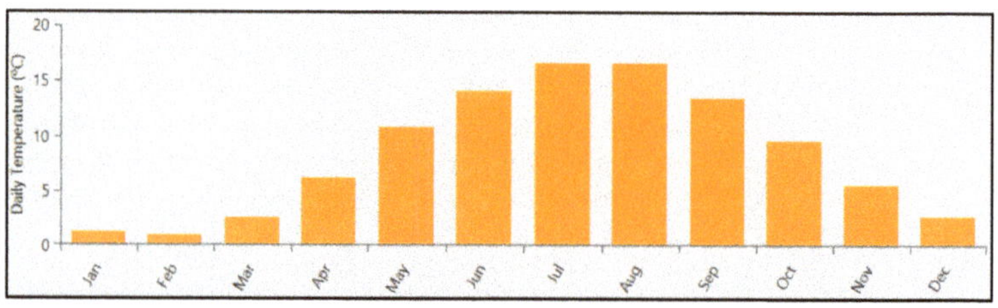

Figure 5. Average monthly temperature data of the target site in one year.

A. Haidar et al. [33] in their article highlighted that the generated power from PV modules is directly related to solar irradiance, temperature, the capacity of PV array, and its derating factor. This can be explained by Equation (1) that is used to determine the optimal capacity of the solar PV at time t.

$$P_{pv}(t) = P_{pv}^{ra}(t) \times f_{pv} \times \left(\frac{I}{I_e}\right) \times [1 + T_c\,(C_T - C_{Tc})] \quad (1)$$

where $P_{pv}^{ra}(t)$ is the PV array rated capacity (kW), f_{pv} is derating factor, I is irradiance incident on PV plate (kW/m^2), I_e is irradiance at the standard test condition (kW/m^2), T_c is temperature coefficient, C_T is cell temperature, and C_{Tc} is cell temperature at the standard test condition.

3.3.2. Wind Resources

The power generation of WTs relies heavily on weather conditions such as wind speed and wind direction [34]. A common linear mathematical model for wind energy generation estimation used to find the dynamical power curve of WTs output power is defined in Equation (2) [35]:

$$P_w(v) = \begin{cases} 0, & 0 \le v \ge v_{in} \\ P_r \cdot \dfrac{v - v_{in}}{v_r - v_{in}}, & v_{in} \le v \ge v_r \\ P_r, & v_r \le v \ge v_{out} \\ 0, & v \ge v_{out} \end{cases} \quad (2)$$

where $P_w(v)$ is the power output of the WT at wind speed v, P_r is the WT rated power, v_r is the rated wind speed, v_{in} is cut-in wind speed, and v_{out} is cut-out wind speed.

Figure 6 illustrates the monthly average wind speed at the height of 50 m above the surface of the earth over one year period. The annual average wind speed is 8.17 (m/s), with the highest value of 9.62 (m/s) in January and the lowest value of 6.81 (m/s) in July. Due to the wind speed variation with height, it is plotted on a logarithmic scale. As can be seen in this figure, the wind speed increases with height because there are fewer obstacles and turbulence.

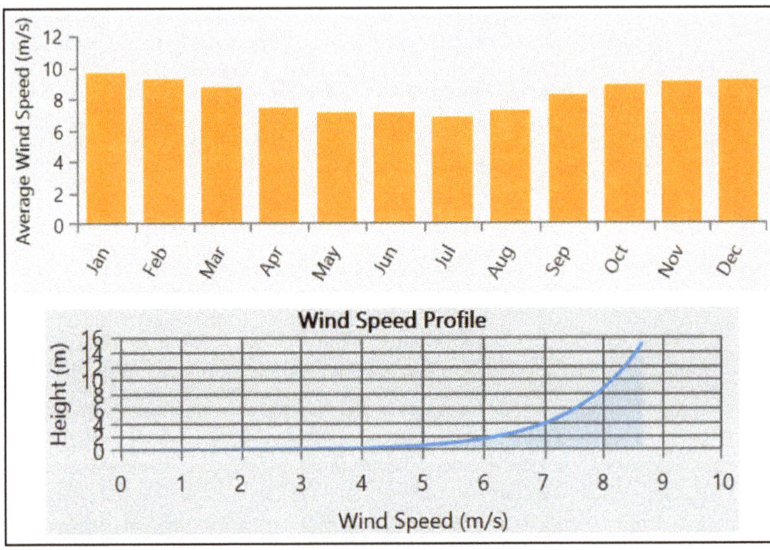

Figure 6. Wind speed profile of the target site in one year.

3.4. HOMER Optimizer

3.4.1. Schematic and Design Parameters of the Proposed Seaport Microgrid

Hybrid Optimization of Multiple Energy Resources (HOMER) is a simulation-based software used to optimize any integrated system by finding the right size of the equipment and the best possible system configuration while minimizing the net present cost (NPC). It stimulates the designed electric power system hour by hour for a year in the specific region considering the available energy resources present at the target location. For each time step, HOMER searches for many different configurations that satisfy the technical constraints at the lowest life cycle cost to meet the electrical load. Users can simulate their proposed power system and HOMER delivers the optimal candidate design.

Furthermore, it can perform sensitivity analysis to examine the impact of uncontrollable variables to see how they may affect the designed system costs. These variables include the price of fuel, which is always volatile, the price of components, lifetime data, efficiency, and other parameters that their change in the future is not known. Sensitivity analysis is critical for understanding the design's robustness. It enables users to be aware of how changing the input parameters affects the system architecture and cost.

Based on the proposed system in Section 3.1, the required components are added to the software. The schematic diagram of the system is shown in Figure 2. All the technical parameters and economic features are either from the software database or real data from commercial datasheets. Replacement cost is assumed to be 5–10% less than the capital cost.

3.4.2. Optimization Algorithm

HOMER has two optimization algorithms, which are: (1) the derivative-free algorithm and (2) the search algorithm. During the design process, the user may encounter a problem with determining the appropriate sizing or capacity for components that are compatible with the system. For the derivative-free algorithm, the software will automatically choose the appropriate sizing aiming for the least system cost. On the other hand, the search algorithm will simulate every possible system configuration by the quantity defined in the search space. The input value of the 'search space' varies depending on the peak load measurement. The selected size has a portion that is at the upper or lower end of the search space. It then specifies the best realizable system configuration capable of meeting the electric demand and finalizes it into a few categories. Table 4 shows the type of optimization option used for each of the components in this simulation.

Table 4. Optimization option used to determine the size for each component in the proposed design.

Component	HOMER Optimizer	Search Space
Diesel generator	✓	-
Wind turbine	-	5, 10, 15, 20, 25
PV	-	250, 500, 1000, 1500, 2000, 2500, 3000, 3500
Battery	✓	-
Converter	-	0~2880

The evaluation indices that are used in the software are net present cost (NPC), levelized cost of energy (COE), operating cost, and renewable fraction. The NPC or the life-cycle cost of a component is the present value of all the costs of establishing and operating every component in a system over the project lifetime, minus the present value of all the revenues that are earned over the project lifetime as shown in (3).

$$C_{NPC} = C_P - C_R \tag{3}$$

where C_P and C_R are the present values of all costs and revenues earned over the project lifetime, respectively.

Costs calculated in the algorithm include capital costs, replacement costs, O&M costs, fuel costs, emissions penalties, and the costs of buying power from the grid. Meanwhile, salvage value and grid sales income are included in the revenue. Overall, NPC is the HOMER's main economic output which is used to rank all system configurations in the optimization results and is the basis from which the total annualized cost and the levelized cost of energy (*COE*) are calculated as follows:

$$COE = \frac{C_{ann,tot} - C_{boiler} H_{served}}{E_{served}} \tag{4}$$

where $C_{ann,tot}$ is the total annualized cost of the system [USD/yr], C_{boiler} is the boiler marginal cost [USD/kWh], H_{served} and E_{served} are the total thermal and electrical loads served [kWh/yr], respectively. Meanwhile, to calculate the operating cost of the system, Equation (5) is utilized.

$$C_{operating} = C_{ann,tot} - C_{ann,capital} \tag{5}$$

where $C_{ann,capital}$ is the total annualized capital cost [USD/yr]. To have an efficient system, it is highly recommended to include the RESs in the architecture. The renewable fraction used in the software refers to the percentage of the energy supplied to the loads that come from RESs. It can be calculated using the following equation:

$$f_{ren} = 1 - \frac{E_{nonren} - H_{nonren}}{E_{served} - H_{served}} \tag{6}$$

where E_{nonren} is nonrenewable electrical production [kWh/yr], H_{nonren} is nonrenewable thermal production [kWh/yr].

3.4.3. Mathematical Formulation of the Objective Function and Constraint

The objective of the optimization problem is to minimize the operation cost, as follows: Objective function:

$$\min(C_{total}) = \sum P_{grid}C_{grid} + P_{gen}C_{gen} + P_{WT}C_{WT} + P_{PV}C_{PV} + P_{bat}C_{bat} + P_{con}C_{con} \tag{7}$$

where the cost of each components includes capital cost, operation, maintenance, replacement, and fuel costs as given below:

$$C_{element} = \sum C_{capital} + C_{O\&M} + C_{replacement} + C_{fuel} \tag{8}$$

The objective function is subject to the power balance constraint between generation and demand and also the percentage of renewable energy.

Constraints:

$$P_{grid} + P_{gen} + P_{WT} + P_{PV} + P_{bat} + P_{con} \geq P_{load} \tag{9}$$

$$P_{WT} + P_{PV} \geq 60\% \tag{10}$$

3.5. *Energy Management System (EMS) in the Design Phase*

The complexity of multiple power resources coordination, varying electrical loads, fluctuation in economic variables, and assessing the environmental impact require the use of an EMS in both design and planning phases. EMS focuses on the use and coordination of energy sources over a specific time frame and is often combined with future forecasting and projection systems [20]. It is strongly related to both energy efficiency as well as cost savings. However, different functionalities and requirements can be defined for the EMS depending on the situation and the microgrid application.

In this design, three key concerns that are taken into account are the minimization of fuel consumption, environmental impact, and economic investment. The big question is what kind of system configuration and sizing specification should be considered. To satisfy this, RESs and storage elements are included in the design process. PVs and WTs in the schematic diagram represent the RESs; meanwhile, battery represents the storage component. The use of RESs will help to sustain the energy supply and its environmentally friendly characteristic will protect the environment from harmful emissions. Since the energy sources are from nature (sun and wind), RESs also provide the microgrids with a cost-effective power generation solution. Furthermore, any excess energy generated by RESs can be sold back to the grid bringing further economic benefits.

Utilization of the battery as an ESS is beneficial for both having an energy backup during critical periods (energy shortage and peak hours) and storing the excess energy when energy generation is abundant. RESs such as PV and WT have some limitations for power production. PVs are only able to generate power during the day, depending on solar radiation and temperature. PV power output is also reduced on cloudy and rainy days. The same situation exists for the WTs, where wind speed varies depending on the site location and seasonal wind patterns. ESS is a vital component to store excess energy from both solar and wind sources when they have high power generation and supply the loads during periods of low energy availability, such as peak hours and power shortage intervals. Figure 7 illustrates how EMS plays part in the seaport microgrid system.

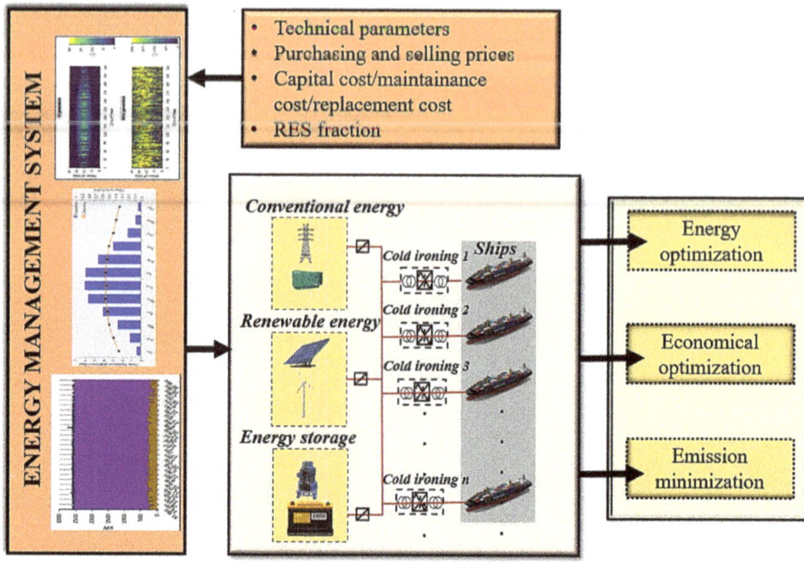

Figure 7. Energy Management System (EMS) in seaport microgrid.

4. Optimal Sizing of Seaport Microgrid

In this section, three different models, namely, Model 1, Model 2, and Model 3, are investigated for the optimal hybrid seaport microgrid, where Model 1 is the proposed system, Model 2 is the best system recommended by HOMER, and Model 3 is a conventional system without a battery and RESs. Table 5 shows the detailed comparison between optimization results for Model 1, Model 2, and Model 3. All models are simulated during the 25-year lifetime of the project.

Table 5. Optimization results (All the data are retrieved from Homer Pro, accessed on 5 May 2021. Abbreviations: net present cost (NPC), levelized cost of energy (COE), renewable energy source (RES), Operation and Maintenance (O&M)).

Parameter		Proposed System (Model 1)	Optimal System (Model 2)	Conventional System (Model 3)
Architecture	Configuration	Grid/Generator/PV/ WT/Converter/Battery	Grid/PV/WT/Converter	Grid/Generator
	Difference	-	-no battery -no diesel generator	-no RESs -no battery
	PV (kW)	250	250	-
	PV-MPPT (kW)	500	500	-
	Wind turbine	25	25	-
	Generator (kW)	3100	-	2800
	Battery	1	-	-
	Grid (kW)	999,999	999,999	999,999
	Converter (kW)	90	90	-

Table 5. *Cont.*

Cost	NPC (USD)	5,920,870	4,718,247	26,981,960
	COE (USD)	0.0309	0.0246	0.2385
	Operating cost (USD/yr)	−42,392	−14,748	1,978,877
	Initial capital (USD)	6.47 M	4.91 M	1.4 M
System	RES frac (%)	85.2	85.2	-
	Total fuel (L/yr)	0	0	0
PV	Capital cost (USD)	300,000	300,000	-
	Production (kWh/yr)	315,315	315,315	-
Wind turbine	Capital cost (USD)	4,500,000	4,500,000	-
	Production (kWh/yr)	12,435,239	12,435,239	-
	O&M	125,000	125,000	-
Battery	Autonomy (h)	0.0801	-	-
	Annual throughput (kWh/yr)	320	-	-
	Nominal capacity (kWh)	100	-	-
	Usable Nominal Capacity (kWh)	80	-	-
Converter	Rectifier Mean Output (kW)	0.0385	0	-
	Inverter Mean Output (kW)	24.4	24.3	-
Grid	Energy purchased (kWh)	2,189,430	2,189,312	8,751,839
	Energy sold (kWh)	6,085,820	6,085,762	-
Emission	Carbon Dioxides (kg/yr)	1,383,720	1,383,645	5,531,162
	Sulfur Dioxide (kg/yr)	5999	5999	23,980
	Nitrogen Oxides (kg/yr)	2934	2934	11,727

Optimization results from the proposed design indicate that the best architecture for the target hybrid seaport microgrid is the grid/PV/WT/power converter configuration (Model 2) with the optimally sized components featuring the lowest NPC. The best system is selected based on the lowest value of NPC, which is USD 4,718,247 for Model 2, USD 5,920,870 for Model 1, and USD 26,981,960 for Model 3. The significant change that can be seen from Model 2 compared with the proposed system concerns its components where it has no diesel generator and battery.

Conventionally, without microgrid technology, the seaport sector relied only on the supply of electricity from the main grid and diesel generators to run their daily operation. According to Table 5, this architecture (Model 3) has the lowest initial capital cost, which is USD 1.4M, as the system does not need to invest in any other component such as RESs, ESSs, and power converters. Unfortunately, the system has the highest NPC value with USD 26M. The conventional system is highly dependent on the raw material of the power resources such as coal and diesel that produce very high greenhouse gas emissions. In terms of emission, this system produces 5,531,162 kg/yr of CO_2, 23,980 kg/yr of SO_2, and 11,727 kg/yr of NO_X. Meanwhile, the hybrid microgrid system with optimal sizing in

Model 2 offers a big amount of emission reduction with a total reduction of 82.51% in CO_2, 74.98% in SO_2 and 74.98% in NO_x compared with the conventional model. Significant emission reduction in Model 2 gives a positive impact to the environment toward achieving the green port goal. Besides producing hazardous pollutants, the conventional system also faces the problem of fossil resource depletion [36,37]. That explains, the reason why microgrid technology utilization is important for the seaport sector. The advantage of hybrid RESs and ESSs from microgrids will increase energy efficiency and provide sustainable energy while offering a long-term cost-effective solution for port electrification and decarbonization. This is supported by the significant amount of emission reduction in Model 1 and Model 2 that will certainly result in a better port environment.

According to the simulation results, the best architecture for the seaport microgrid in this project which is Model 2 consists of a 250 kW-PV, 25 100 kW-WTs, and a 90 kW-converter with a connection to the main grid. This microgrid requires 40,665 kWh/day and has a peak power demand of 2734 kW. The amount of excess electricity is 93,388 kWh/yr. Figure 8 shows the graph of energy production from both the utility grid and RESs that serve the electrical load. According to this figure, a big share of the produced power is related to RESs where 12,435,239 kWh/year is produced by the WTs, 315,315 kWh/year from PV, and only 2,189,312 kWh energy is purchased from the main grid. This indicates a RES fraction of 85.3%, whereas 14.7% of the power is from the main grid. The total energy production from PV, WT, and the main grid is equal to 14,939,866 kWh, from which 59% is used to supply the required power of the electrical loads (ships) and the remaining 41% of excess energy is sold back to the main grid. The amount of sold energy is 6,085,762 kWh, which is much higher than the energy purchased from the grid. In terms of emission, 1,383,645 kg/yr of CO_2, 5999 kg/yr of SO_2, and 2934 kg/yr of NO_x are produced. The opportunity to gain extra money from selling back the energy to the main grid and reduce emissions compared with the conventional model is due to the deployment of RESs in the microgrid. Elimination of diesel generator and ESS in this model compared with the proposed system is due to the fact that power generation from RESs is more than enough to supply the current load. Thus, energy from the diesel generator and battery is not necessary in this case. Moreover, including a battery will increase the capital cost.

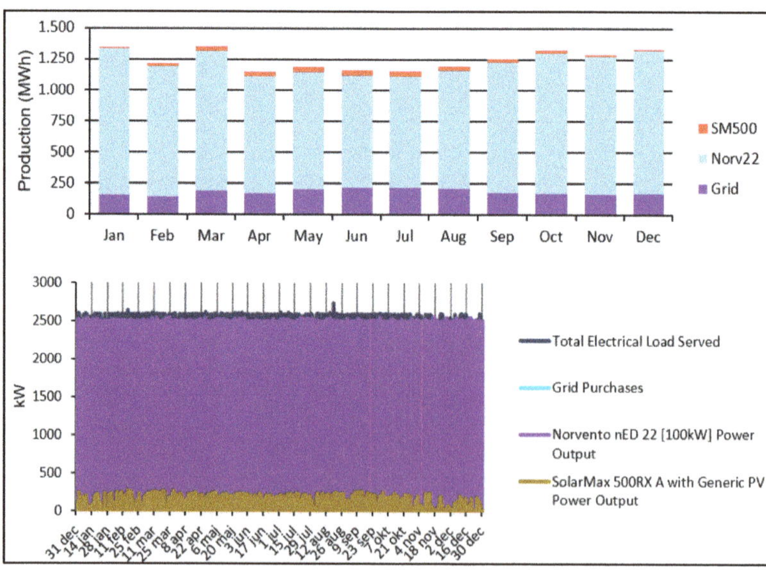

Figure 8. Energy generation and load consumption of Model 2.

Figure 9 shows both the PV and WTs power generation in Model 2 for a one-year duration. From the 85.3% of renewable energy fraction, 2.11% is from solar PV SM500 whereas the remaining 83.2% comes from Norv22 WT. The amount of energy produced by RESs and its reliability depend on the site location and the available natural resources. In this case study, Denmark as a seasonal country experiences high irradiation during the warm months (May, June, and July) due to the sunny climate and long hours of daylight. In those particular months, the output from the PV system is be higher compared with the power produced in other months with a high chance of having cloudy and rainy days. This limitation explains the reason for the small contribution of the PV system.

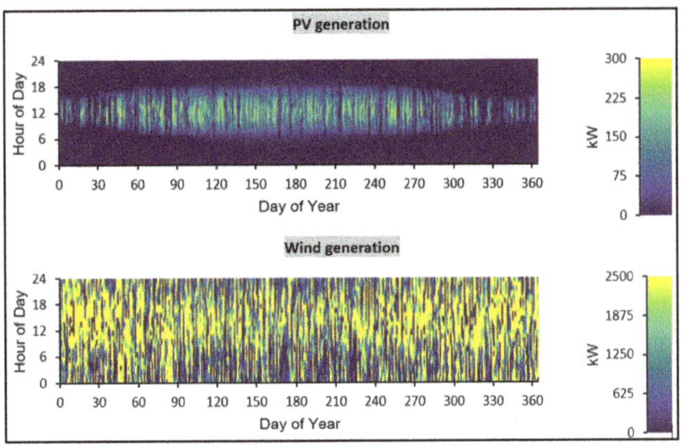

Figure 9. Solar and Wind generation of Model 2.

In the meantime, Denmark is a rich country in terms of wind resources as this country has relatively average wind speeds of 4.9–5.6 m/s measured at a height of 10 m. Denmark has vast offshore and onshore wind resources, as well as large swaths of sea territory with shallow water depths of 5–15 m, ideal for sitting WTs. Higher wind speeds, in the range of 7.08–9.62 m/s at 50 m height, are also available at these locations. There have been no big issues from wind variability.

A sensitivity analysis has been performed to evaluate the impact of the electricity price on the optimal system design. Four values of 0.20 USD/kWh, 0.4 USD/kWh, 0.6 USD/kWh, and 0.8 USD/kWh were evaluated. Table 6 summarizes the results obtained for various electricity rates loaded into the software. There is an increasing value of the NPC as the electricity rate increase. This is due to the growing share of the RES fraction and the requirement for more battery units to store the access energy from renewable sources. This makes sense considering that the optimizer is attempting to reduce the expenditure in electricity cost, hence increasing the utilization of power from RESs. In addition, as the RES fraction increases, a significant amount of emission is also reduced for CO_2, SO_2, and NO_x gases.

Table 6. Sensitivity analysis of electricity price (all the data are retrieved from Homer Pro, accessed on 18 October 2021).

Electricity Price (USD/kWh)	Battery (Unit)	NPC USD	RES Fraction %	Emissions Production		
				CO_2 kg/yr	SO_2 kg/yr	NO_x kg/yr
0.2	-	4.72 M	85.2	1,383,645	5999	2934
0.4	38	8.71 M	91.8	746,512	3236	1583
0.6	51	10.8 M	92.9	641,640	2782	1360
0.8	128	11.4 M	97.1	251,339	1090	533

5. Conclusions

This paper provided an optimal architecture for a seaport microgrid to support cold ironing services at ports. The design goal was to minimize the net present cost by optimizing the size of the components. A comparative discussion between a conventional seaport power system (grid/diesel generator) and two other hybrid designs of seaport microgrids was provided highlighting their difference in terms of cost, energy production, and emissions. The conclusion was that a grid/PV/wind/converter configuration with a 250 kW-PV, 25,100 kW-WTs, and a 90 kW converter with an NPC of USD 4,718,247 is the best seaport microgrid design to serve the given ship's load at the target port. More than 80% of electricity was generated from RESs, whereas only 14.7% of energy was purchased from the main grid. The proposed seaport microgrid was also able to sell a considerable amount of energy back to the main grid. In terms of emissions, the optimal design could offer a significant reduction in CO_2, SO_2, and NO_x emissions, with a total reduction of 82.51%, 74.98%, and 74.98%, respectively. Emission reduction and cost-effectiveness of the optimal design of the seaport microgrid results in a promising solution for electrification of the future seaports following green maritime goals and the sustainable energy systems paradigm.

Author Contributions: Conceptualization, N.N.A.B.; software, N.N.A.B.; validation, N.B. and M.O.; formal analysis, N.N.A.B., visualization, N.N.A.B., writing—original draft preparation, N.N.A.B.; writing—review and editing, N.N.A.B. and N.B.; data provision, B.D.R.; supervision, J.M.G. and J.C.V.; project administration, J.M.G.; funding acquisition, Y.A.A.-T. All authors have read and agreed to the published version of the manuscript.

Funding: The Deanship of Scientific Research (DSR) at King Abdulaziz University, Jeddah, Saudi Arabia has funded this project, under grant no. (RG-49-135-40). The research work was also funded by a Villum Investigator grant (no. 25920) from The Villum Fonden.

Institutional Review Board Statement: Not applicable.

Informed Consent Statement: Not applicable.

Data Availability Statement: Not applicable.

Acknowledgments: This research work was supported by a Villum Investigator grant (no. 25920) from The Villum Fonden, University Malaysia Perlis, and Ministry of Education Malaysia. The Deanship of Scientific Research (DSR) at King Abdulaziz University, Jeddah, Saudi Arabia, has funded this project under grant no. (RG-49-135-40).

Conflicts of Interest: The authors declare no conflict of interest.

References

1. Ministry of Foreign Affairs of Denmark. Pioneers in Clean Energy. Available online: https://denmark.dk/innovation-and-design/clean-energy (accessed on 13 May 2021).
2. Danish Port Targets 70 Percent Carbon Reduction Using Wind Energy. The Maritime Executive. 2021. Available online: https://www.maritime-executive.com/article/danish-port-targets-70-percent-carbon-reduction-using-wind-energy (accessed on 10 May 2021).
3. Wu, D.; Ma, X.; Huang, S.; Fu, T.; Balducci, P. Stochastic optimal sizing of distributed energy resources for a cost-effective and resilient Microgrid. *Energy* **2020**, *198*, 117284. [CrossRef]
4. Fathi, M.; Khezri, R.; Yazdani, A.; Mahmoudi, A. Comparative study of metaheuristic algorithms for optimal sizing of standalone microgrids in a remote area community. *Neural Comput. Appl.* **2021**, *6*, 1–19. [CrossRef]
5. Wang, Y.; Member, S.; Rousis, A.O.; Strbac, G. A Three-Level Planning Model for Optimal Sizing of Networked Microgrids Considering a Trade-Off Between Resilience and Cost. *IEEE Trans. Power Syst.* **2021**, *36*, 5657–5669. [CrossRef]
6. Gong, K.; Member, S.; Wang, X.; Jiang, C. Security-Constrained Optimal Sizing and Siting of BESS in Hybrid AC / DC Microgrid Considering Post-Contingency Corrective Rescheduling. *IEEE Trans. Sustain. Energy* **2021**, *12*, 2110–2122. [CrossRef]
7. Khezri, R.; Member, S.; Mahmoudi, A.; Member, S.; Haque, M.H.; Member, S. Sizing of Standalone Renewable-Battery Systems. *IEEE Trans. Sustain. Energy* **2021**, *12*, 2184–2194. [CrossRef]
8. Takano, H.; Hayashi, R.; Asano, H.; Goda, T. Optimal Sizing of Battery Energy Storage Systems Considering Cooperative Operation with Microgrid Components. *Energies* **2021**, *14*, 7442. [CrossRef]

9. Xie, C.; Wang, D.; Lai, C.S.; Wu, R.; Wu, X.; Lai, L.L. Optimal sizing of battery energy storage system in smart microgrid considering virtual energy storage system and high photovoltaic penetration. *J. Clean. Prod.* **2020**, *281*, 125308. [CrossRef]
10. Amini, M.; Khorsandi, A.; Vahidi, B.; Hosseinian, S.H.; Malakmahmoudi, A. Optimal sizing of battery energy storage in a microgrid considering capacity degradation and replacement year. *Electr. Power Syst. Res.* **2021**, *195*, 107170. [CrossRef]
11. Mueller, D.; Uibel, S.; Takemura, M.; Klingelhoefer, D.; Groneberg, D.A. Ships, ports and particulate air pollution—An analysis of recent studies. *J. Occup. Med. Toxicol.* **2011**, *6*, 31. [CrossRef]
12. Abu Bakar, N.N.; Guerrero, J.M.; Vasquez, J.C.; Bazmohammadi, N.; Yu, Y.; Abusorrah, A.; Al-Turki, Y.A. A Review of the Conceptualization and Operational Management of Seaport Microgrids on the Shore and Seaside. *Energies* **2021**, *14*, 7941. [CrossRef]
13. Sciberras, E.A.; Zahawi, B.; Atkinson, D.J. Electrical characteristics of cold ironing energy supply for berthed ships. *Transp. Res. Part D Transp. Environ.* **2015**, *39*, 31–43. [CrossRef]
14. Sciberras, E.A. Shipboard Electrification—Emission Reduction and Energy Control. Ph.D. Thesis, Newcastle University, Newcastle, UK, 2016; p. 240.
15. Colarossi, D.; Principi, P. Technical analysis and economic evaluation of a complex shore-to-ship power supply system. *Appl. Therm. Eng.* **2020**, *181*, 115988. [CrossRef]
16. Abu Bakar, N.N.; Hassan, M.Y.; Sulaima, M.F.; Nasir, M.N.M.; Khamis, A. Microgrid and load shedding scheme during islanded mode: A review. *Renew. Sustain. Energy Rev.* **2017**, *71*, 161–169. [CrossRef]
17. Reusser, C.A.; Pérez, J.R. Evaluation of the Emission Impact of Cold-Ironing Power Systems, Using a Bi-Directional Power Flow Control Strategy. *Sustainability* **2020**, *13*, 334. [CrossRef]
18. Roy, A.; Auger, F.; Olivier, J.; Schae, E.; Auvity, B. Design, Sizing and Energy Management of Microgrids in Harbor Areas: A Review. *Energies* **2020**, *13*, 5314. [CrossRef]
19. Odun-ayo, T. Scholars' Mine Impact of Stochastic Loads and Generations on Power System Transient Stability. Ph.D. Thesis, Missouri University, Columbia, MO, USA, 2011; p. 58.
20. Xie, P.; Guerrero, J.M.; Tan, S.; Bazmohammadi, N.; Vasquez, J.C.; Mehrzadi, M.; Al-Turki, Y. Optimization-Based Power and Energy Management System in Shipboard Microgrid: A Review. *IEEE Syst. J.* **2021**, 1–13. [CrossRef]
21. Rousis, A.O.; Tzelepis, D.; Konstantelos, I.; Booth, C.; Strbac, G. Design of a Hybrid AC/DC Microgrid Using HOMER Pro: Case Study on an Islanded Residential Application. *Inventions* **2018**, *3*, 55. [CrossRef]
22. Yenalem, M.G.; Hinga, P. Modelling and Optimal Sizing of Grid-Connected Micro grid System using HOMER in Bahir Dar City, Ethiopia 2 Literature Review. *Int. J. Power Syst.* **2020**, *5*, 1–12.
23. Shahinzadeh, H.; Moazzami, M.; Fathi, S.H.; Gharehpetian, G.B. Optimal sizing and energy management of a grid-connected microgrid using HOMER software. In Proceedings of the 2016 Smart Grids Conference (SGC), Kerman, Iran, 20–21 December 2016; pp. 13–18. [CrossRef]
24. Kumar, P.; Pukale, R.; Kumabhar, N.; Patil, U. Optimal Design Configuration Using HOMER. *Procedia Technol.* **2016**, *24*, 499–504. [CrossRef]
25. Krishna, K.M. Optimization analysis of microgrid using HOMER—A case study. In Proceedings of the 2011 Annual IEEE India Conference, Hyderabad, India, 16–18 December 2011. [CrossRef]
26. Motjoadi, V.; Adetunji, K.E.; Joseph, P.K.M. Planning of a sustainable microgrid system using HOMER software. In Proceedings of the 2020 Conference on Information Communications Technology and Society (ICTAS), Durban, South Africa, 11–12 March 2020; pp. 1–5. [CrossRef]
27. Yasin, A.; Alsayed, M. Optimization with excess electricity management of a PV, energy storage and diesel generator hybrid system using HOMER Pro software. *Int. J. Appl. Power Eng. (IJAPE)* **2020**, *9*, 267–283. [CrossRef]
28. Lata-Garcìa, J.; Reyes-Lopez, C.; Jurado, F.; Fernández-Ramírez, L.M.; Sanchez, H. Sizing optimization of a small hydro/photovoltaic hybrid system for electricity generation in Santay Island, Ecuador by two methods. In Proceedings of the 2017 CHILEAN Conference on Electrical, Electronics Engineering, Information and Communication Technologies (CHILECON), Pucon, Chile, 18–20 October 2017; pp. 1–6. [CrossRef]
29. Ahamad, N.B.; Othman, M.; Vasquez, J.C.; Guerrero, J.M.; Su, C.-L. Optimal sizing and performance evaluation of a renewable energy based microgrid in future seaports. In Proceedings of the 2018 IEEE International Conference on Industrial Technology (ICIT), Lyon, France, 20–22 February 2018; pp. 1043–1048. [CrossRef]
30. Kharrich, M.; Kamel, S.; Alghamdi, A.; Eid, A.; Mosaad, M.; Akherraz, M.; Abdel-Akher, M. Optimal Design of an Isolated Hybrid Microgrid for Enhanced Deployment of Renewable Energy Sources in Saudi Arabia. *Sustainability* **2021**, *13*, 4708. [CrossRef]
31. Innes, A.; Monios, J. Identifying the unique challenges of installing cold ironing at small and medium ports—The case of aberdeen. *Transp. Res. Part D Transp. Environ.* **2018**, *62*, 298–313. [CrossRef]
32. Ahamad, N.B.B.; Guerrero, J.M.; Su, C.L.; Vasquez, J.C.; Zhaoxia, X. Microgrids Technologies in Future Seaports. In Proceedings of the 2018 IEEE International Conference on Environment and Electrical Engineering and 2018 IEEE Industrial and Commercial Power Systems Europe (EEEIC/I&CPS Europe), Palermo, Italy, 12–15 June 2018; pp. 1–6. [CrossRef]
33. Haidar, A.M.; Fakhar, A.; Helwig, A. Sustainable energy planning for cost minimization of autonomous hybrid microgrid using combined multi-objective optimization algorithm. *Sustain. Cities Soc.* **2020**, *62*, 102391. [CrossRef]
34. Ciulla, G.; D'Amico, A.; Di Dio, V.; Brano, V.L. Modelling and analysis of real-world wind turbine power curves: Assessing deviations from nominal curve by neural networks. *Renew. Energy* **2019**, *140*, 477–492. [CrossRef]

35. Wang, W.; Peng, Y.; Li, X.; Qi, Q.; Feng, P.; Zhang, Y. A two-stage framework for the optimal design of a hybrid renewable energy system for port application. *Ocean Eng.* **2019**, *191*, 106555. [CrossRef]
36. Zhou, Z.; Benbouzid, M.; Charpentier, J.F.; Scuiller, F.; Tang, T. A review of energy storage technologies for marine current energy systems. *Renew. Sustain. Energy Rev.* **2013**, *18*, 390–400. [CrossRef]
37. Brando, G.; Dannier, A.; Del Pizzo, A.; Di Noia, L.P.; Pisani, C. Grid connection of wave energy converter in heaving mode operation by supercapacitor storage technology. *IET Renew. Power Gener.* **2016**, *10*, 88–97. [CrossRef]

energies

Article

Participation of Aggregated DERs to the Ancillary Services Market: A Monte Carlo Simulation-Based Heuristic Greedy-Indexing Model

Davide Falabretti [1,*], Francesco Gulotta [1] and Lorenzo Spinelli [2]

1 Department of Energy, Politecnico di Milano, Via Lambruschini 4, I-20133 Milan, Italy; francesco.gulotta@polimi.it
2 Tesla International BV, Burgemeester Stramanweg 122, 1101 EN Amsterdam, The Netherlands; lspinelli@tesla.com
* Correspondence: davide.falabretti@polimi.it

Abstract: In an effort to improve the stability and secure operation of the grid, regulatory bodies are opening Ancillary Services Markets participation to Distributed Energy Resources (DERs), energy storage systems, and demand response. Within this framework, this study proposes a model that simulates the coordinated operation of an aggregate of power plants, including non-dispatchable DERs and, as regulating units, Combined Heat and Power (CHP) generation and electrochemical energy storage systems. A Monte Carlo procedure is adopted to realistically create a population of aggregation scenarios. The real-time operation of the DER portfolio is managed through a Heuristic Greedy-Indexing logic, which allows the Aggregator to select the optimal control action to implement according to the technical and economic quantities characterizing the market and the grid. The techno-economic performance of the proposed algorithm is evaluated by simulating its interaction with the electricity markets. Finally, a sensitivity analysis is performed to analyze the profitability in different scenarios. The novel mathematical model proposed showed to be effective in managing a complex problem like the one at hand with an acceptable computational effort. The numerical results obtained confirmed that the aggregated participation in the market could provide interesting economic returns, especially if a CHP unit is involved as regulating unit, while the feasibility of the batteries adoption is still limited by the actual cost of the technology.

Keywords: aggregator; ancillary services; distributed resources; enabled virtual unit; energy storage systems; heuristic greedy-indexing; Monte Carlo; virtual power plant

Citation: Falabretti, D.; Gulotta, F.; Spinelli, L. Participation of Aggregated DERs to the Ancillary Services Market: A Monte Carlo Simulation-Based Heuristic Greedy-Indexing Model. *Energies* **2022**, *15*, 1037. https://doi.org/10.3390/en15031037

Academic Editor: Miguel Jiménez Carrizosa

Received: 29 December 2021
Accepted: 27 January 2022
Published: 29 January 2022

1. Introduction

The formidable growth of Distributed Energy Resources (DERs) in the last decades was both a boon and a major challenge for the safe and economical dispatch of electric power systems. Larger adoption of DER raises concerns regarding the reliability of transmission and distribution systems, which must improve their flexibility to face the issues introduced by the variability and unpredictability of renewable energy sources [1]. Designs aimed to improve the fitness of the electricity markets for a high DER penetration have been thoroughly researched in recent years, with a focus on two matters:

1. Counteracting the problem of number and small size of involved power plants, which causes DERs to be often unmanageable and "invisible" to the Transmission System Operator (TSO), by aggregating them into bigger market entities, able to deliver electricity with dispatchable and controllable production schedules.
2. Reducing the costs arising on the Ancillary Services Market (ASM) for the collection of the regulation services required to ensure to the power system an adequate degree of reliability (operation performed, for example, by relaxing the restrictions on technical

requirements to get access to the market, with the purpose to involve new subjects, e.g., DERs) [2].

In this framework, in Italy, Resolution 300/2017/R/eel [3] started a pilot regulation, consisting of a reformation of the dispatching discipline in place, aimed to enable the participation of DERs in the ASM, in an aggregated form and on a voluntary basis, for the provision of balancing services to the power system. This new entity, named Enabled Virtual Unit (EVU), can include small power plants, storage systems, controllable loads, and electric vehicles, managed and coordinated by an Aggregator (commonly known also as Balance Service Provider).

DER aggregation has been proven to yield several improvements to grid operation and market performance: not only does it help to hedge against the risks of imbalance fees within the Day-Ahead Market (DAM) by the sheer effect of technological diversification (initially the first driver for market aggregates), but it also provides better controllability of small scale generators [4], increased visibility [5], and allows the participation in the ASM by increasing their flexibility when coupled to controllable units (e.g., combined heat and power, storage systems and demand response) [6]. As far as previous studies addressing the operation and optimization of DERs aggregates are concerned, the use of Stochastic Linear Programming to model the uncertainties associated with power output and market prices is very common [7]. These studies focus on bidding decisions [8,9], as well as on power control [7,10] and minimization of imbalances [11]. Non-Linear Programming strategies are employed to determine bidding strategies for dispersed generation, storage, and demand response participating in an aggregated form in energy and secondary reserve markets in [12], while the optimal sizing of DER for minimization of operational costs is studied in [13,14].

However, these methods require large computational effort to reach an optimal solution, with high modeling complexity; heuristic or meta-heuristic methods, on the other hand, are easier to implement and can give more feasible results for real-time operation. An example of these techniques is seen in [15], where hill-climbing algorithms are used to evaluate the optimal dispatch of a virtual power plant, or in [16], where a large fleet of electric vehicles is aggregated and managed by a centralized heuristic scheduler to provide Ancillary Services (ASs) to the power system.

According to the issues at hand, the problem of DER aggregation for ASM participation is tackled in this work through a Heuristic Greedy-Indexing model (similar to the one proposed in [17] to evaluate the e-vehicles charging dispatch), which defines, by means of a centralized architecture, the optimal control actions to be performed by a Combined Heat and Power (CHP) unit and an Energy Storage System (ESS) to make programmable an aggregate of Non-Dispatchable Renewable Energy Sources (NDRES) and provide ASs (secondary and replacement reserve services) to the power system. In the literature, greedy algorithms are widely used when computation time is crucial, but the solutions do not have to be optimal [18]. In the study, uncertainties related to NDRES production are reproduced through a Monte Carlo procedure, generating stochastically a suitable number of aggregation scenarios in which the performance of the Heuristic Greedy-Indexing model is assessed. Technical and economic analyses are developed in compliance with the features of the Italian power system and electricity market. The purposes of this work are: (i) to evaluate the effectiveness of the coordination strategy proposed based on the Heuristic Greedy-Indexing; (ii) to evaluate the business opportunities and technical challenges that arise from the participation of a DER portfolio in the ASM; (iii) to determine the most profitable and effective configuration for the aggregate and the best regulating units size.

In the following of the present work, in Section 2, the framework of the model is set up and the algorithms and approaches developed are presented. In Section 3, the case study taken as reference is described. Results are then assessed and commented in Section 4, and a sensitivity analysis is performed in Section 5. Finally, some conclusions are drawn in Section 6.

2. The Proposed Approach

The proposed architecture, based on a Heuristic Greedy-Indexing approach [19], controls the real-time EVU operation, making the aggregate of units dispatchable and enabling the provision of Secondary Reserve and Replacement Reserve services (hereinafter SR and RR). Different EVU scenarios are created through a Monte Carlo procedure, which defines the number, technology, and size of the NDRES plants involved. The configuration of choice for the aggregate comprises PhotoVoltaic (PV) plants, Wind Turbines (WTs), and run-of-river Hydro Power (HP) generators. Then, CHP generators and ESS are introduced in the EVU as regulating units, providing the system with imbalance correction, as well as with reserve capabilities that can be offered on the ASM. The model runs through the simulation period (up to one year), determining in advance, with the information available, the market commitments, based on historical market data and the operational constraints of the power plant. Real-time decisions enacted by the Heuristic Greedy-Indexing algorithm are based both on the technical characteristics of the generation (such as their production unpredictability, available flexibility, and ramp limits) and on economic figures (fuel cost and market prices). The overarching logic that governs the model is presented in Figure 1.

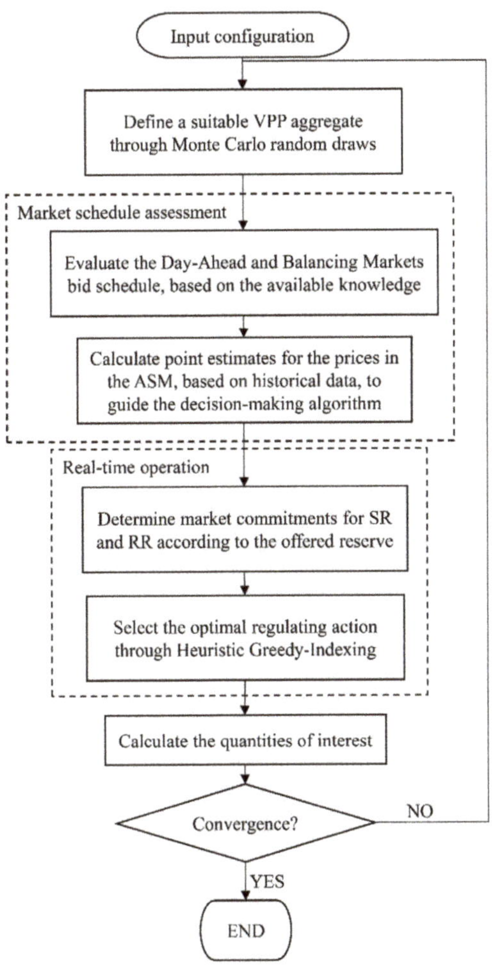

Figure 1. Flowchart of the simulation procedure adopted in the study.

2.1. Enabled Virtual Unit Portfolio Definition

The EVU configuration, for each Monte Carlo run, is chosen through a sequence of random selections that define the technology and size of each NDRES unit i within the aggregate: first, the technology j ($j \in PV, WT, HP$) is picked from a stepwise distribution with weights $w_{tech,j}$, then a power class is selected from $w_{power,j}$ and the nominal power $P_{nomDG,i}$ of the considered unit i is randomly drawn within that class. This process is repeated until the cumulative power of the current EVU scenario ($\sum_i P_{nomDG,i}$) reaches a fixed total nominal power (P_{totEVU}).

As already mentioned, once having defined each EVU configuration (Monte Carlo scenario) as a set of NDRES power plants, a regulating unit is introduced in the aggregate. In the study, ESSs and CHP plants are considered for this purpose. ESSs are assumed to be totally devoted to respecting the market commitments (imbalance correction, secondary and replacement reserve regulations). This assumption is made by taking as reference the scenario in which an Aggregator, managing an NDRES portfolio, decides to adopt an ESS to make the portfolio programmable enough to enable its market participation. The ESS size, in terms of both rated power and energy capacity, respectively P_{nomESS} and E_{nomESS}, is set as a parameter of the simulation. To avoid economic penalties, the ESS must guarantee the provision of the service requested on the market, which in the worst situation could last even several hours (for example, according to the prescriptions in place in Italy for the secondary frequency control, a power equal to the full band offered could be required for 2 h). Therefore, to allow the exchange of power for the time needed, ESSs with an energy/power ratio of some hours must be adopted for this application. For this reason, electrochemical ESSs are considered in this study.

Concerning the CHP unit, on the other hand, it is assumed aimed to follow the electricity demand (hereinafter p_{ind}) of the industry it serves. Therefore, the Aggregator, to supply ASs to the system, can only exploit the remaining CHP flexibility ($P_{flexCHP}$): i.e., the available up/downward power production margins determined with respect to the working point of the generator (i.e., power production required to cover at a given instant the industrial load). In particular, starting from the production profile of the considered CHP plant, the remaining up/downward flexibility is calculated considering the number of engines in operation (n), and their technical limits (max/minimum power).

Each EVU scenario generated by the Monte Carlo procedure is then processed on a given time frame (e.g., one year) through the model described in the following of the work; timesteps of $\Delta t = 15$ min are adopted, coherently with the time resolution used in the Italian ASM. In order to arrest the generation of new scenarios, a converge criterion is applied based on the error evaluated on the estimates of the quantities of interest (e.g., revenues, costs, imbalances, etc.): when the uncertainty relevant to the selected quantities becomes lower than a given threshold (e_{min}), the algorithm is stopped, and the results are deemed satisfactory.

2.2. DAM Commitments and Schedule Forecasts

Due to the heuristic nature of the model, all decisions regarding the control actions to be performed on the EVU by the Heuristic Greedy-Indexing logic must be taken with the knowledge and information available at the time of closure of the relevant market session. This is assumed, for the DAM, 12:00 of day $D - 1$ (where D is the day of dispatch). Regarding the ASM, the market structures in the EU are very different, even if a general trend is ongoing to reduce the distance to the delivery time of market gate closure [2]. For example, in Italy, it has been recently set for some services one hour in advance. Despite this, in the present study, for the sake of simplicity and to keep the approach as far as possible general, a gate closure 4 h before the delivery time has been considered. For the same reasons and the minor impact on the Aggregator operation, the commitments arising from the intra-day market are neglected.

During the participation in the DAM, the EVU defines the binding schedule of the commitments for the sale of the energy produced by the NDRES power plants. Indeed, the CHP and the ESS will not participate in the DAM, since they are assumed just devoted to, respectively, supply the relevant industrial load and provide flexibility on the ASM to the Aggregator.

To forecast the NDRES production, a persistence model is adopted. This prediction technique is based on the assumption that meteorological phenomena vary slowly enough that the most likely outcome for the weather in day $D + 1$ is to be the same as D. A simplistic forecasting approach has been purposefully used, being the development of ad-hoc designed forecasting methods beyond the scope of the study. In real-life, the Aggregator will be free to choose the preferred forecasting method and utilize, for instance, data collected from weather forecast providers. In this case, the forecasting error committed will be lower, thus it will be easier for the EVU to follow the power schedule agreed on the market. According to the persistence model technique, the NDRES production profile used in the DAM ($P_{expected}$) is computed considering the actual power profile measured ($P_{measured}$) the day before:

$$P_{expected}(t) = P_{measured}(t - d{\cdot}24\text{ h}) \tag{1}$$

Therefore, the schedule of commitments for the electricity offered on the DAM ($P_{expected}$) is obtained by time-shifting the available production data forward by 1 day or 2 days (d), for hours respectively before and after noon time. On the other hand, the updated prediction used in the ASM bidding process (P_{updt}) is entirely based on the previous day's data, as described in the next section.

2.3. Reserve Bidding

Bids for the ASM are defined through the available flexibility of CHP and ESS, in both up/downward directions, suitably adjusted to take into account the imbalance correction with respect to the forecast (updated at the time of ASM sessions) and the deviation of the ESS State of Charge (SoC). Indeed, the formulation of bids on the ASM exploits the EVU flexibility, taking into account also the requirements of:

- minimizing the imbalances with respect to the DAM schedule, defined in Equation (1);
- maintaining at a predetermined reference value the battery SoC (E_{setESS}), assumed equal to 50% of ESS rated capacity.

In the Italian market framework, both the bids for SR and RR must include a quantity-price pair, and—if accepted—they are remunerated on a pay-as-bid basis. To compute the most adequate bidding price to be used by the Aggregator on the ASM, in this work, a point estimate based on historical price data is implemented. The same approach is also used to determine, in advance, the most convenient service to bid (RR or SR).

The bidding price for service X ($X \in \{SR; RR\}$), calculated for both for upward and downward reserve (respectively, $e_{X,up}$ and $e_{X,dn}$), is determined considering the mean value of the n_{av}-day historical price data from the same market session:

$$\begin{aligned} e_{X,up}(t) &= \frac{K_{pup}}{n_{av}} \sum_{d=D-1-n_{av}}^{D-1} p_{X,up,d}(t) \\ e_{X,dn}(t) &= \frac{K_{pdn}}{n_{av}} \sum_{d=D-1-n_{av}}^{D-1} p_{X,dn,d}(t) \end{aligned} \tag{2}$$

In Equation (2), $K_{p_{up}}$ and $K_{p_{dn}}$ are parametric coefficients that modulate the aggressiveness of bid prices as a compromise between earnings and likelihood of acceptance (the greater the value of coefficients, the greater the aggressiveness of bid prices), whilst $p_{X,up,d}/p_{X,dn,d}$ are the historical price data from the market session of day d for up/downward reserve.

Once price estimations are calculated, the capacity available for the reserve should be assessed considering: (i) the energy required to bring storage to the set level, (ii) the CHP flexibility, and (iii) the requirement to reduce the imbalance (i.e., the gap between the actual power profile and the power schedule offered on the DAM).

Firstly, the SoC of the ESS is verified, to determine the amount of energy in excess or deficit to reach the set level (E_{setESS}) in the next 4 h. The power required to bring the ESS to the set level (P_{reqESS}) can be calculated as:

$$P_{reqESS}(t) = min\left[\frac{1}{4} \cdot (E_{setESS} - SoC(t)) \cdot \eta_{ESS}, P_{lim,ch/dis}(SoC)\right] \tag{3}$$

where:

- $SoC(t)$ is the State of Charge of the ESS at time t;
- $P_{lim,ch/dis}(SoC)$ is the charge/discharge technical limitation on power, calculated from the ESS characteristic curve presented in Figure 2 [20];
- E_{setESS} is the ESS set level, assumed equal to half of its capacity to consider symmetric up/downward regulating bands;
- Finally, the ESS round trip efficiency is defined as:

$$\eta_{ESS} = \begin{cases} \eta_{ch} & \Delta SoC > 0 \\ \frac{1}{\eta_{dis}} & \Delta SoC < 0 \end{cases} \tag{4}$$

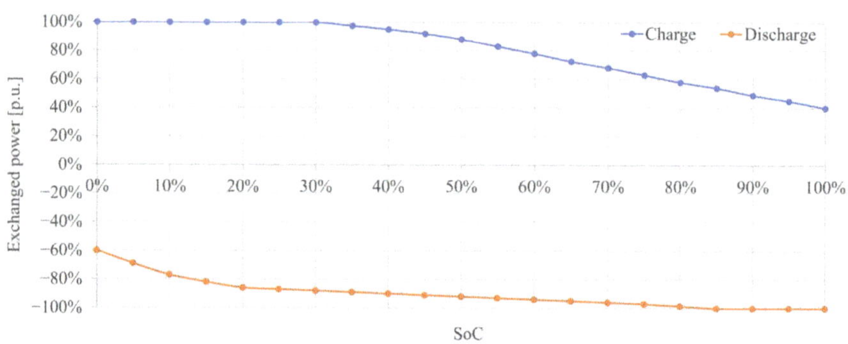

Figure 2. Maximum power exchangeable by the ESS, as a function of SoC.

Please note that, according to Equation (3), to allow the battery to charge or discharge to a given power P_{reqESS}, it is necessary that both the conditions about the technical limitation on power and the current SoC are satisfied. Therefore, in practice, when the ESS is almost fully charged or discharged, it will be able to absorb/release an amount of energy lower than the one reported in the chart in Figure 2.

Secondly, the CHP flexibility ($P_{CHPflex}$) in both directions is assessed. To this purpose, a real CHP power plant, having two internal combustion engines, is taken as a reference. As already introduced, the power plant operates at the power needed to cover the industrial load it serves $p_{ind}(t)$. At time t, the number of combustion engines in operation is defined as the lowest integer quantity n ($0 \div 2$), making true the inequality $p_{ind}(t) \leq \sum_n (n \cdot P_{nomCHP})$. Consequently, the up/down flexibilities are defined, respectively, as $P_{CHPflex,up}(t) = n \cdot P_{nomCHP} - p_{ind}(t)$ and $P_{CHPflex,dn}(t) = p_{ind}(t) - n \cdot P_{minCHP}$, where P_{nomCHP} and P_{minCHP} is the maximum (nominal) and minimum admissible power of each combustion engine.

Finally, the expected error (P_{err}) between the DAM schedule ($P_{DAM} = P_{expected}$) and updated forecasts available at the time of ASM bidding (P_{updt}) is evaluated as:

$$P_{err}(t) = P_{updt}(t) - P_{DAM}(t) \tag{5}$$

Therefore, P_{err} is an estimate of the expected imbalance, which the EVU will attempt to fix later through the bid in the ASM.

From these quantities, the available band for up/downward regulation ($P_{avEVUup/dn}$) for each period (15 min) of the session can be calculated as:

$$\begin{aligned} P_{avEVU,up}(t) &= P_{CHPflex,up}(t) + P_{err}(t) - P_{reqESS}(t) \cdot [max(e_{RR,up}(t), e_{SR,up}(t)) - p_{DAM}(t) > c_{wear}] \\ P_{avEVU,dn}(t) &= P_{CHPflex,dn}(t) - P_{err}(t) + P_{reqESS}(t) \cdot [p_{DAM}(t) - min(e_{RR,dn}(t), e_{SR,dn}(t)) > c_{wear}] \end{aligned} \quad (6)$$

In Equation (6), p_{DAM} is the DAM clearing price, while $e_{RR,up/dn}$ and $e_{SR,up/dn}$ are defined according to Equation (2). P_{reqESS} is multiplied by a logic operator, in square brackets, which models the economic profitability of the ESS employment for the ASM bid. In particular, the logic operator is defined as inequality between the quantities:

- $(max(e_{RR,up}, e_{SR,up}) - p_{DAM})$ and $(p_{DAM} - min(e_{RR,dn}, e_{SR,dn}))$ are the expected net earnings from up and downward regulation, respectively;
- c_{wear} is the cost of wearing down the batteries, calculated as

$$c_{wear} = \frac{c_{invESS}}{2 \cdot n_{cycles,tot} \cdot DoD} \quad (7)$$

where c_{invESS} $\left[\frac{€}{\text{kWh}}\right]$ is the investment cost of the ESS, and $n_{cycles,tot}$ is the number of expected cycles with a Depth of Discharge (DoD) equal to 80%.

Therefore, the logic condition is introduced in Equation (6) to enable the ESS regulation only when the ASM estimated revenues at least cover the battery wear costs.

Note that P_{err} has been included in Equation (6) because if an imbalance is predicted when the production forecast is updated, the best option is to take advantage of the expected imbalance by offering the excess or deficit of production on the ASM as SR/RR reserve.

Given the available regulation bands ($P_{avEVU,up}(t)$ and $P_{avEVU,dn}(t)$), in each time interval, the most profitable service (i.e., RR or SR) shall be identified. If $(e_{RR,up} > e_{SR,up}) \vee (e_{RR,dn} < e_{SR,dn})$, the supply of RR is more convenient than the SR one. Therefore, the quantity quantities bid on the ASM are computed as:

$$\begin{aligned} bid_{RR,up}(t) &= min\left(P_{avEVU,up}(t), \overline{q}_{RR,up}\right) \\ bid_{RR,dn}(t) &= min\left(P_{avEVU,dn}(t), \overline{q}_{RR,dn}\right) \end{aligned} \quad (8)$$

where $\overline{q}_{RR,up/dn}$ is the average amount of RR service acquired by the TSO in the previous ASM sessions. The minimum value between the available regulating band and the historically requested service is offered on the market. With the purpose to maximize the revenues from the ASM participation, the remaining regulating band is offered for the SR service:

$$\begin{aligned} bid_{SR,up}(t) &= K_{q_{up}} \cdot (P_{avEVU,up}(t) - P_{avRR,up}(t)) \\ bid_{SR,dn}(t) &= K_{q_{dn}} \cdot (P_{avEVU,dn}(t) - P_{avRR,dn}(t)) \end{aligned} \quad (9)$$

In Equation (9), $K_{p_{up/dn}}$ is a parameter that modulates the quantities committed to SR, based on the accepted level of risk, since the SR control signal delivered by the TSO ($s_{SR,ctrl}$), which will modulate in real-time the amount of secondary frequency control requested to the EVU, often does not command the whole of the auctioned band (i.e., typically $s_{SR,ctrl} < 1$, so $K_{q_{up/dn}}$ can be >1).

In the opposite case, when SR is deemed more profitable than RR $(e_{SR,up} > e_{RR,up}) \vee (e_{SR,dn} < e_{RR,dn})$, the opposite approach is adopted: most of the available flexibility is offered for the SR regulation and the remaining one for the RR service.

2.4. Market Commitment

After the definition of the flexibility offered on the market and the relevant bidding prices, it is determined whether bids for reserve services are actually accepted, checking if the price presented by the Aggregator for the X-th service (with $X \in \{SR; RR\}$) is more profitable than the other market players' offers (i.e., $e_{X,dn} > p_{X,dn}$ and $e_{X,up} < p_{X,up}$). Accordingly, the accepted quantities (q_{accX}) are defined as:

$$\begin{aligned} q_{accX,up}(t) &= min(bid_{X,up}(t), q_{X,up}(t)) \\ q_{accX,dn}(t) &= min(bid_{X,dn}(t), q_{X,dn}(t)) \end{aligned} \tag{10}$$

In Equation (10), $q_{X,up}$ and $q_{X,dn}$ are the quantities that, according to the historical data available, have been accepted by the TSO in the ASM for the service X. Then, the amount of AS offered by the EVU that can be accepted on the market has to be lower than, or equal to, the quantity required in real-life by the system operator at that time. The regulation needed by the TSO (energy measured in the quarter-hour Δt, $E_{regSR,up/dn}$) is calculated using Equation (11) for SR and Equation (12) for RR.

$$\begin{aligned} E_{regSR,up}(t) &= \sum_{>0} s_{SR,ctrl}(t) \cdot q_{accSR,up}(t) \cdot \frac{1}{60} \\ E_{regSR,dn}(t) &= \sum_{<0} s_{SR,ctrl}(t) \cdot q_{accSR,dn}(t) \cdot \frac{1}{60} \end{aligned} \tag{11}$$

$$E_{regRR,up/dn}(t) = \pm q_{accRR,up/dn}(t) \cdot \frac{1}{4} \tag{12}$$

where, as already mentioned, $s_{SR,ctrl}$ is the SR control signal (1 min. resolution) sent by the TSO to the EVU to modulate the real-time regulation. This quantity has been determined by using the actual control signal delivered by the Italian TSO to power plants on the European continental power system [21]. Whereas for RR regulation, following the Italian market rules, the quantity of service that the EVU is requested to provide is considered equal to the accepted one.

Therefore, the final commitment toward the market, resulting from the combination of DAM and ASM, is calculated as:

$$E_{sold}(t) = E_{DAM}(t) + E_{regSR,up}(t) + E_{regSR,dn}(t) + E_{regRR,up}(t) + E_{regRR,dn}(t) \tag{13}$$

and the real-time regulation ($E_{reg}(t)$) required to the EVU to fulfill the market obligation is:

$$\begin{aligned} E_{reg}(t) = \; & E_{sold}(t) - E_{prod}(t) = E_{sold}(t) - \sum_i P_i(t)/4 \\ E_{reg,up}(t) = \; & E_{sold}(t) - E_{prod}(t) \quad s_r > 0 \\ E_{reg,dn}(t) = \; & E_{prod}(t) - E_{sold}(t) \quad s_r < 0 \end{aligned} \tag{14}$$

In Equations (13) and (14), E_{DAM} is the energy exchanged on the DAM resulting from the power forecast performed the day-ahead ($P_{expected}$), E_{prod} is the sum of the energy produced in the current timestep by all RES power plants of the EVU, s_r is the sign of E_{reg}, and $E_{reg,up}$ and $E_{reg,dn}$ is the energy required respectively for upward and downward regulation.

Once the commitments to the market for SR and RR services are defined, the control actions requested in real-time to regulating units have to be defined. To this purpose, as explained in the following section, the scheduling logic evaluates, by means of the proposed Heuristic Greedy-Indexing, the costs and benefits of each corrective action that could be taken to supply the ASs and to extinguish imbalances. To quantify the contribution to the AS provision admissible for each regulating unit, the relevant operative limits must also be considered.

2.5. Real-Time Operation

The final step in the logic is the decision on the regulation to be performed in real-time, which is performed as shown in Figure 3. At time t, the algorithm receives in input the current state of the EVU and performs proper evaluations on the costs and earnings related to the regulating options considered admissible. Then, a list of feasible options is defined, sorted in order of increasing cost, among which the most cost-efficient one is selected. In particular, four possible actions are considered, aiming, when possible, to eliminate the imbalance and respecting the power schedule offered in the DAM and ASM:

1. exploiting the CHP flexibility;
2. activating the ESS;
3. reducing the energy generated by the RES (i.e., energy curtailment);
4. maintaining the residual imbalance.

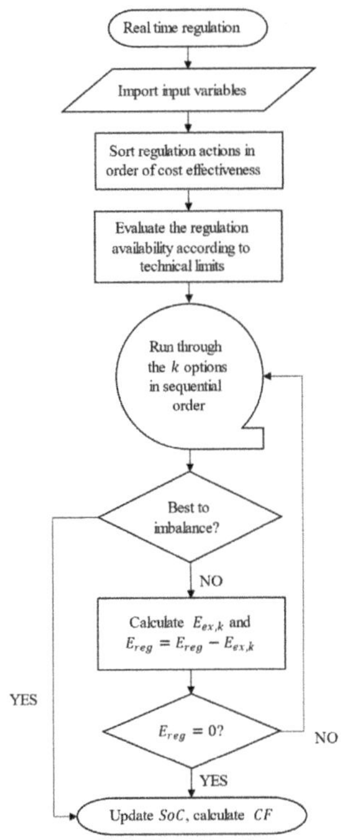

Figure 3. Real-time regulation—Greedy-Indexing Heuristic.

To choose the proper regulating action to be undertaken, the costs of the following options are considered:

1. Costs (earnings) deriving from the regulation performed through the ESS (c_{ESS});
2. Costs (earnings) deriving from the regulation performed through the CHP (c_{CHP});
3. Cost of NDRES curtailment (c_{curt});
4. Costs of residual imbalance (c_{imb}).

It is relevant pointing out that, as is obvious, the EVU will earn from the provisions of the AS to the power system but, since all the possible regulations performed by the Aggregator are feasible to provide the services sold, all control actions are expected to bring the same profit. Therefore, to carry out the selection of the optimal action to be implemented, the algorithm evaluates the costs relevant to each one rather than the net earnings. In the following, the procedure adopted for the evaluation of costs and availability of each action is reported.

2.5.1. Cost and Availability of ESS Regulation

In order to evaluate the cost of intervention of the ESS (c_{ESS}), different aspects need to be taken into account. First of all, due to the large investment cost associated with the ESS, and the comparatively low number of cycles that batteries can withstand [22], the batteries wearing has to be considered. To this purpose, the penalty factor c_{wear} is used, as defined in Equation (7), evaluated as investment cost for a unit of ESS capacity spread out on the ESS lifetime (expected number of charge/discharge cycles).

Moreover, the decision on the regulation of the ESS at the present time depends also on the price of the services expected in the following hours, because a control action performed at time t determines a deviation of the battery's SoC from the reference value, which needs to be adjusted in the future (otherwise, the ESS would tend to saturate to the max/minimum SoC).

Therefore, the cost assigned by the real-time logic to the ESS regulation (c_{ESS}) is evaluated as:

$$c_{ESS}(t) \quad = [\bar{c}_{ASM} \cdot s_r(t) + c_{wear}] \cdot \eta_{ESS} \tag{15}$$

where $\bar{c}_{ASM}$ is the average cost associated with the requirement of restoring the battery SoC during the next ASM sessions, evaluated by a point estimate method similar to the one in Equation (2), c_{wear} are the ESS wearing costs, and η_{ESS} is the efficiency of the storage depending on the direction of the power flow. In particular, $\eta_{ESS} = \eta_{ch}$ if $s_r > 0$ and $\eta_{ESS} = 1/\eta_{dis}$ if $s_r < 0$ (i.e., the regulation requires the battery to discharge).

After the evaluation of the cost of ESS regulation (c_{ESS}), it is necessary to quantify the availability of down/upward regulation (E_{avESS}). To this purpose, both the amount of energy stored in the batteries and their charge/discharge power limits ($P_{lim,ch/dis}$) must be taken into account. In particular, at time step t, the energy available for the regulation is:

$$\begin{array}{ll} E_{avESS,up}(t) = min\left(\frac{SoC(t)}{\eta_{dis}}, P_{lim,dis}{\cdot}\Delta t\right) & s_r > 0 \\ E_{avESS,dn}(t) = min((E_{nomESS} - SoC(t)){\cdot}\eta_{ch}, P_{lim,ch}{\cdot}\Delta t) & s_r < 0 \end{array} \tag{16}$$

where $P_{lim,ch}$ and $P_{lim,dis}$ are, respectively, the charge and discharge power limits of the ESS, shown in Figure 2.

2.5.2. Cost and Availability of CHP Regulation

The cost of the regulation performed through the CHP power plant is mostly dependent on the savings/losses related to the reduction/increase of fuel required to power the internal combustion engines. As a consequence, it can be calculated as $c_{CHP} = c_f \cdot s_r$, where $s_r = sign(E_{reg})$ and c_f is the fuel cost, meaning that if an upward control action is required ($E_{reg} > 0$), this will entail burning more fuel; vice versa, in case of a downward regulation, the fuel saving will allow for a reduction in the working costs of the generator. The availability of CHP regulation is evaluated according to its flexibility, defined in Section 2.3. Therefore, in a quarter-hour, the energy that can be exchanged for the regulation is $E_{avCHP,up/dn}(t) = P_{CHPflex,up/dn}(t)/4$.

2.5.3. Cost and Availability of NDRES Curtailment

A further option to consider for providing downward regulation is the possibility to curtail NDRES production. The cost of the curtailment (c_{curt}) is set equal to 0 €/kWh, assuming that NDRES power plants can reduce the power injected into the grid without additional costs with respect to the investment costs covered for the power plant realization and without an increase of the expenses compared to the standard operation. At time slot Δt, the energy available to be curtailed can not exceed the total energy produced by the NDRES units (E_{prod}).

Please consider that, usually, this option is less convenient than performing the same regulation by the ESS or CHP, since in both cases a downward regulation usually brings to savings: in the former case (ESS), earnings are given by the energy stored in the battery that can be used to supply ASs in the next timesteps; in the latter one (CHP), they are given instead by a reduction in the fuel costs.

2.5.4. Cost of Residual Imbalance

If the EVU flexibility is not enough to fulfill the system operator's dispatching orders, an energy imbalance occurs. To evaluate the costs that the Aggregator must cover in this case, a dual pricing mechanism is applied [23]: therefore, the imbalance cost depends on the sign and amplitude of the error (respectively s_r and E_{imb}), but also on the zonal imbalance sign s_z, that is the direction of the overall imbalance in the market zone where the EVU is located. If s_z is opposite to the imbalance registered by the EVU ($s_r \cdot s_z < 0$), the energy gap is not penalized, since such an error actually helps the system stability; thus, the unitary imbalance cost is set equal to the DAM price ($c_{imb} = s_r \cdot p_{DAM}$). In case of concordant signs, the imbalance caused by the EVU is detrimental for the grid, hence it shall be penalized:

$$\begin{array}{ll} c_{imb}(t) = max(p_{DAM}(t), p_{maxASM,up}) & s_r > 0 \\ c_{imb}(t) = -min(p_{DAM}(t), p_{minASM,dn}) & s_r < 0 \end{array} \quad (17)$$

where $p_{maxASM,up}$ and $p_{minASM,dn}$ are the max/minimum prices for the up/downward regulation, estimated considering the data of the previous market sessions. Since the sign of the zonal imbalance is not known in real-time by the Aggregator (only the TSO has all the information to evaluate it), a conservative approach is assumed to estimate c_{imb} for the real-time decisions performed by the Heuristic Greedy-Indexing, in particular by taking the worst-case scenario (i.e., concordant signs $s_r \cdot s_z > 0$).

2.5.5. Selection of the Best Control Action

Once technical and economic quantities associated with each control action are known, a list of options, sorted in order of increasing cost, is produced by the Heuristic Greedy-Indexing algorithm. The logic runs through all options in order, calculating $E_{ex,k,\ up/dn}$, that is the energy exchangeable through the k-th option ($k \in \{ESS, CHP, curtailment\}$) for up/downward regulation:

$$E_{ex,k,\ up/dn}(t) = min\left(E_{av,k,up/dn}(t), E_{reg,up/dn}(t)\right) \quad (18)$$

After each regulation action, the residual regulation required $E_{reg,up/dn}$ to fulfill the market commitments is updated:$E_{reg,up/dn} = E_{reg,up/dn} - E_{ex,k,up/dn}$, until either $E_{reg,up/dn} = 0$, or the selected option is more expensive than keeping the imbalance (in this case $E_{imb} = E_{reg}$).

As the last step of the real-time regulation algorithm, the battery SoC is updated ($SoC(t + \Delta t) = SoC(t) - E_{exESS}(t)/\eta_{ESS}$) and cash flows in the current timestep t are calculated as:

$$\begin{array}{l} CF(t) = E_{DAM}(t) \cdot p_{DAM}(t) + E_{regSR,up}(t) \cdot e_{SR,up}(t) + E_{regSR,dn}(t) \cdot e_{SR,dn(t)} + E_{regRR,up}(t) \cdot e_{RR,up}(t) \\ \qquad + E_{regRR,dn}(t) \cdot e_{RR,dn}(t) - c_{CHP}(t) \cdot E_{exCHP}(t) - CF_{imb}\ (t) \end{array} \quad (19)$$

In Equation (19), a pay-as-bid approach has been adopted to calculate the revenues from the AS provision (i.e., $E_{regX} \cdot e_{X,up}$) and cash flows related to imbalance are evaluated as $CF_{imb} = s_r|c_{imb}||E_{imb}|$.

3. Case Study

The proposed method runs through each timestep in the simulated period (one year), defining the best control action to perform through the Heuristic Greedy-Indexing algorithm. To achieve statistically significant outcomes, the proposed method has been tested through the Monte Carlo procedure described in Section 2.1, which randomly generates a population of different EVUs. In each scenario, the production of NDRES power plants is modeled through weather data (solar radiation and wind speed, respectively for PV and WT generation) from 135 stations scattered around Northern Italy. These profiles are suitably transformed to per-unit power and then fed to the Monte Carlo algorithm. The technology and size of each power plant involved in the EVU are defined according to the real distribution of NDRES in Italy (see Table 1) [24,25]. Rated power of DER units ranges from $P_{min} = 100$ kW to $P_{max} = 6$ MW, assuming the latter as maximum rated power of dispersed generation units that can be connected to medium voltage networks according to the Italian regulation framework [26]. Regarding the CHP power plant usage, it has been derived from the real power profile measured in 2016 on an industrial user in Northern Italy, equipped with 2 × 1605 kW natural gas-powered generators.

Table 1. Distribution of technology (w_{tech}) and power (w_{power}) classes.

	Class I	Class II	Class III	Class IV	Class V	Class VI	Class VII	Class VIII	Class IX	Class X	Class XI	Class XII	Class XIII	Class XIV	
P_{class_min} [kW]	100.0	134.0	179.5	240.5	322.1	431.6	578.2	774.6	1037.7	1390.3	1862.6	2495.3	3342.9	4478.6	
P_{class_max} [kW]	134.0	179.5	240.5	322.1	431.6	578.2	774.6	1037.7	1390.3	1862.6	2495.3	3342.9	4478.6	6000.0	
Photovoltaic	16.18%	10.47%	17.90%	5.78%	6.22%	8.19%	6.59%	24.21%	0.57%	0.91%	0.94%	0.86%	0.57%	0.63%	**91.93%**
Wind	2.87%	3.58%	48.03%	0.72%	0.36%	0.00%	2.51%	34.41%	0.36%	1.43%	1.43%	1.08%	2.51%	0.72%	**1.50%**
Hydroelectric	10.59%	6.65%	9.03%	10.34%	9.77%	10.76%	11.00%	8.78%	7.96%	4.93%	3.37%	2.96%	2.05%	1.81%	**6.57%**

The real prices and quantities registered on the Italian market in 2016 [27] are used to determine the accepted average, maximum, and minimum prices and quantities for each session of DAM and ASM. An example of the price trend in a given day is shown in Figure 4, where maximum (Max), average (avg), minimum (min) prices for upward (Up) and downward (Dn) Secondary (SR) and Replacement Reserve (RR) are shown.

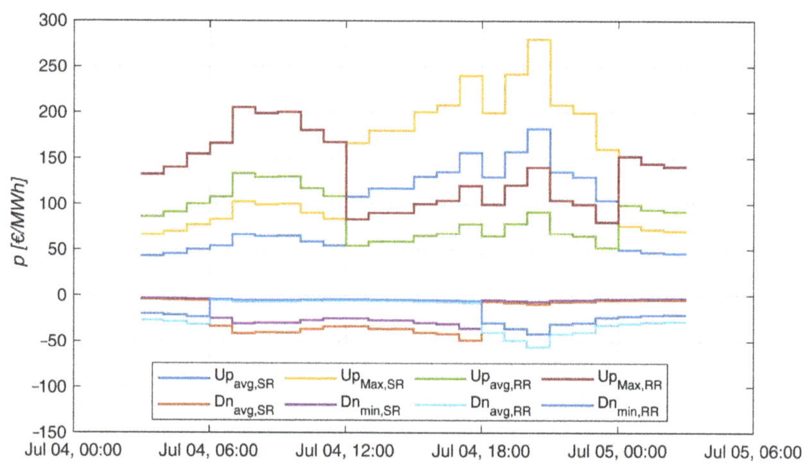

Figure 4. Prices trend in the ASM for a day of example.

Zonal sign (required to determine imbalance settlements in the ex-post economic analysis) and secondary regulation control signal—which modulates the regulation over the bands accepted in the ASM—have been downloaded from the website of the Italian TSO (Terna) [21].

The performance of the strategy proposed in optimizing the Aggregator's revenue is evaluated by simulating the EVU operation during the entire year, for a $P_{totEVU} = 10$ MW sized configuration. Table 2 shows the technical characteristics adopted for the ESS, such as the number of expected cycles ($n_{cycles,tot}$), charge/discharge efficiency ($\eta_{ch/dis}$) and relevant investment costs (c_{invESS}), and the parameters that modulate the aggressiveness of bid prices (K_q,K_p). For what concerns the ESS and CHP sizing, their power is defined as a percentage of the EVU total rated power. A detailed sensitivity analysis on these parameters is provided in Section 5. For the batteries, an energy/power ratio equal to 4 is assumed.

Table 2. Simulation parameters.

h_{eq}	4 MWh/MW
$\eta_{ch/dis}$	0.85
c_{invESS}	500 €/kWh
$n_{cycles,tot}$	4000
K_q	1.2–1.4
K_p	0.95–1.05

About capital and operating expenditures, proper assumptions are needed, defined in compliance with the scenario under investigation. Since the CHP plant is in charge of supplying the relevant industrial load, and the Aggregator only uses its residual flexibility for market participation, the investment cost for the power plant realization and the fuel expenditure needed to cover the local consumption are considered in charge to the industrial user. Hence, they are excluded from the economic analysis of the present work (developed from the Aggregator's perspective). However, the fuel costs/savings caused by the AS provision are included in the economic evaluation. Concerning the ESS, it is assumed entirely devoted to supplying AS to the power system (e.g., it is installed and managed by the Aggregator), then the investment costs are included in the economic analysis.

4. Numerical Results

To assess the validity of the mathematical model developed, it is crucial to verify its performance in providing the regulation required by the power system and the economic viability of the investment made by the Aggregator. To this purpose, the Heuristic Greedy-Indexing control logic is tested in the case study described in Section 3.

4.1. EVUs Generation

As a first step, a suitable number of EVU scenarios is generated to test the proposed method. To this purpose, the Monte Carlo procedure is used to randomly create an adequate number of EVU scenarios able to represent the typical NDRES distribution in Italy. The random generation procedure is stopped when the uncertainty measured on the quantities of interest (Q) compared to the expected value becomes lower than a predefined threshold (e_{min}).

$$e_Q = \frac{1}{N \cdot Q_N} \cdot \sqrt{\sum_{n=1}^{N} (Q_n - E[Q])^2} \tag{20}$$

The following quantities (Q) are taken into account:

- total energy produced by NDRES power plants of the EVU over the year (E_{totEVU});
- total imbalances (E_{imb_tot}) and average nonzero ($t^\star = t : E_{imb\_tot} \neq 0$) imbalance to grid $E_{imb\_avg} = mean(|E_{imb}(t^\star)|)$;
- average relative imbalance calculated with respect to the energy sold on the DAM $E_{imb\_rel} = mean\left(\frac{|E_{imb}(t^\star)|}{E_{DAM}(t^\star)}\right)$;
- total revenues in the considered period $R = \sum_t CF(t)$;
- total revenues in the ideal scenario R_{id}, i.e., assuming a perfect day-ahead forecasting of EVU production.

According to the convergence criterion identified, the Monte Carlo procedure has been stopped after the generation of 15 EVU scenarios, reaching an adequate trade-off between the uncertainty of results and computational effort. In particular, for what concerns the former aspect, Table 3 presents the results in terms of expected values ($E[Q]$) and uncertainty (e_Q) for each quantity of interest. It is worth noting that 15 EVU scenarios allow keeping, for almost all the quantities, $e_Q \leq 3\%$.

Table 3. Results relevant to the base case scenario.

	$E[Q]$	e_Q
E_{totEVU} [MWh/y]	13,611	±1.50%
E_{imb_tot} [MWh/y]	4341	±0.69%
E_{imb_avg} [MWh/y]	129.67	±0.33%
E_{imb_rel} [p.u.]	0.241	±3.72%
R [€/y]	343,496	±2.49%
R_{id} [€/y]	486,509	±2.38%

The distribution of the NDRES power plants as a function of the relevant rated power units, resulting from the Monte Carlo simulations, is shown in Figure 5, where continuous lines represent the mean value evaluated over the whole simulated scenarios and bars the max/minimum values. It is possible to observe that, despite the low probability of occurrence, also some large power plants are selected in the EVU scenarios generation process (the largest one is a PV plant of 4.9 MW). Even if the limited number of EVU scenarios did not allow getting samples over the whole admissible range (since, as mentioned before, the maximum NDRES rated power according to the probability distributions in Table 1 was equal to 6 MW), the NDRES plants actually selected are quite near to its limits. Therefore, the numerical simulations performed are considered able to provide a satisfactory representation of the situation in place in Italy. Figure 6 shows an example of EVU production profile obtained in output to the Monte Carlo procedure.

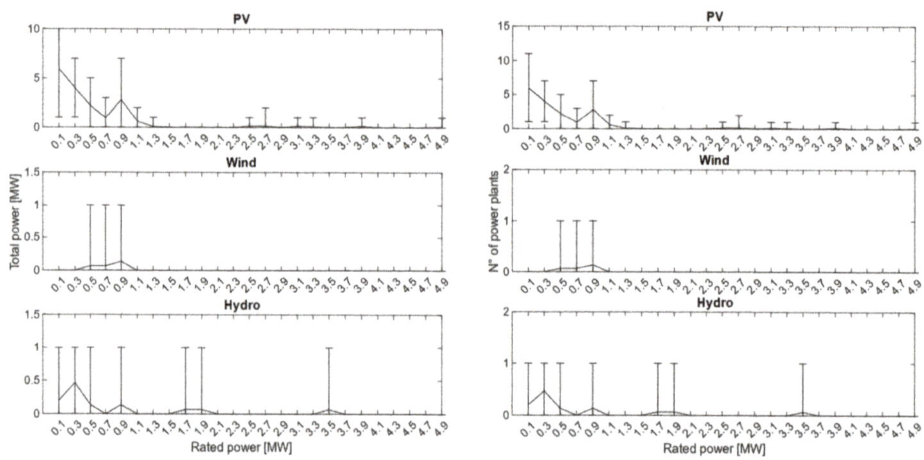

Figure 5. Distribution of NDRES power plants in the EVU: overall generation (**left**) and number of NDRES units (**right**) per category of plant's rated size (200 kW).

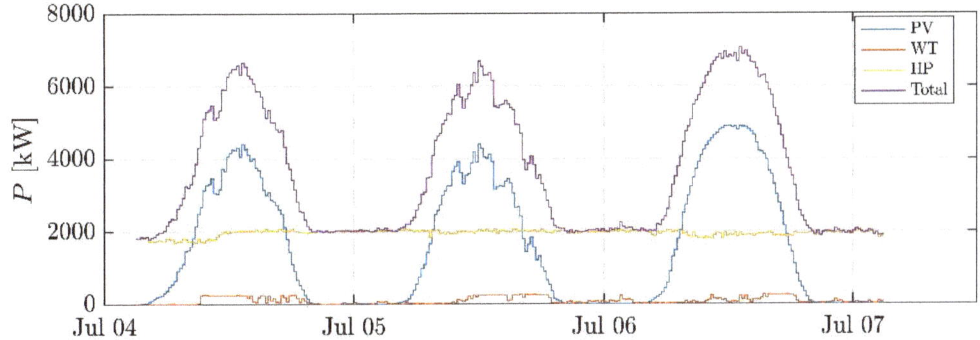

Figure 6. Example of EVU production profile.

4.2. Discussion on Test Results

To validate the model and compare the techno-economic results obtained in the different possible EVU configurations, several setups of the aggregate under analysis are considered:

- *No Reg* scenario: in this configuration, no regulation is performed by the Heuristic Greedy-Indexing; this scenario is used as a benchmark to compare results.
- *C* scenario: only the CHP power plant (sized $P_{nomCHP} = 1605$ kW $\times$ 2) is considered as regulating unit in the EVU, providing imbalance relief, secondary and replacement frequency reserve.
- *C(RR)* scenario: CHP regulation is limited to imbalance relief and replacement reserve.
- *S* scenario: ESS (sized 20% of EVU total rated power: 2 MW/8 MWh) is used to adjust the EVU power profile and to supply ancillary services to the power system.
- *S(RR)* scenario: only energy imbalance correction and replacement reserve are provided through the ESS (20% of EVU's rated power).
- *C+S* scenario: the regulation is supplied by both CHP and ESS units, each one sized according to the above-mentioned criteria.
- *C+S-half* scenario: CHP and ESS are used, each one sized half than in the previous scenarios ($P_{nomCHP} = 802.5$ kW $\times$ 2 and $P_{nomESS} = 1$ MW).
- *C+S(RR)* scenario: ESS regulation is limited to imbalance correction and replacement reserve, while CHP also supplies secondary reserve.

Figure 7 shows, for the different EVU configurations listed before, the average value of revenues *R* and costs *C*, categorized by source. The net average profit is plotted in red with the relative Monte Carlo uncertainty, which is always $\leq$3% (maximum admitted value, adopted as the threshold to arrest the Monte Carlo procedure for the scenario generation). As a reference, the average revenue obtained in the *No Reg* configuration is shown in dashed green. The bar diagram reports as positive values the average yearly revenues from the energy sold on the DAM (R_{DAM}) and from the supply of SR and RR services on ASM (respectively R_{SR} and R_{RR}), as well as the imbalance costs avoided thanks to the CHP/ESS regulation ($R_{err,corr}$), and as negative ones the costs incurred by the Aggregator due to energy imbalance (C_{imb}) and ESS wearing (C_{wear}).

In general, configurations with only the CHP as regulating unit, i.e., *C* and *C(RR)* scenarios, show promising economical returns, as highlighted by the red continuous line in the figure. The Heuristic Greedy-Indexing logic applied to a portfolio of NDRES power plants and a CHP unit improves the EVU returns compared to the non-regulating case (*No Reg*) by 27.93% in the case of both SR and RR services provision (C scenario), or even more (32.71%), if only RR is offered (*C(RR)* scenario).

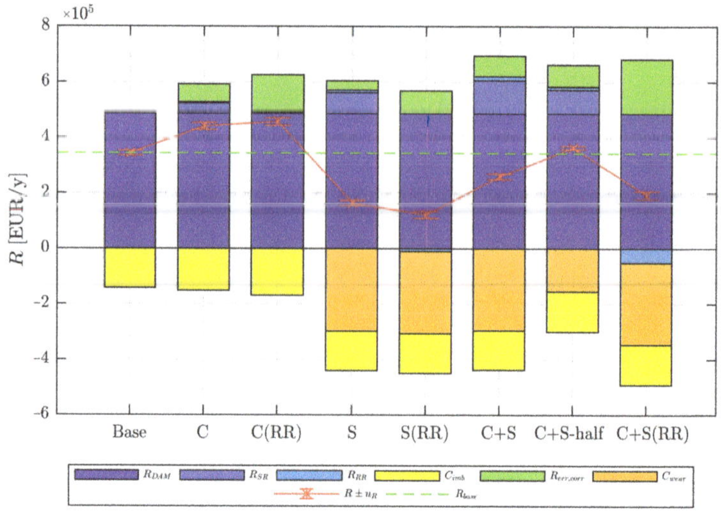

Figure 7. Economic results of the proposed EVU configurations.

On the other hand, ASM participation and imbalance correction seem to provide economic yields insufficient to justify investments in large electrochemical storages (*S* configuration). This is mainly due to the current cost of the technology, despite its high potential in terms of technical capability. In particular, the ESS wearing costs, compared to the prices rewarded on the market for the AS provision, cause limited employment of storage capacity. Regarding the scarce affordability of the ESS, it is worth noticing that in perspective other remunerated services may be implemented through the batteries, such that Uninterruptible Power Supply and Fast Reserve [28], which have been neglected in this study. Their coordinated provision (service stacking) could further justify from an economic point of view the adoption of electrochemical storage technologies for behind-the-meter and front-of-the-meter services. The configuration with both CHP and batteries undersized compared to the benchmark case (*C+S-half* scenario) shows promising results, but it achieves marginally better results than the *No Reg* case.

Figure 8 displays the energy exchanges resulting from market commitments: positive values represent an increase in the energy injected into the network (or a decrease in the power absorption), vice versa for negative values. As a consequence of the modulating effect of SR control signal ($s_{SR,ctrl}$), the configurations in which only RR is considered (e.g., *C(RR)* and *C+S(RR)* scenarios) show much larger energy quantities. In general, CHP regulation is biased toward the downward direction, due to the available flexibility of the power plant resulting from the working point to cover the local load.

Figure 9 provides an analysis about the use of the ESS, in the configurations in which it is adopted: the average number of yearly cycles performed, with a DoD of 80% (n_{cycles}), is plotted in solid blue line against the number of hours per year in which the storage is empty ($SoC = 0$) and the number of hours at which the storage is full ($SoC = 1$), drawn respectively in orange solid and dotted lines. The corresponding Monte Carlo uncertainty is also displayed with the two horizontal lines. The trends of these quantities provide a useful figure to assess the proper sizing of ESS: high n_{cycles} and a few hours of saturation are index of a correct sizing of the storage system for the required operation. This is far from being the case when only Replacement Reserve is offered (*S(RR)* scenario), since, for the small average profit margins, the ESS usage is rarely justified.

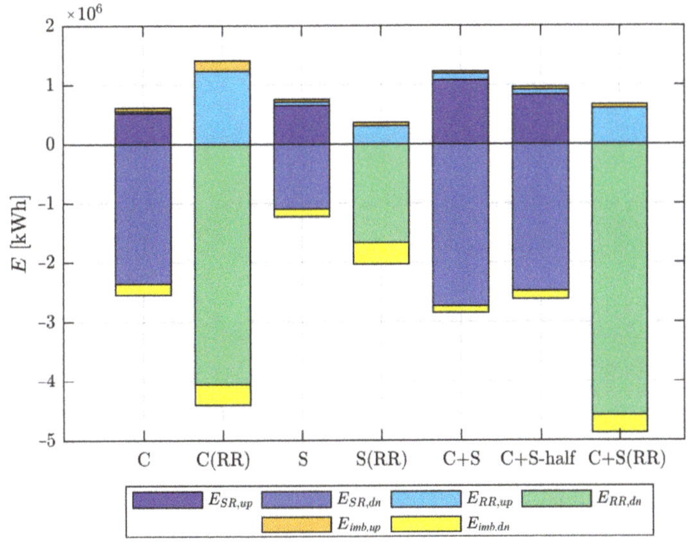

Figure 8. Enabled Virtual Unit's market exchanges.

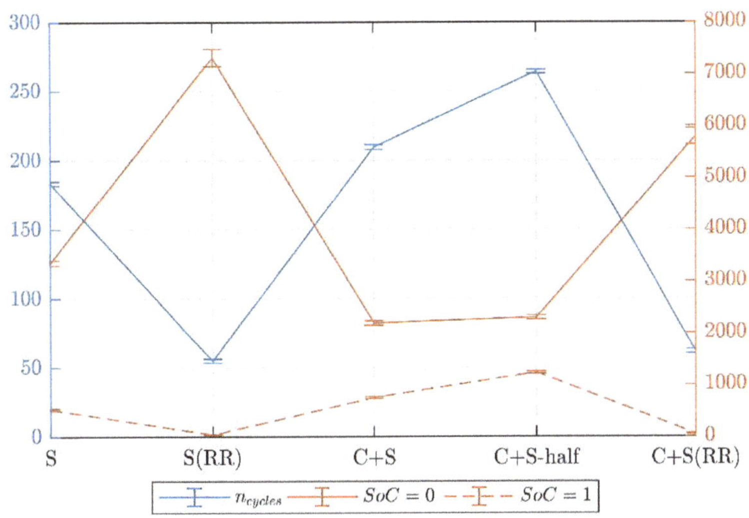

Figure 9. Usage of the ESS: charge/discharge cycles per year (n_{cycles}), axis on the left, and hours/year at a given SoC, on the right.

Table 4 reports more in detail the EVU market exchanges (Figure 7) and ESS usage (Figure 8) with the statistical uncertainties over the different Monte Carlo scenarios related to each quantity, assessed through the Equation (20). According to the approach explained in Section 4.1, the Monte Carlo procedure adopted to define the EVU scenarios is designed to keep the uncertainty affecting the quantities of interest (reported in Table 3) below a predefined threshold (3%). Other quantities involved in the analysis could also be affected by greater uncertainty, but this only happens for those having little impact on the outcomes of interest of the simulation, i.e., very small compared to the others (as one can observe in Table 4).

Table 4. Detail of the EVU market exchanges and ESS usage, and related statistical uncertainties.

	C	C(RR)	S	S(RR)	C+S	C+S-Half	C+S(RR)
$E_{SR,up}$ [MWh]	523.30 ± 0.14%	-	649.88 ± 0.24%	-	1078.26 ± 0.21%	836.02 ± 0.44%	-
$E_{SR,dn}$ [MWh]	−2354.39 ± 0.06%	-	−1090.71 ± 0.18%	-	−2728.71 ± 0.09%	−2476.71 ± 0.43%	
$E_{RR,up}$ [MWh]	29.11 ± 0.19%	1231.00 ± 0.16%	68.01 ± 0.50%	307.18 ± 0.96%	112.97 ± 0.49%	81.80 ± 0.62%	605.04 ± 0.70%
$E_{RR,dn}$ [MWh]	−3.26 ± 0.02%	−4057.72 ± 0.06%	0.00 ± 3.23%	−1664.26 ± 0.20%	−3.26 ± 0.02%	−3.15 ± 0.57%	−4574.25 ± 0.09%
$E_{imb,up}$ [MWh]	58.09 ± 0.92%	172.26 ± 0.76%	42.89 ± 6.99%	55.84 ± 16.36%	34.33 ± 3.36%	47.90 ± 4.45%	63.39 ± 10.38%
$E_{imb,dn}$ [MWh]	−185.90 ± 0.77%	−347.92 ± 1.76%	−132.83 ± 4.26%	−365.28 ± 4.22%	−112.73 ± 1.04%	−137.65 ± 1.11%	−298.69 ± 1.38%
n_{cycles}	-	-	176.25 ± 0.83%	52.68 ± 2.65%	201.49 ± 0.86%	253.93 ± 0.54%	59.19 ± 2.60%
$h@SoC = 0\%$	-	-	3358.38 ± 1.51%	7316.98 ± 2.33%	2193.18 ± 2.01%	2302.77 ± 1.72%	5791.02 ± 2.70%
$h@SoC = 100\%$	-	-	528.45 ± 3.35%	37.22 ± 13.38%	751.68 ± 2.73%	1232.78 ± 1.84%	52.44 ± 11.97%

5. Sensitivity Analysis

In this section, a sensitivity analysis is performed by parametrically changing the CHP and ESS size, and the costs in charge to the Aggregator for their use. In particular, the relationship between revenues of the EVU and size of the regulating unit is evaluated by iteratively changing the CHP/ESS size (respectively, P_{nomCHP} and P_{nomESS}) from 500 kW up to 4 MW (from 5 to 40% of the total rated power of the aggregate of NDRES units considered in the study, equal to 10 MW).

The effect of different values of P_{nomCHP} on yearly revenues R are reported in Figure 10. An increase in CHP rated power results in an almost linear proportional improvement of revenues (linear correlation coefficient equal to = 0.99). This fact occurs because the investment costs of the CHP plant are assumed totally in charge to the industrial user where the power plant is deployed. As already introduced, this hypothesis corresponds to a scenario in which the power unit is installed by the user to supply the local load; therefore, no investment cost is required to the Aggregator. However, to develop an economic analysis sticking to reality, it should also be considered that, to exploit the CHP regulation in the EVU, the Aggregator should acknowledge to the CHP owner a proper wage. The amount of this remuneration is very hard to estimate, since it depends on both the Aggregator and user's business plans and strategies. Therefore, in this work, a parametric analysis is provided to highlight its effects on the economic viability of the EVU. To this purpose, assuming as a raw estimation of investment costs required for the realization of the CHP unit 1000 k€/MW and a lifespan of the power plant of 20 years, a yearly depreciation expense is evaluated (for example, for a CHP unit with rated power 2 MW, the yearly depreciation is equal to $1000\frac{\text{k€}}{\text{MW}}\cdot 2\ \text{MW}/20\ \text{y} = 100\ \text{k€/y}$). Then, a percentage of that amount (from 0 to 50%) is assumed in charge to the Aggregator. With the increase of the share of investment costs applied to the Aggregator, as expected, the revenues of the EVU reduce. As one can observe in Figure 10, the Aggregator can easily cover expenses up to $30 \div 40\%$ of the overall CHP investment costs. Over that threshold, revenues rapidly decline with the increase of the CHP size. This fact is somewhat related to the hypotheses adopted in the study: with the increasing of CHP size, investment costs are assumed to increase proportionally (even if, in real-life, economies of scale cause the cost per unit of power to decrease with the increasing of the power plant size), whilst revenues from ASM participation growth less than proportionally.

Concerning the ESS, a similar approach has been adopted to set up the parametric analysis, iteratively changing the system rated power (P_{nomESS}) as a percentage of the EVU nominal power (P_{nomEVU}). In all cases, for the sake of simplicity, the energy/power ratio has been kept equal to 4 h. Moreover, since with a cost of the technology of 500 €/kWh batteries have shown not to be yet competitive compared to other more consolidated solutions, the revenues of the Aggregator have been reported for different costs per unit of energy capacity (from 100 to 500 €/kWh) [29].

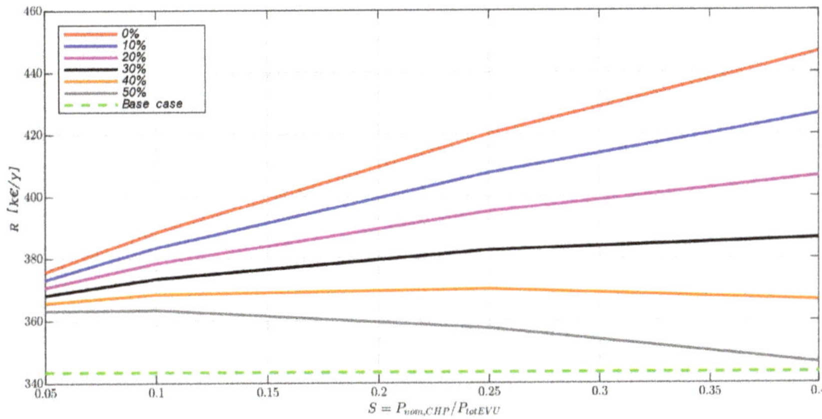

Figure 10. CHP power plant size and cost sensitivity analysis.

As shown in Figure 11, increasing the ESS size results in a general reduction of yearly revenues of the Aggregator, because the investment costs tend to outweigh the benefits of the regulation. In this regard, it is important to recall that the cost for the adoption of the ESS has been supposed in this study totally in charge to the Aggregator, because batteries are only used to supply ASs to be sold on the market and provide imbalance correction. Therefore, with the increasing of ESS size, the expenses related to the wearing and depreciation of batteries (supposed to be replaced in any case after 15 years) also grow, making the investment rapidly unprofitable. In particular, in Figure 11, it is possible to observe that, even if a cost of the technology $\leq$ 400 €/kWh yields a positive net result over the case with no regulation, a significant reduction of the price of batteries (for example, up to 100 €/kWh) is still required to make the adoption of large ESSs (e.g., 100 $\div$ 200 kW) interesting from the Aggregator's perspective. Finally, even if the results of numerical analyses seem to suggest that a very small ESS is the best opinion for the Aggregator, it should be recalled again that in practice, as for the CHP power plant, the price of the ESS per unit of power usually increases with the reduction of its size (for the effect, for example, of fixed costs, such as of installation and connection to the network). Thus, even if the price of electrochemical storage technologies is expected to drop significantly in the near future, the lowest prices reported in the chart (100 $\div$ 200 €/kWh) will be most probably a target feasible only for large ESSs (hundreds or thousands kW/kWh).

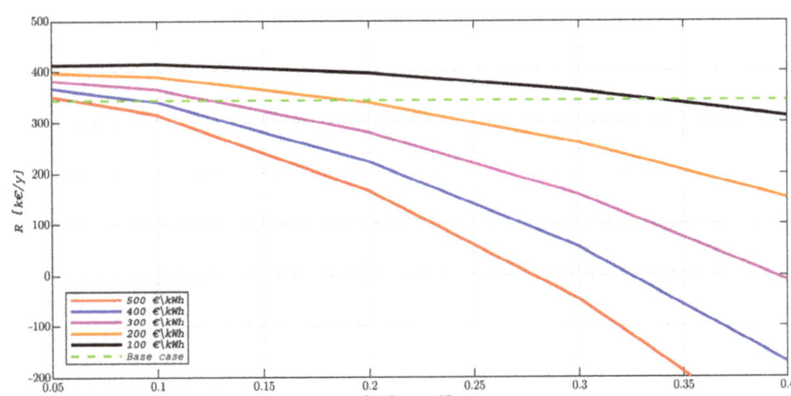

Figure 11. Storage size and cost sensitivity analysis.

6. Conclusions

In this work, an approach based on a Heuristic Greedy-Indexing logic has been proposed to make controllable a portfolio of DER power plants including, as regulating unit, a CHP generator and an ESS. The purpose of the numerical methods developed is twofold. On the one hand, the logic and approaches designed will enable in perspective an Aggregator to participate in the Ancillary Service Market, opening to the possibility of making profits by selling regulation services to the network and limiting possible penalties related to power imbalances. On the other hand, the value of this work stands in the simulation framework proposed, having modeled with a Monte Carlo procedure an aggregated of NDRES power plants, coupled with a controllable unit, and the economic variables affecting its operation. This allowed to simulate, in a comprehensive and realistic way, the behavior of the portfolio of power plants managed by the Aggregator and to evaluate under which conditions the participation in the ASM market can be economically viable.

Author Contributions: Conceptualization, D.F. and L.S.; methodology, D.F. and L.S.; software, L.S.; validation, D.F. and L.S.; resources, D.F. and L.S.; data curation, L.S.; writing—original draft preparation, D.F. and L.S.; writing—review and editing, D.F., F.G. and L.S.; visualization, D.F., F.G. and L.S.; supervision, D.F.; project administration, D.F.; funding acquisition, D.F. All authors have read and agreed to the published version of the manuscript.

Funding: F. Gulotta's position is partially funded by RSE S.p.A. within the National Research Fund for Electric Systems in compliance with the Decree of the Italian Minister of Economic Development of 16 April 2018.

Institutional Review Board Statement: Not applicable.

Informed Consent Statement: Not applicable.

Data Availability Statement: Not applicable.

Acknowledgments: Authors gratefully acknowledge the support of Enel X in developing this study.

Conflicts of Interest: The authors declare no conflict of interest.

Nomenclature

Acronyms:

AS	Ancillary Services
ASM	Ancillary Services Market
C	Costs
CF	Cash Flow
CHP	Combined Heat and Power
DAM	Day-Ahead Market
DER	Distributed Energy Resources
ESS	Energy Storage System
EVU	Enabled Virtual Unit
HP	Hydro Power
NDRES	Non-Dispatchable Renewable Energy Sources
PV	PhotoVoltaic
R	Revenues
RR	Replacement Reserve
SoC	State of Charge
SR	Secondary Reserve
TSO	Transmission System Operator
WT	Wind Turbine

Parameters and Variables:

$bid_{RR/SR}$	Quantities bid in a RR/SR market session
$c_{CHP/curt/ESS}$	Cost of the regulation performed by the CHP/NDRES curtailment/ESS
$\bar{c}_{ASM}$	Average cost expected on future ASM sessions

c_f	Fuel cost of the CHP generator
CF	Total cash flow of the EVU
CF_{imb}	Cash flows related to imbalance
C_{imb}/c_{imb}	Cost of residual imbalance (total/per unit of energy)
c_{invESS}	Investment cost per unit of power capacity of the ESS
C_{wear}/c_{wear}	Battery wearing cost (total/per unit of energy)
Δt	Timestep of the simulation
$E_{av,k,up/dn}$	Energy available by the *k*-th regulation action (up/downward regulation)
$E_{avCHP/ESS,up/dn}$	Energy available by the CHP/ESS unit for up/downward regulation
E_{DAM}	Energy sold on the DAM
$E_{exCHP/ESS}$	Energy exchanged by the CHP/ESS unit
$E_{ex,k,up/dn}$	Energy exchanged through the *k*-th regulation action (up/downward regulation)
E_{imb}	Residual imbalance
$E_{imb\_avg/rel/tot}$	Residual imbalance (average/relative/total value)
e_{min}	Threshold of the Monte Carlo stopping criterion
E_{nomESS}	Nominal ESS capacity
E_{prod}	Energy produced by the NDRES
$E_{reg,up/dn}$	Total energy required to fulfill the market commitments (up/downward regulation)
$E_{regRR/SR,up/dn}$	Energy required to fulfill the commitments for RR/SR service (up/downward regulation)
$e_{RR/SR,up/dn}$	Bidding price of the RR/SR service (up/downward regulation)
E_{setESS}	Set level for the ESS state of charge
$\eta_{ch/dis}$	ESS charge/discharge efficiency
η_{ESS}	ESS round trip efficiency
$K_{p_{up/dn}}$	Factor modulating bid price aggressiveness (up/downward regulation)
$K_{q_{up/dn}}$	Factor modulating the aggressiveness of the SR service offering process (up/downward reserve)
$n_{cycles,tot}$	Number of expected storage cycles at DoD = 80%
$P_{avEVU,up/dn}$	Up/downward band available for the ASM bidding
$P_{CHPflex,up/dn}$	CHP residual up/downward flexibility bands
P_{DAM}	DAM power schedule
p_{DAM}	DAM clearing price
P_{err}	Error correction in bidding
$P_{expected}$	Power forecast used in the DAM
P_{ind}	Power absorbed by the industrial load supplied by the CHP generator
$P_{lim,ch/dis}$	Technical limit for charging/discharging power of the ESS
$p_{maxASM,up}$	Maximum price accepted in the ASM for upward service
$p_{minASM,dn}$	Minimum price accepted in the ASM for downward service
P_{minCHP}	CHP minimum admissible power production
$P_{nomCHP/DG/ESS}$	Nominal power of the CHP/NDRES/ESS unit
P_{reqESS}	Power required to bring the ESS to the set level
P_{totEVU}	Total rated power of the EVU
P_{updt}	Updated power forecast used in the ASM
$p_{X,up/dn}$	Historical AS prices for the *X* service
$q_{accRR/SR,up/dn}$	Quantities accepted on the market for RR/SR service (up/downward regulation)
$\overline{q}_{RR/SR,up/dn}$	Average quantity accepted on the market for RR/SR service (up/downward regulation)
R_{DAM}	Revenues from DAM
$R_{err,corr}$	Cost avoided for the imbalance correction
R_{id}	Revenues relevant in the ideal scenario (perfect forecast)
$R_{RR/SR}$	Revenues from the provision of RR/SR service
s_r	Sign of E_{reg}
$s_{SR,ctrl}$	Secondary regulation control signal
s_z	Imbalance zonal sign
w_{power}	Power statistical distribution of NDRES
w_{tech}	Technology statistical distribution of NDRES

References

1. Steele, A.J.H.; Burnett, J.W.; Bergstrom, J.C. The impact of variable renewable energy resources on power system reliability. *Energy Policy* **2021**, *151*, 111947. [CrossRef]
2. Rancilio, G.; Rossi, A.; Falabretti, D.; Galliani, A.; Merlo, M. Ancillary services markets in europe: Evolution and regulatory trade-offs. *Renew. Sustain. Energy Rev.* **2021**, *154*, 111850. [CrossRef]
3. Italian Energy Authority (ARERA). Resolution 300/2017/r/eel. May 2017. Available online: https://www.arera.it/it/docs/17/300-17.htm (accessed on 28 December 2021).
4. Orths, A.G.; Eriksen, P.B. European test field: VPP Denmark. In Proceedings of the 2009 IEEE Power & Energy Society General Meeting, Piscataway, NJ, USA, 26–30 July 2009; Volume 9, pp. 1–5.
5. Mashhour, E.; Tafreshi, S.M.M. Bidding Strategy of Virtual Power Plant for Participating in Energy and Spinning Reserve Markets—Part I: Problem Formulation. *IEEE Trans. Power Syst.* **2011**, *26*, 949–956. [CrossRef]
6. Trovato, V.; Kantharaj, B. Energy storage behind-the-meter with renewable generators: Techno-economic value of optimal imbalance management. *Int. J. Electr. Power Energy Syst.* **2020**, *118*, 105813. [CrossRef]
7. Kardakos, E.G.; Simoglou, C.K.; Bakirtzis, A.G. Optimal Offering Strategy of a Virtual Power Plant: A Stochastic Bi-Level Approach. *IEEE Trans. Smart Grid* **2015**, *7*, 794. [CrossRef]
8. Nezamabadi, H.; Nazar, M.S. Arbitrage strategy of virtual power plants in energy, spinning reserve and reactive power markets. *IET Gener. Transm. Distrib.* **2016**, *10*, 750–763. [CrossRef]
9. Yang, D.; He, S.; Wang, M.; Pandzic, H. Bidding Strategy for Virtual Power Plant Considering the Large-Scale Integrations of Electric Vehicles. *IEEE Trans. Ind. Appl.* **2020**, *56*, 5890–5900. [CrossRef]
10. Yuan, Y.; Wei, Z.; Sun, G.; Sun, Y.; Wang, D. A real-time optimal generation cost control method for virtual power plant. *Neurocomputing* **2014**, *143*, 322–330. [CrossRef]
11. Zapata, J.; Vandewalle, J.; D'Haeseleer, W. A comparative study of imbalance reduction strategies for virtual power plant operation. *Appl. Therm. Eng.* **2014**, *71*, 847–857. [CrossRef]
12. Mashhour, E.; Tafreshi, S.M.M. Bidding Strategy of Virtual Power Plant for Participating in Energy and Spinning Reserve Markets—Part II: Numerical Analysis. *IEEE Trans. Power Syst.* **2010**, *26*, 957–964. [CrossRef]
13. Logenthiran, T.; Srinivasan, D.; Khambadkone, A.; Raj, T.S. Optimal sizing of Distributed Energy Resources for integrated microgrids using Evolutionary Strategy. In *2012 IEEE Congress on Evolutionary Computation*; Institute of Electrical and Electronics Engineers (IEEE): Piscataway, NJ, USA, 2012; pp. 1–8.
14. Maity, S.; Paul, S.; Karbouj, H.; Rather, Z.H. Optimal Sizing and Placement of Wind Farm in a Radial Distribution Network Considering Reliability, Operational, Economic and Environmental Factors. *IEEE Trans. Power Deliv.* **2021**, *36*, 3043–3054. [CrossRef]
15. Petersen, M.K.; Hansen, L.H.; Bendtsen, J.D.; Edlund, K.; Stoustrup, J. Heuristic Optimization for the Discrete Virtual Power Plant Dispatch Problem. *IEEE Trans. Smart Grid* **2014**, *5*, 2910–2918. [CrossRef]
16. Gulotta, F.; Rancilio, G.; Blaco, A.; Bovera, F.; Merlo, M.; Moncecchi, M.; Falabretti, D. E-Mobility Scheduling for the Provision of Ancillary Services to the Power System. *Int. J. Electr. Electron. Eng. Telecommun.* **2020**, *9*, 349–355. [CrossRef]
17. Ke, X.D.; Wu, D.; Lu, N. A Real-Time Greedy-Index Dispatching Policy for Using PEVs to Provide Frequency Regulation Service. *IEEE Trans. Smart Grid* **2019**, *10*, 864–877. [CrossRef]
18. Shi, H.; Blaauwbroek, N.; Nguyen, P.; Kamphuis, R.I.G. Energy management in Multi-Commodity Smart Energy Systems with a greedy approach. *Appl. Energy* **2016**, *167*, 385–396. [CrossRef]
19. Lagodimos, A.; Leopoulos, V.I. Greedy heuristic algorithms for manpower shift planning. *Int. J. Prod. Econ.* **2000**, *68*, 95–106. [CrossRef]
20. Falabretti, D.; Lindholm, M.; Merlo, M.; Scapeccia, G. Energy storage coupling in a high efficiency household scenario: A real life experimental application. *J. Energy Storage* **2018**, *17*, 496–506. [CrossRef]
21. Terna. Secondary Reserve Control Signal. Available online: https://myterna.terna.it/SunSet/Public (accessed on 28 December 2021).
22. Eyer, J.; Iannucci, J.; Butler, P.C. *Estimating Electricity Storage Power Rating and Discharge Duration for Utility Transmission and Distribution Deferral. A Study for the DOE Energy Storage Systems Program*; SANDIA Report SAND2005-7069; Sandia National Laboratories: Albuquerque, NM, USA, 2005.
23. Italian Energy Authority (ARERA). Annex A to Resolution ARG/Let 111/06. Available online: https://www.arera.it/it/docs/06/111-06.htm (accessed on 28 December 2021).
24. GSE. AtlaImpianti. 2021. Available online: https://atla.gse.it/atlaimpianti (accessed on 28 December 2021).
25. GSE. Rapporto Statistico—Energia da Fonti Rinnovabili—Anno 2019. 2021. Available online: https://www.gse.it/dati-e-scenari/statistiche (accessed on 28 December 2021).
26. Italian Electrical Committee (CEI). CEI 0-16. Reference Technical Rules for the Connection of Active and Passive Consumers to the HV and MV Electrical Networks of Distribution Company. 2019. Available online: https://mycatalogo.ceinorme.it/cei/item/0000016796 (accessed on 28 December 2021).
27. GME Website. 2021. Available online: http://www.mercatoelettrico.org/it (accessed on 28 December 2021).

28. Italian Energy Authority (ARERA). Resolution 200/2020/r/eel. 2020. Available online: https://www.arera.it/it/docs/20/200-20.htm (accessed on 28 December 2021).
29. Penisa, X.N.; Castro, M.T.; Pascasio, J.D.A.; Esparcia, E.A.; Schmidt, O.; Ocon, J.D. Projecting the Price of Lithium-Ion NMC Battery Packs Using a Multifactor Learning Curve Model. *Energies* **2020**, *13*, 5276. [CrossRef]

energies

Article

Synthetic Theft Attacks and Long Short Term Memory-Based Preprocessing for Electricity Theft Detection Using Gated Recurrent Unit

Pamir [1], Nadeem Javaid [1,2,*], Saher Javaid [3,*], Muhammad Asif [1], Muhammad Umar Javed [1], Adamu Sani Yahaya [1] and Sheraz Aslam [4,*]

1 Department of Computer Science, COMSATS University Islamabad, Islamabad 44000, Pakistan; pamirshams2011@yahoo.com (P.); muhammad.asif.comsat@gmail.com (M.A.); umarkhokhar1091@gmail.com (M.U.J.); asyahaya.it@buk.edu.ng (A.S.Y.)
2 School of Computer Science, University of Technology Sydney, Ultimo, NSW 2007, Australia
3 Graduate School of Advanced Science and Technology, Japan Advanced Institute of Science and Technology, 1-1 Asahidai, Nomi City 923-1292, Ishikawa, Japan
4 Department of Electrical Engineering, Computer Engineering and Informatics, Cyprus University of Technology, Limassol 3036, Cyprus
* Correspondence: nadeemjavaidqau@gmail.com (N.J.); saher@jaist.ac.jp (S.J.); sheraz.aslam@cut.ac.cy (S.A.)

Abstract: Electricity theft is one of the challenging problems in smart grids. The power utilities around the globe face huge economic loss due to ET. The traditional electricity theft detection (ETD) models confront several challenges, such as highly imbalance distribution of electricity consumption data, curse of dimensionality and inevitable effects of non-malicious factors. To cope with the aforementioned concerns, this paper presents a novel ETD strategy for smart grids based on theft attacks, long short-term memory (LSTM) and gated recurrent unit (GRU) called TLGRU. It includes three subunits: (1) synthetic theft attacks based data balancing, (2) LSTM based feature extraction, and (3) GRU based theft classification. GRU is used for drift identification. It stores and extracts the long-term dependency in the power consumption data. It is beneficial for drift identification. In this way, a minimum false positive rate (FPR) is obtained. Moreover, dropout regularization and Adam optimizer are added in GRU for tackling overfitting and trapping model in the local minima, respectively. The proposed TLGRU model uses the realistic EC profiles of the Chinese power utility state grid corporation of China for analysis and to solve the ETD problem. From the simulation results, it is exhibited that 1% FPR, 97.96% precision, 91.56% accuracy, and 91.68% area under curve for ETD are obtained by the proposed model. The proposed model outperforms the existing models in terms of ETD.

Keywords: theft attacks; long short term memory; gated recurrent unit; deep learning techniques; machine learning techniques; electricity theft detection; smart grids

Citation: Pamir; Javaid, N.; Javaid, S.; Asif, M.; Javed, M.U.; Yahaya, A.S.; Aslam, S. Synthetic Theft Attacks and Long Short Term Memory-Based Preprocessing for Electricity Theft Detection Using Gated Recurrent Unit. *Energies* **2022**, *15*, 2778. https://doi.org/10.3390/en15082778

Academic Editors: Miguel Jiménez Carrizosa and Abu-Siada Ahmed

Received: 31 December 2021
Accepted: 30 March 2022
Published: 10 April 2022

1. Introduction

From global perspective, the traditional metering system is still a commonly applied system, especially in the residential sector [1]. However, the electrical meters used in the traditional metering system, i.e., electromechanical meters, do not perform as accurately as expected and their measurement ability is mostly affected by the waveform distortion, operating temperature, and other factors. Moreover, this category of meters allows unidirectional communication [2]. In addition, the consumed electricity measurement needs manual reading by the electric utility personnel, in which there are many chances of measurement errors. Furthermore, it should also be considered that manual meter readings lead the utilities to incur high operational cost, which is in fact charged from the energy users. Therefore, the advanced metering infrastructure [3] is introduced to overcome the issues caused by the conventional metering systems.

The electricity theft detection (ETD) is one of the major issues and a trending research area in the current era. It lies in the category of non technical losses (NTLs). Generally, NTLs along with technical losses (TLs) are the categories in which losses of electricity are grouped [4]. Majorly, TLs occur in power system's equipment owing their resistance against the power flow. While NTLs arise because of the electricity theft in terms of meter hacking, bypassing, or tampering. The theft of electricity leads to many dangerous issues like huge economic loss, operational inefficiency in electric grids, public safety hazards, etc. The economic loss arises due to the electricity theft that amounts around 100 million Canadian dollars per year according to the British Columbia power and hydro authority [5]. In addition, the revenue loss incurred due to the NTL throughout the world is around 96 billion USD yearly [6]. Hence, it is very crucial to have an efficient and effective ETD approach in the smart grids (SGs).

The detection approaches of the electricity theft used in the literature are grouped into three main categories: the hardware based approaches, the classification based approaches, and the game-theory based approaches. The ETD approaches based on the hardware devices [7,8] employ different hardware equipment to obtain higher theft detection accuracy. However, these techniques require high monetary cost for the installation and maintenance of the hardware equipment. ETD is referred as a game, in game-theory based approaches, between two players [9,10] where both players try to optimize their utility functions. Furthermore, a zero-sum game is introduced among the power entities for achieving the equilibrium state. These methods do not require extra payment. However, they are still not a suitable and optimal remedy to minimize electricity theft because the formulation of a suitable utility function in a real environment is a tiresome job in hand.

Several deep learning (DL) and machine learning (ML) based classification models are developed in [5,6,11–16] for ETD and they use energy consumption data stored in smart meters (SMs). Therefore, their costs are reasonable. However, there are some issues with the existing DL and ML based classification models, which negatively affect the classifiers in terms of false positive rate (FPR) and true positive rate (TPR). One crucial problem that causes the DL and ML based classifiers to perform poorly in detecting the electricity theft is imbalanced class problem. The imbalanced class or data imbalanced issue denotes that the count of the data points related to the abnormal consumers is not equal to the count of the normal consumers' present in a dataset. The data related to the normal electricity consumers is easily available in comparison with abnormal class data because the abnormal data are gathered in a limited amount from the real environment. Hence, the problem of data imbalance leads the DL and ML classifiers to be biased towards the majority class (normal users) when performing classification, which results in high FPR. In addition, another crucial issue that negatively affects the classification algorithms in terms of TPR, FPR, and overfitting is the curse of dimensionality. It occurs while dealing with the data of high dimensionality. In other words, the curse of dimensionality refers to a principle in which the increment in number of features (dimensions) is directly proportional to the increment in the classification error. Furthermore, another crucial issue that affects the ML and DL classification algorithms negatively is ignoring the drift. It refers to the irregular consumption of the electricity that occurs from non theft factors like changes in the number of family members, seasonal changes, changes in electric appliances in terms of their type or number, etc.

This paper presents the extended version of the work already published in [17]. This work uses six theft attacks (TAs) to produce theft data samples for balancing the data. The combined model having TAs, long short-term memory (LSTM) [18], and gated recurrent unit (GRU) [19], termed as TLGRU, is proposed for efficient ETD. Moreover, the proposed approach learns and pinpoints the real abnormal consumers instead of pinpointing the drift as theft. In this way, it reduces the FPR. The research article comprises the following major contributions.

- A TLGRU model is proposed for effective and reliable ETD in SGs. In the proposed model, the synthetic TAs are implemented to generate theft samples in the dataset acquired from state grid corporation of China (SGCC) for tackling imbalance problem. Moreover, LSTM is employed to efficiently extract and maintain the vital characteristics from the huge time series data, which handle curse of dimensionality problem.

- A powerful recurrent memory network, termed as GRU, is utilized to initially investigate the electricity consumption (EC) profiles of consumers and then tackle the problem of drift accordingly.
- An efficient regularization technique, known as dropout, is integrated in the proposed TLGRU model to avoid overfitting and increase the convergence speed.

The remaining sections of the article are arranged as discussed. Section 2 comprises the study of the existing literature. While, the subject matter of Section 3 is the proposed system model. The outcomes obtained after performing extensive simulations are elaborated in Section 4. Finally, the concluding remarks are given in Section 5 .

2. Literature Review

Tuning the hyperparameters, data imbalance, ensuring privacy, and the dimensionality curse problems are the four broad categories in which the existing literature is divided and studied into four groups in this section. The research articles that deal with the hyperparameters' tuning of the ML techniques are given in initial category. In [4–6,11–14], those ML and DL techniques are under consideration that deal with efficient tuning of hyperparameters. In [4], a stacked sparse denoising autoencoder (SSDAE) is proposed to extract the most effective features to deal with the FPR and generalization issues. The low value for FPR, high detection rate, high robustness, and important feature extraction are achieved by introducing the noise and sparsity parameters into SSDAE. The hyperparameters of SSDAE are tuned using particle swarm optimization algorithm. Moreover, the SGCC hourly data are used for analysis in this work. Furthermore, in [5], to detect electricity theft, a wide and deep convolutional neural network (WDCNN) is proposed. The theft detection is performed by using both one dimensional and two dimensional data for model training and model testing. The wide component is used to process the global features and the deep component is utilized to find whether the periodicity exists between EC patterns or not. Moreover, the behavior of the data are checked using a statistical technique, i.e., pearson correlation coefficient. The proposed classifier is validated using area under the curve (AUC) and mean average precision (MAP).

However, manual hyperparameter tuning is done in the work, which reduces the accuracy and efficiency of the classifier.

To ensure correct identification and detection of the theft patterns, a hybrid model of two deep neural networks is proposed in [6]. The hybrid model comprises of multi layer perceptron (MLP) and LSTM. LSTM is responsible for analyzing EC data while MLP deals with the exogenous variables. Clearly, it is proved that the hybrid model outperforms the benchmark models. However, the proposed hybrid model is prone to overfitting issue due to the full connectivity of neurons. Furthermore, the class imbalance problem is not tackled, which reduces the model's generalization ability.

The adjustment of the hyperparameters' values and the imbalanced data problem are tackled by the studies conducted in [11–13]. The authors in [11] propose a consumption pattern based ETD (CPBETD) model. The data imbalance problem is resolved by generating a synthetic attacks' dataset. In the model, low sampling rate of EC values is considered to preserve customers' privacy. Furthermore, a sustainable energy authority of Ireland (SEAI) dataset is employed to check the performance of the model. The proposed model is validated using FPR, DR, and bayesian detection rate (BDR) performance metrics. However, the feature extraction step is not considered, which increases the computational complexity. One of the main performance indicators, i.e., accuracy is ignored in this paper. The authors in [12] utilize ensemble learning models for ETD. The ensemble ML models employed in the study are extreme gradient boosting (XGBoost), random forest (RF), adaptive boosting (AdaBoost), light gradient boosting (LGB), extra trees, and categorical boosting (CatBoost). The commission for energy regulation (CER) data are used for models' evaluation. The data preprocessing step is also performed for the TPR improvement. The class imbalance problem is tackled using SMOTE. However, it requires high computational cost for training and testing processes. Another issue with synthetic minority oversampling technique (SMOTE) is that it lacks in capturing probability distribution curve from the complex EC data, which degrades the generalization ability of the classifiers. Similarly, the authors in [13] employ a variant of generative adversarial network (GAN) [20], named as wasserstein GAN (WGAN), and K-nearest neighbor for balancing imbalanced data and classifying data points that are near to support vector machine (SVM's) hyperplane, respectively. Moreover, a combined technique of supervised and unsupervised methods is proposed, i.e., decision tree combined with the K-nearest neighbor SVM (DT-KSVM) and WGAN techniques. In [14], the authors propose an XGBoost technique for detecting anomalies present in the dataset of SMs. The hyperparameters' values are adjusted using a grid search method in this paper. However, the grid search method is computationally expensive. Moreover, feature extraction is not efficiently performed in this work.

The second group of literature review focuses on imbalanced problem in SM data. In [15,21–25], the data imbalance problem is resolved by the authors using several sampling techniques. In [15], the authors tackle the data imbalance problem using a hybrid of k-means SMOTE (K-SMOTE) technique while RF is used for theft classification. The proposed model is evaluated and validated using the EC data obtained from Hebei province of China. The authors in [21] propose a framework that consists of maximal overlap discrete wavelet packet transform (MODWPT) and random undersampling boosting (RUSBoost) for feature extraction and theft classification, respectively. The data imbalance problem is tackled using a random undersampling (RUS) technique. The EC data of commercial and industrial users from Honduras is employed for the evaluation of the framework. Further, the proposed framework is validated using AUC, Matthews correlation coefficient (MCC), accuracy, precision, recall, F_β score, and specificity. However, some important information is lost because of random removal of observations from the majority class using RUS. This information loss results in a high FPR. In addition, the authors in [22] propose a combined model of CNN and LSTM for detecting energy theft in SGs. Furthermore, the class imbalance problem in the EC data is resolved using an oversampling technique, known as SMOTE. Eventually, validation of the proposed model is performed using MCC, recall, F1-score, accuracy, and precision. In addition, the EC data of Multan electric power company (MEPCO) is considered for conducting simulations. However, SMOTE based synthetic data generation leads to a class overlapping problem. Furthermore, in [23], the authors propose a methodology based on ensemble bagged tree (EBT) for detecting the fraudulent electricity consumers.

The research articles in the third group consider the electricity users' privacy preservation problem. The authors in [16,26–28] focus on dealing with the maintenance of the consumers' privacy. The study in [16] proposes a semi supervised auto encoder (SSAE) for extracting significant features from the SM data of the industrial consumers. The proposed SSAE model, for addressing the overfitting issue, makes use of the data that is unlabeled. Along with that, an adversarial module is also used. The proposed model's exceptional performance with a small set of samples and preservation of consumers' privacy are exhibited through simulations. In addition, the research conducted in [26] proposes an efficient ETD model. Privacy preservation is ensured using a functional encryption (FE) method used by the SMs of the users to encrypt their readings. The issue of imbalance between classes is tackled through adaptive synthetic (ADASYN) and by creating a malicious attack dataset. Similarly, in [27], the authors propose a privacy preserving based ETD (PPETD) algorithm. The generalized CNN (GCNN) classifier is employed for ETD using the encrypted small-sized consumption slots data. However, it has high computational complexity due to training of the excessive parameters of GCNN. Furthermore, in [28], a multiple linear regression model (MLRM) is proposed for NTL detection in SGs. A significant benefit of this method is that it detects electricity theft after performing comparison between the data recorded by an SM and the collector (a sensing device attached with the SM) without violating the customers' privacy.

The research articles in the final category deal with the curse of dimensionality issue using different feature generation, feature selection, and feature extraction strategies. The authors in [29] employ feature generation using mean, standard deviation, and minimum and maximum values of features. They develop a gradient boosting theft detector (GBTD) for electricity theft identification in SGs. A framework of practical feature engineering is proposed in [30]. The framework is the combination of the finite mixture model (FMM) based clustering for segmentation of the customers and genetic algorithm (GA) based feature generation from one or more already available features to improve prediction accuracy. A gradient boosting machine is used for classification. The model's validation is done using various performance measures. Furthermore, the authors in [31] focus on the important features' selection for pinpointing the anomalous consumers. They practice binary black hole algorithm (BBHA), which is a metaheuristic technique, for feature selection. The proposed algorithm is validated using accuracy and execution time metrics, which are not sufficient for a fair evaluation. Moreover, many novel optimization algorithms are developed that can be used for selecting the most suitable features in order to achieve better performance as compared to BBHA in terms of accuracy and execution time. In [32], a probabilistic technique is put forward for classifying patterns along with the mathematical formulation on the levenberg-marquardt technique's basis. Feature extraction is performed using the encoding algorithm. However, the model input parameters' tuning is ignored, which becomes the reason for overfitting.

The summarized form of the available literature review is given in Table 1. Moreover, the problems addressed from the aforementioned literature and motivation of this work are finalized as follows.

Table 1. Literature review summary.

Methodology	Objectives	Dataset	Performance Metrics	Limitations
SSDAE [4]	To tackle NTLs	SGCC hourly data	FPR, TPR and AUC	Inadequate evaluation metrics
WDCNN [5]	To secure SGs by detecting electricity theft	SGCC daily data	AUC and MAP	Data imbalance issue
LSTM-MLP [6]	To overcome NTLs	Endesa	AUC, precision, recall and precision-recall-AUC	Data imbalance issue
CPBETD [11]	To improve ETD performance	SEAI	TPR, FPR and BDR	No feature extraction is performed
RF, AdaBoost, XGBoost, LGB, ensemble tree and CatBoost [12]	To detect energy theft in power grids	CER	Precision, AUC and accuracy	Ensemble techniques are computationally complex
DT-KSVM [13]	To decrease power losses	SEAI	AUC and accuracy	Inadequate performance metrics
XGBoost [14]	To enhance ETD performance	Endesa	AUC, precision-recall and execution time	High computational time
RF [15]	To detect NTL behavior	Hebei province	AUC and accuracy	No feature extraction is done

Table 1. Cont.

Methodology	Objectives	Dataset	Performance Metrics	Limitations
SSAE [16]	To reduce NTLs by employing semi-supervised data	SGCC daily data	Accuracy, TPR, FPR, precision, recall and F1-score	Inappropriate hyperparameter tuning
MODWPT, RUSBoost [21]	To reduce NTLs	Honduras	F1-score, MCC, precision, recall, AUC and accuracy	Important information is lost due to RUS
CNN-LSTM [22]	To detect abnormal EC profiles of consumers	SGCC daily data	F1-score., MCC, precision, recall and accuracy	Classes overlap due to SMOTE
EBT [23]	To minimize NTLs	MEPCO	Accuracy, sensitivity, specificity, F1-score and FPR	Curse of dimensionality problem is not tackled
ETDFE [26]	To detect ET by preserving consumers' privacy	CER	Highest difference (HD), FPR, DR and Accuracy	High computational complexity due to improper hyperparameter optimization
PPETD [27]	To perform ETD while maintaining consumers' privacy	CER	HD, DR and FPR	Improper hyperparameter tuning
MLRM [28]	To overcome NTLs	Neighborhood area network dataset	Accuracy, sensitivity and specificity	Curse of dimensionality problem is not handled

The authors of [4] introduce SSDAE for ETD in SGs. The principal challenge they considered is high value of FPR because of the low generalization of classification techniques. The CPBETD technique is presented in [11]. CPBETD uses the data of the electric users as well as the transformer meters for ETD. The imbalanced data issue is handled by creating a synthetic TA dataset. Finally, motivated from [4,11], we focused on anomalies' prevention and detection due to nonmalicious intermediaries (drift), curse of dimensionality, and dealing with the data imbalance problem. The authors in [5,6] introduce new WDCNN [33] and LSTM-MLP based techniques for electricity theft classification, respectively. However, the unavoidable issue of data imbalance is not considered that leads to classifier's biasness with respect to the normal class that causes high FPR. In addition, the high value of the FPR is neglected by these research articles. In addition, in the wide module of the WDCNN model, only one layer, i.e., fully connected layer, is employed, which causes the model to be trapped into the local optima. Furthermore, the authors in [11] present CPBETD for efficient detection of NTL as previously metioned. Moreover, CPBETD obtained better performance even with a low granularity of EC data, which assists to maintain consumers' privacy. However, the curse of dimensionality issue is ignored, which leads to high FPR and overfitting issues.

3. Proposed System Model

The model designed for the underlying work is made of two main units and some subunits. The main units are (1) preprocessing unit and (2) theft classification unit. The details of the units and their relevant subunits are given in the subsequent sections. Figure 1 exhibits the graphical view of the system model.

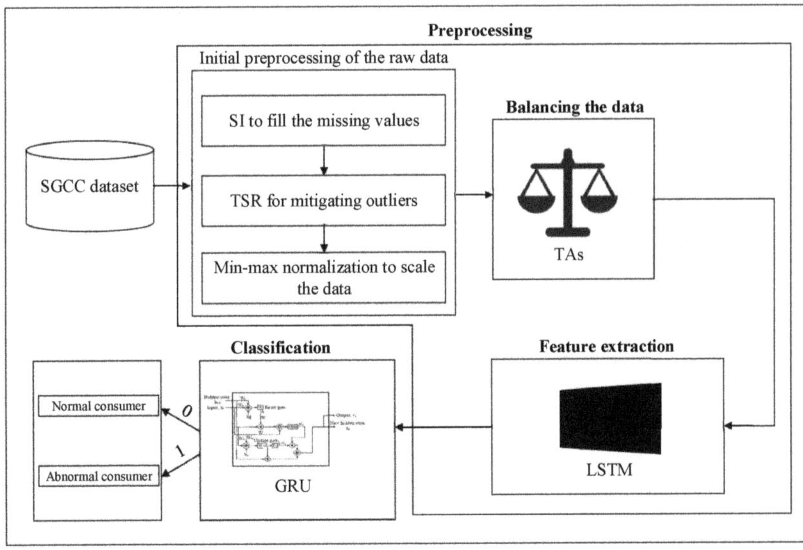

Figure 1. Overview of the proposed ETD model.

3.1. Dataset Description

The data of energy consumption acquired from SGCC (which is both realistic and easily accessible) is employed for the proposed model's validation via different performance metrics [5]. The SGCC possesses imbalanced EC data.

The total number of consumers' data records is 42,372 of which 38,752 are the normal users' records and 3615 are theft consumers' records. The sampling frequency of the dataset is set to daily. In the dataset, the overall EC values are represented in terms of rows while the EC value on a specific day is given in terms of a column. Furthermore, the data are collected after conducting onsite inspections. So, it contains NaN values, outliers, and data being dispersed on a huge scale. These abnormalities should be treated before proceeding to the development of the ETD model. In this regard, to recover the missing values, mitigate the outliers, and scale the data in a specific range, the preprocessing step is required, which is discussed in detail in the next subsection. Before preprocessing, the total number of consumers' data is 42,372, whereas, after the preprocessing, 5 rows are dropped by the simple imputer (SI) method because all the values in these rows are NaN values. In such cases, the SI does not know what value is to be imputed. The imputer will impute some values instead of deleting a record if it finds atleast one non-NaN value in the targeted record. It is also important to note that SI method works column-wise, so you need to take the transpose of your data before applying the imputer method. After imputation, again take the transpose of the data to revert it to the original shape. Table 2 presents details of the used dataset.

Table 2. Dataset detail.

Dataset Description	Values
Dataset acquisition intervals	2014–2016
Total abnormal users count before the data balancing	3615
Total benign users count before the data balancing	38,752
Total abnormal users count after the data balancing	21,183
Total benign users count after the data balancing	21,184
Total users count before the initial preprocessing of raw data	42,372
Total users count after the initial preprocessing of raw data	42,367

3.2. Preprocessing Unit

The data balancing, initial preprocessing and the feature extraction are the subunits of the preprocessing unit. These subunits are elaborated below.

3.2.1. Initial Preprocessing of the Raw Data

The preprocessing of the raw data is very important as the performance of any model is not only limited to the classification of electricity theft using ML models, but related to the data quality as well. Generally, the consumption data stored by the electric meters mostly has the missing values or the outliers. In our case, we considered the real EC data from a Chinese dataset, i.e., SGCC [5], that also contains the outliers and the missing values. The values exist because of various causes like the unreliable dispatch of the consumption data, the faulty meters, the storage related problems, etc., [5]. We utilized an SI technique for computing the mean of the consumption data present in the previous and the next cells to deal with the missing data by. The SI working mechanism is taken from [34] and is given in Equation (1).

$$f(x_{i,s}) = \begin{cases} \frac{x_{i,s-1}+x_{i,s+1}}{2}, & x_{i,s} \in NaN, x_{i,s-1}, \\ & x_{i,s+1} \notin NaN \\ 0, & x_{i,s} \in NaN, x_{i,s-1} \text{ or } \\ & x_{i,s+1} \in NaN, \\ x_{i,s}, & Otherwise, \end{cases} \tag{1}$$

where, i and s show a specific electricity customer and a time slot (day), respectively. $x_{i,s-1}$ and $x_{i,s+1}$ denote the EC data of a consumer for the previous day and the next day, respectively. Not a number (NaN) represents the missing values.

The availability of the outliers in the dataset negatively affects the classifier's performance and maximizes the FPR value. Hence, the outliers need to be removed from the dataset. Therefore, we use the three-sigma rule (TSR) of thumb method to remove the outliers from the dataset. The mathematical representation of TSR is taken from [5] and is given in Equation (2).

$$f(x_{i,s}) = \begin{cases} avg(X) + 2 \cdot \sigma(X), & x_{i,s} > avg(X) + 2 \cdot \sigma(X), \\ x_{i,s}, & Otherwise, \end{cases} \tag{2}$$

where, X shows a vector that is made of multiple $x_{i,s}$ values. The term $avg(X)$ and $\sigma(X)$ represent the average and standard deviation of X, respectively.

As up to now, the NaN and outlier values are successfully dealt with, so now, the dataset normalization is required because DL techniques are sensitive to the sparsed, diversed, and unscaled data. Therefore, we use min-max data scaling method in order to scale or normalize the data. The data scaling is performed using the Equation (3) [5].

$$f(x_{i,s}) = \frac{x_{i,s} - min(X)}{max(X) - min(X)} \quad (3)$$

where, $min(X)$ and $max(X)$ functions return the minimum and the maximum values of vector X, respectively.

3.2.2. Data Balancing by TAs' Implementation

The synthetic TAs are employed for balancing the data in the SGCC dataset. All the six TAs are introduced in [11] while the updated and revised version of the attacks are introduced in [29]. We select the updated TAs to create more practical and real abnormal consumption patterns to balance the dataset. The real consumption of a user is denoted by e_t, where, ($t \in [0, 1034]$). In this study, the employed dataset contains the total of 1035 days' consumption data. The mathematical representations of all TAs are taken from [29] and are presented in Equations (4)–(9).

$$t1(x_t) = x_t * random(0.1, 0.9), \quad (4)$$

$$t2(x_t) = x_t * r_t, r_t = random(0.1, 1), \quad (5)$$

$$t3(x_t) = x_t * random[0, 1], \quad (6)$$

$$t4(x_t) = mean(X) * random(0.1, 1), \quad (7)$$

$$t5(x_t) = mean(X), \quad (8)$$

$$t6(x_t) = x_{1034-t}, \quad (9)$$

where, $X = \{x_1, x_2, \ldots, x_{1034}\}$. In theft attack 1, $t1(.)$ multiplies the complete row (actual reading) by the same randomly generated value between 0.1 and 0.9. It is argued in [29] that it is not necessary that a theft might occur continuously in the real world but some discontinuous values may also be reported by the theft. Therefore, in theft attack 2, $t2(.)$ multiples each timestamp in a row with a different random value ranges from (and including) 0.1 to 1. Here, when the upper limit (1) of the random number is generated, the actual reading of that particular timestamp will be reported as theft. In theft attack 3, the theft consumer sometimes reports the real EC value and sometimes reports a 0 value. In theft attacks 4 and 5, the mean value of the actual readings is involved. Where, in attack 4, the mean value is multiplied with the same randomly generated value. Finally, theft attack 6 mimics the behaviour of a theft consumer when it sends the actual readings in a reverse order.

We applied the TAs on benign users' consumption data to establish an appropriate balance between the theft and honest data points. The SGCC dataset has 3615 abnormal and 38,757 normal consumers' data instances out of total of 42,372 records. The ratio of the normal and abnormal consumers in the dataset is 1:9. There are large number of data points in the dataset and it is difficult to use all of them for analysis due to the higher computational complexity problem. So, as a sample, 9999 records out of the 42,372 are employed for analysis. In this way, the last 900 real theft records out of 3615 (2714–3615) are selected and the other real theft records (0–2713) are ignored. Moreover, the remaining deficient abnormal records (4098) are synthetically created using the TAs, generation over the benign consumers' data starting from 901th and ending at 4998th record. Attacks 1–6 are implemented on the benign consumers' data ranges 901–1583, 1584–2266, 2267–2949, 2950–3632, 3633–4315, and 4316–4998, respectively. The TAs' pattern and the normal user's consumption pattern are shown in Figures 2 and 3, respectively. Furthermore, the data from 4999–9998 is the benign consumers' data. So now, the data are balanced (where 0–4998 is theft consumers' data and 4999–9998 is normal consumers' data) and is forwarded to LSTM for extracting the necessary features from them, which improves the classification performance. Finally, the GRU classifier is trained using the extracted data received for efficient and effective electricity theft classification. The dataset contains 1035 days' energy consumption data and the attacks are implemented on the entire data. However, in the figures referred above, only 30 days' synthetic TAs and normal patterns are given as an example.

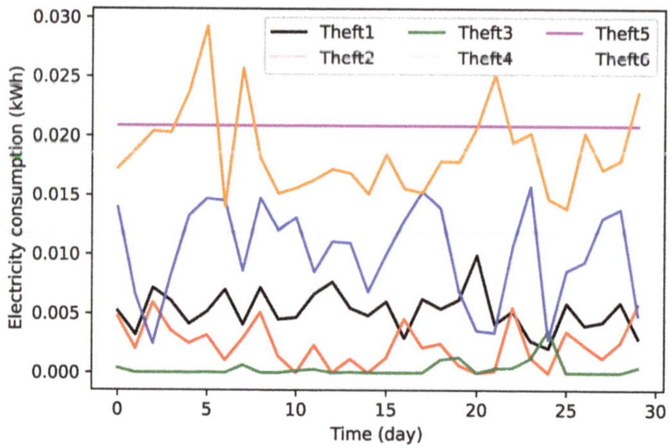

Figure 2. Synthetic theft attack patterns.

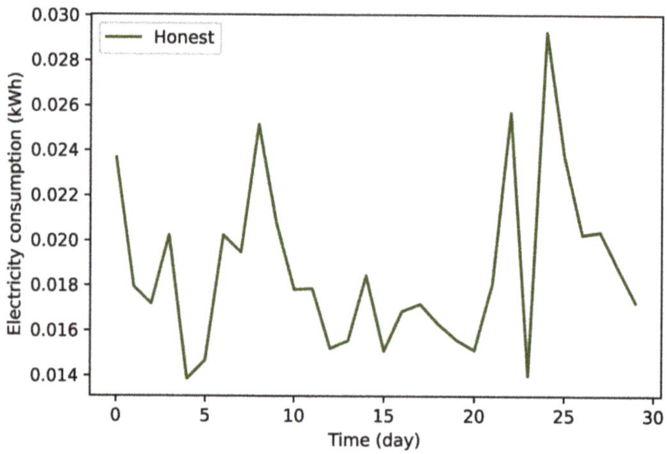

Figure 3. Normal pattern.

3.2.3. Long Short Term Memory Based Feature Extraction

Once the initial preprocessing and balancing of the data are performed, LSTM [35] is used for extracting important features. SGCC dataset contains huge and high dimensional (features) data and we need to perform dimensionality reduction. In such situations, the conventional recurrent neural network (RNN) [36] can not be employed as it faces the problems of the vanishing gradient and exploding gradient while dealing with the huge amount of data for dimensionality reduction. LSTM, an advanced RNN variant, is employed widely for successfully dealing with exploding and vanishing gradients issue. In training process, the temporal correlation between the current input and the previous state is found via using the previous data by RNN, and the output is finalized. However, because of its short and temporary memory, it fails to re-achieve and re-gain the previous information for the huge time series data and therefore, it fails to capture the temporal correlation between the current and the previous states. On the other hand, LSTM is able to easily and smoothly capture the temporal correlation. It also helps in dimensionality reduction of the huge time series data. The reason is that it has special and unique memory cells, which make use of the historical information. Furthermore, the significant features from the huge time series data are retained and memorized using the cells. This information is kept and maintained by the cell state (long term memory) in the LSTM. The significant features enfold the necessary information of the whole dataset.

The LSTM comprises the forget gate (f_t), the input gate (i_t), and the output gate (o_t). The determination and decision that whether the information taken from the current input (x_t) and the previous hidden state (h_{t-1}) should be retained or discarded from the cell state is made by the forget gate. In this way, the information from h_{t-1} and x_t are passed through the sigmoid (σ) activation function and it will decide either to keep or discard the previous output from the cell state by generating 1 or 0, respectively. The input gate determines that which values or data are employed for updating memory state or cell state, denoted by (C_t). Again the details from h_{t-1} and x_t are passed from the second sigmoid and decision will be made that what to do with the information; either discard it or save it. Similarly, in the cell state (C_t), the $tanh$ activation function is applied on h_{t-1} and x_t, and the output is stored into the cell state. The results from the cell state and the input gate are multiplied and added with the multiplication result of the forget gate and cell state. It is finally stored into the current cell state C'_t, which is now an updated cell state. Similarly, the final output gate (o_t) takes the x_t and h_{t-1} as the inputs, applies the sigmoid operation on them and the final result is stored in o_t. In order to calculate the next hidden state (h_t), the multiplication of $tanh(C'_t)$ and o_t is performed, sigmoid is applied on their multiplication result and the final output is stored into h_t. Moreover, the mathematical formulations of the forget, input, and output gates are given in Equations (10)–(15) [37].

$$f_t = \sigma(W_f(x_t, h_{t-1}) + b_f), \tag{10}$$

$$i_t = \sigma(W_i(x_t, h_{t-1}) + b_i), \tag{11}$$

$$C_t = tanh(W_c(x_t, h_{t-1})), \tag{12}$$

$$C'_t = (f_t * (C_t)) + (i_t * (C_t)), \tag{13}$$

$$o_t = \sigma(W_o(x_t, h_{t-1}) + b_o), \tag{14}$$

$$h_t = \sigma(o_t * tanh(C'_t)). \tag{15}$$

where, b_o, b_i, and b_f are the biases for the output, input, and forget gates, respectively. The W_o, W_i, and W_f denote the weights of the output, input, and forget gates, respectively. Moving ahead, for denoting previous hidden state information along with the updated cell state information, C_t and C'_t are used, respectively.

The optimal adjustments of the hyperparameters' values are very important to attain better results for feature extraction using LSTM. In order to perform better feature extraction, we perform manual parameter tuning. We set 200 and 100 neurons for each LSTM's layer. Whereas, the dense or fully connected layer has only one neuron. The dropout layer value is set to be 0.2 in order to protect the proposed TLGRU model from overfitting. The more detailed picture of the hyperparameters' values are given in Table 3. In our proposed LSTM feature extractor, we employ six layers, i.e., two LSTM, two LeakyReLU, one BatchNormalization, and one Dropout layer. We use 200 neurons in the first LSTM layer, the learning rate (Alpha) for both LeakyReLU layers is chosen to be 0.001, 100 neurons are used for the second LSTM layer, and finally, 0.2 is selected as the dropout probability for Dropout layer.

Table 3. Optimal settings of the hyperparameters' values.

Hyperparameters	Values
Units	200 and 100
Alpha	0.001
Dropout	0.2

3.3. Theft Classification Unit

TLGRU based classification is put forward for detecting the thefts in in electricity usage in SG. Furthermore, the popular benchmark models, LR, DT, SVM and GRU, are used for performance comparison with the proposed classifier. The details of these classifiers are given in the following subsections.

3.3.1. GRU

It was developed in 2014 [38]. GRU is a sub module of our proposed TLGRU model. GRU is used in two ways: (1) as a sub module of our proposed TLGRU model to tackle drift, ovefitting, high FPR, and local minima trap problems, (2) benchmark method for comparison purpose. GRU is faster in comparison with RNN and LSTM with respect to training time. It is employed widely in other research areas [39–42]; but, it is rarely employed for efficient ETD.

GRU is mainly developed to deal with the vanishing gradient problem with which the RNN fails to deal. The GRU merges the forget gates and the input gate of the LSTM into one gate known as the update gate. Moreover, the combination of both cell state and hidden state is made in GRU. The GRU's basic architecture is illustrated in Figure 4 [39,40].

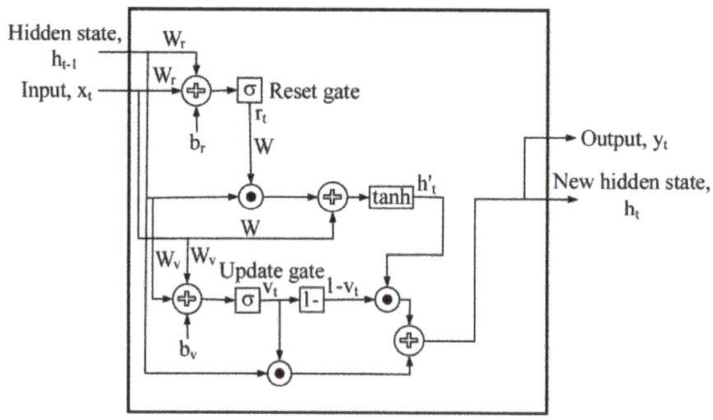

Figure 4. Gated recurrent unit architecture.

GRU consists of two gates: update gate (long term memory) and reset gate (short term memory), which are used to solve the gradient vanishing issue of RNN. These gates are two different vectors that finalize and determine that which information must be passed and proceeded to the output. One special and unique property about them is that the gates have the capability to be trained in order to retain information for a long duration, without discarding it with time or discarding information that is irrelevant and insignificant for prediction. By this feature GRU can clearly differentiate between the irregularity because of nonmalicious intermediaries (drift) and the irregularity due to the real theft factors, which consequently minimizes the FPR value. Therefore, we select the GRU classifier as the theft detection module of our proposed TLGRU classifier. The complete mathematical formulation of reset gate (r_t) and update gate (v_t) is given in Equations (16) and (17), respectively [38].

$$r_t = \sigma(W_r x_t + W_r h_{t-1} + b_r), \tag{16}$$

$$v_t = \sigma(W_v x_t + W_v h_{t-1} + b_v), \tag{17}$$

where, W_r, b_r, W_v, b_v, and x_t denote the weight related to reset gate, the bias value related to the reset gate, the weight related to update gate, bias related to the update gate, and the input vector, respectively. The reset gate is used in GRU to decide the amount of historical information to forget. Whereas, the update gate assists GRU to decide about previous information to be copied or passed to future. To compute the current or new hidden state (h_t), two steps are need to be performed. The first step is to calculate the candidate hidden state (h'_t) while the second step is to compute the h_t. The mathematical forms for h'_t, and h_t are depicted in Equations (18) and (19), respectively [38].

$$h'_t = tanh(Wx_t + Wr_t \odot h_{t-1}), \tag{18}$$

$$h_t = v_t \odot h_{t-1} + (1 - v_t) \odot h'_t. \tag{19}$$

where, W and h_{t-1} denote the weights' values and the hidden state at the previous time-step, respectively. In addition, the symbol $\odot$ denotes the Hadamard product. The *tanh* and σ show the hyperbolic and sigmoid activation functions, respectively. In addition, the value of r_t is used to know about the role or influence of h_{t-1} on h'_t. If r_t value is 0, it means that the information from h_{t-1} is totally ignored or discarded. Likewise, if its value is 1, it means that all the information from h_{t-1} is considered. Furthermore, in the final h_t equation, unlike LSTM, instead of employing an independent or separate gate, in GRU only one update gate is employed to control both the previous information from h_{t-1} as well as the new information from the h'_t. Now, assume that the value of v_t is 0 or near to 0 in h_t's equation, the first term of h_t equation is going to vanish. It means that h_t will not contain a good amount of information from h_{t-1} and the h_t will have information from h'_t only. Similarly, if the value of v_t is 1 or close to 1 in h_t's equation, the second part of the equation will be 0; it means h_t will totally be dependent on the first part of the equation, i.e., h_{t-1}. Hence, it is proved that the value of v_t has a great importance and it ranges from 0 to 1.

In the GRU module of TLGRU, four layers are employed: GRU layer, Flatten layer, Dropout layer, and Dense layer. In the first layer, 50 neurons are used, 20% of dropout probability is selected, and only one neuron for dense layer is employed. In the second layer, the conversion of data from multi dimensional to single dimensional is performed. Furthermore, the 20% of the dropout probability value is used to randomly deactivate the 20% of neurons, so that the overfitting issue can be avoided.

3.3.2. Support Vector Machine

SVM is one of the most popular and widely used classification techniques that is employed by many researchers as their basic proposed or benchmark model in the existing literature [43]. In [11], SVM based CPBETD is proposed and in [35], SVM is employed as an existing or benchmark classifier for ETD. In our case, we also select SVM as the benchmark classifier for performance evaluation of our proposed TLGRU classifier. The SVM can be employed for classification as well as regression. The support vector regression (SVR) and the support vector classification (SVC) are the two classes of SVM that are used for regression and classification, respectively. However, as our task is theft detection (a classification task), so we use the SVC class of the SVM for classification. For classification purpose, the SVM finds a hyperplane that maximizes the margin of the hyperplane to support vectors. This is done in order to separate the benign data and the abnormal data from each other so that the data of both the classes can be more clearly classified. Furthermore, as the EC data in SGCC dataset is not linearly-separable, therefore, we need to use the kernel SVM. In this way, the radial basis function (RBF) kernel is utilized for the classification of the non linearly-separable data (SGCC). The C and γ are the hyperparameters of SVM and their values are selected by default. The curliness of the decision boundary in SVM is decided by γ. Whereas, C is used to control the misclassification error.

3.3.3. Logistic Regression

LR is the simplest supervised ML binary classification algorithm [44]. It is an extensively utilized classifier for ETD in SGs (binary classification problem) [35,45,46]. LR follows the same principles as neural network. Therefore, we can surely say that LR used for binary classification problems (binary LR) is analogous and similar to the neural network with only one hidden layer and a sigmoid activation function (spans from 0 to 1). Where, the value close to 0 is regarded the normal consumer and vice versa. During the coding stage of LR, we have considered the optimal values for the hyperparameters, i.e., $random\_state = 5$ and $solver = liblinear$, where $random\_state$ is employed to control the random number generator and *solver* specifies that which algorithm to employ for optimization. The values for other hyperparameters are chosen as default.

3.3.4. Decision Tree

DT is also a popular classification method that divides the attributes into classes on the basis of their relevant features. DT is widely used by the researchers as the benchmark [45] and base model [47]. DT prepares a road map for the state of the art and advanced ensembled techniques like gradient boosting classifiers, random forest, and bagging techniques [48].

4. Simulation Results

In this study, simulations are performed using Google Colaboratory. The details of the selected SGCC dataset for validation of our proposed model are given in Section 3.1. Section 4.1 comprises the results obtained after performing extensive simulations. Moreover, the comparison between TLGRU's performance and the performance of the existing models, i.e., GRU, SVM, DT, and LR, is made and the validation is done in terms of ETD. Furthermore, recall, area under the curve (AUC), precision, and F1-score metrics are considered the appropriate measures in order to compute the classifiers' performance using the imbalanced data [35]. Based on the cases mentioned above, the accuracy metric is not an appropriate performance measure [48,49].

4.1. Proposed TLGRU and Other Benchmark Techniques' Results

The epoch variable is employed to control the training process of the proposed model. We run our model for 10 iterations or epochs. The convergence of our proposed TLGRU model with respect to accuracy and loss performance measures is exhibited in Figures 5 and 6, respectively. It is visible in the figures that the training accuracy (accuracy on seen data) of the TLGRU increases moderately at each iteration and finally it reaches to 91.77% at the final iteration. Whereas, using the testing data (accuracy using unseen data), the TLGRU accuracy gradually increases as well and it reaches 91.56% at the final iteration. The SGCC dataset contains some zero values due to which the proposed classifier can not learn it properly at the early iterations, therefore, during the first three epochs, the training accuracy is better than testing accuracy, which means that the overfitting issue has occurred. After the 3rd iteration, the proposed model efficiently learns the zero values and the overfitting issue is removed. The loss of the TLGRU is also computed and noted at different iterations. As shown in Figure 6, the training loss is minimizing at every iteration till it reaches to 0.2068 at the 10th iteration. While, the testing loss also reduces till it reaches to 0.2084 at the final iteration. During the first three iterations, the model overfits because of the zero values that exist in the dataset. After the third iteration, the model learns the dataset as well as the zero values and therefore, the overfitting issue is solved. Now, finally, from training and testing accuracy of the proposed model, it is concluded that the model generalizes well and avoids overfitting. The overfitting problem is tackled using the proper tuning of the dropout probability value and powerful and significant features' extraction ability of the LSTM in TLGRU model.

The accuracy and loss convergence analysis results for the benchmark GRU model are shown in Figures 7 and 8, respectively. It is shown that at the final iteration, the training and testing accuracy value for GRU is 89.99% and 82.65%, respectively. Using the seen data, the accuracy slowly increases, whereas, on testing data, the accuracy fluctuates till the 7th iteration and after 7th iteration, it starts decreasing. Moreover, at the 3rd epoch, the GRU model learns and trains using a batch having some zero values that causes overfitting. However, after the 7th epoch, the overfitting again starts and continues till the final iteration. Hence, it is proved that the existing GRU model overfits. The main reasons for the occurrence of this issue are lack of important features' extraction to reduce the data dimensionality and no proper tuning of the dropout regularization probability value. Furthermore, the training and testing loss values for GRU at the last iteration are 0.2852 and 0.5136, respectively. The same trend continues in the loss as well. The training loss continuously decreases while the testing loss fluctuates till the 7th epoch and it starts increasing after the 7th iteration. At the 3rd epoch, the overfitting issue occurs due to the presence of zero values in the batch. Later on, after the 7th iteration, the overfitting issue occurs and it continues till the end, hence, the model overfits. Finally, from the training and testing accuracy and loss plots, it is concluded that GRU model does not have good generalization ability and overfitting issue occurs because of neglecting the tuning of the dropout regularization probability value and features' extraction. Whereas, our proposed TLGRU model outperforms the benchmark GRU in terms of tackling the overfitting problem.

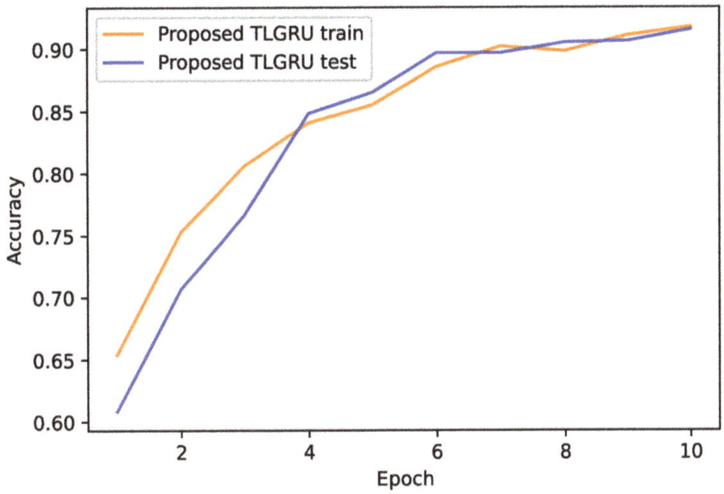

Figure 5. The proposed TLGRU model's training and testing accuracy.

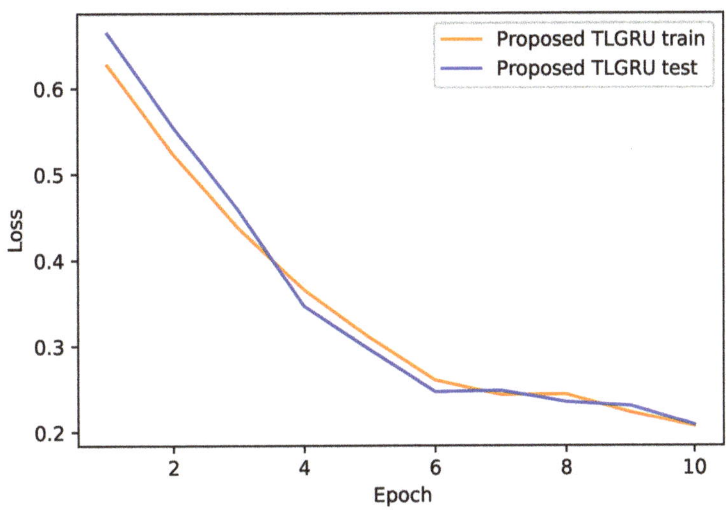

Figure 6. The proposed TLGRU model's training and testing losses.

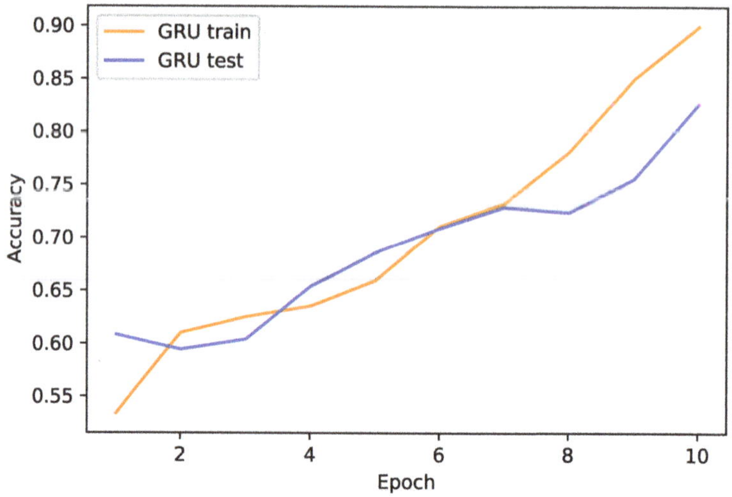

Figure 7. The GRU model's training and testing accuracy.

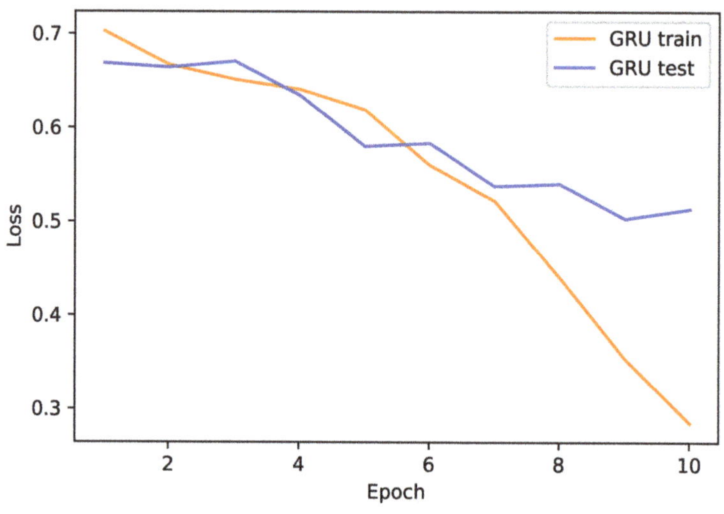

Figure 8. The GRU model's training and testing losses.

For TLGRU's comparison with DT, LR, SVM, and GRU models, the AUC, recall, precision, F1-score, and accuracy are considered, which are the most effective performance parameters. The comparison of the TLGRU's performance with the existing benchmark models is exhibited in Table 4 and Figure 9 with regard to various performance measures. From the results, the superiority of TLGRU model over all the other conventional models with regard to the already discussed indicators of performance due to the following reasons. The first reason is that TLGRU effectively tackles the imbalanced data issue using different TAs, the second reason is that the LSTM module is used to extract the necessary features and solve the high dimensionality issue, and finally, the classification using the GRU module enhances TLGRU's performance. Moreover, in the GRU module, dealing with the drift, overfitting, and local optima trapping issues using the update gate (long term memory), dropout regularization, and Adam optimizer further enhances the performance of the model being put forward our proposed model with respect to the selected evaluators. Contrarily, SVM and LR perform very bad. The reason is that they can not tackle the large and huge time series data and that is why overfitting issue occurs. Whereas, the proposed

TLGRU classifier smoothly handles the large time series data due to the feature extraction ability of the proposed model using LSTM algorithm and solves the issue of overfitting.

Table 4. Performance comparison of the proposed and existing schemes.

Classifier	Accuracy	AUC Score	Precision	Recall	F1-Score	FPR
DT	0.6701	0.6702	0.7019	0.6585	0.6795	0.1485
LR	0.5379	0.5365	0.5207	0.7097	0.6736	0.4370
SVM	0.6433	0.6423	0.4678	0.7162	0.5660	0.2646
GRU	0.8265	0.7552	0.8355	0.7176	0.7721	0.0818
Proposed TLGRU	0.9156	0.9168	0.9796	0.8659	0.9192	0.0100

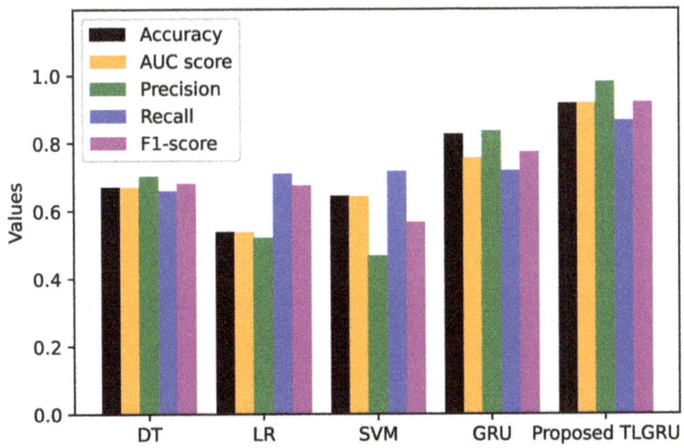

Figure 9. Performance comparison of different classifiers.

FPR is one of the significant performance metrics for ML techniques in which the benign electricity users are classified and shown as theft. The value of FPR is directly proportional to the on field inspection cost. There are many solutions or ways to minimize the FPR value. However, we have only considered and worked on three different ways to reduce the FPR value: data balancing, extraction of the important features from the raw data, and correctly identifying the drift. This article uses the GRU for classification purpose because it has a special quality of keeping the long historical sequence of data using its update gate (long term memory), and the data are then used for analyzing the long relationships among the consumption patterns. The GRU model identifies the anomalies or irregularities in consumption data that arise due to the non-theft reasons. The FPR for DT is 0.1485, for SVM is 0.2646, for LR is 0.4370, for GRU is 0.0818, and for TLGRU is 0.01. Hence, consequences are clear from the FPR numeric values that the TLGRU classifier exhibits the least FPR in comparison with all the other benchmark classifiers. The FPR for our proposed TLGRU and other benchmark models is shown in Figure 10.

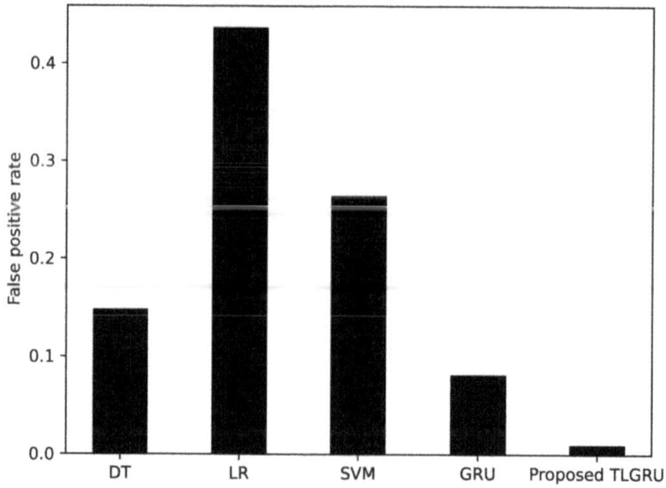

Figure 10. The proposed and benchmark models' FPR analysis.

4.2. Strengths and Weaknesses of the Proposed Work

The fundamental advantage of this work is to provide an efficient ETD model for power utilities, which helps them to reduce economic loss. Furthermore, the accurate and timely detection of energy thieves reduces the line losses in transformers and other grids' components. Besides, the proposed model has some shortcomings. The low-frequency EC data are used to train the model, which limits its performance towards capturing the most granular EC patterns. Consequently, the rate of misclassifying instances increases. Further, it may incur high computational time due to the absence of a hyperparameter tuning technique.

5. Conclusions and Future Work

In this research article, the TLGRU model is presented that consists of two main components and three sub-components. The main components are preprocessing unit and theft classification unit. The preprocessing component is further divided into three sub-components: initial preprocessing of the raw data, data balancing, and feature extraction. In the first sub-component, the NaN values, outliers, and unscaled data are dealt by employing SI, TSR of thumb, and min-max scaler, respectively. In the second sub-component, TAs are employed for creation of the synthetic abnormal data samples to solve the imbalanced data problem. Finally, LSTM classifier is employed for feature extraction and dealing with dimensionality curse problem. Furthermore, the classification component contains the GRU model for theft classification. Moreover, the GRU provides solution for drift identification, overfitting, and trapping in local optima problems. In addition, four popular benchmark models, DT, LR, SVM, and GRU, are implemented for performance comparison with the proposed TLGRU classifier. A realistic EC dataset (SGCC) is employed for simulations. From the simulation results, the superiority of TLGRU over benchmark models in terms of ETD is exhibited. The simulations provide us with 1% FPR, 97.96% precision, 91.92% F1-score, 91.68% AUC, 91.56% accuracy, and 86.59% recall, which are better than the benchmark schemes. Hence, we conclude that the newly developed TLGRU is an efficient model for ETD with minimum FPR. In future works, we will investigate the novel DL models for feature extraction and classification to more efficiently perform the ETD task. Moreover, automated tuning of the hyperparameters' of the models will also be performed using meta-heuristic optimization algorithms. In addition, the aim of the underlying study is the development of a novel deep learning based hybrid model, which helps electric utilities to detect energy frauds in SGs around the globe. Furthermore, the proposed model is trained on a massive EC dataset. So, its real time practicability in terms of identifying the presence of thieves in the SGs is ensured. Further, the model introduced in the underlying work is a quite general solution to detect anomalies and frauds in any time series. It needs only a dataset for its training. So, the EC data collected by the conventional meters can be used to train the proposed model and then it will be applicable in SGs to detect energy frauds.

Author Contributions: Conceptualization, P., N.J. and S.A.; methodology, P.; software, P. and S.J.; validation, N.J., M.A. and M.U.J.; formal analysis, A.S.Y. and S.J.; investigation, N.J.; resources, S.J.; data curation, P.; writing—original draft preparation, P. and N.J.; writing—review and editing, N.J. and S.A.; visualization, N.J. and S.A.; supervision, N.J.; project administration, S.A. and N.J. All authors have read and agreed to the published version of the manuscript.

Funding: This research received no external funding.

Institutional Review Board Statement: Not applicable.

Informed Consent Statement: Not applicable.

Data Availability Statement: The dataset employed in this research is available online at https://github.com/henryRDlab/ElectricityTheftDetection (accessed on 30 December 2021).

Conflicts of Interest: The authors declare no conflict of interest.

References

1. Garcia Deluno, F.; Marafão, F.P.; de Souza, W.A.; da Silva, L.C.P. Power metering: History and future trends. In Proceedings of the 2017 Ninth Annual IEEE Green Technologies Conference (GreenTech), Denver, CO, USA, 29–31 March 2017; pp. 26–33.
2. Weranga, K.S.K.; Kumarawadu, K.; Chandima, D.P. *Smart Metering Design and Applications*; Springer: Singapore, 2014.
3. Foudeh Husam A.; Mokhtar, A.S. Automated meter reading and advanced metering infrastructure projects. In Proceedings of the 2015 9th Jordanian International Electrical and Electronics Engineering Conference, Amman, Jordan, 12–14 November 2015; pp. 1–6.
4. Huang, Y.; Qifeng, X. Electricity theft detection based on stacked sparse denoising autoencoder. *Int. J. Electr. Power Energy Syst.* **2021**, *125*, 106448. [CrossRef]
5. Zheng, Z.; Yatao, Y.; Xiangdong, N.; Dai, H.; Zhou, Y. Wide and deep convolutional neural networks for electricity-theft detection to secure smart grids. *IEEE Transact. Ind. Inform.* **2017**, *14*, 1606–1615. [CrossRef]
6. Buzau, M.-M.; Tejedor-Aguilera, J.; Cruz-Romero, P.; Gómez-Expósito, A. Hybrid deep neural networks for detection of non-technical losses in electricity smart meters. *IEEE Trans. Power Syst.* **2019**, *35*, 1254–1263. [CrossRef]
7. Khoo, B.; Ye, C. Using RFID for anti-theft in a Chinese electrical supply company: A cost-benefit analysis. In Proceedings of the 2011 Wireless Telecommunications Symposium (WTS), New York, NY, USA, 13–15 April 2011; pp. 1–6.
8. McLaughlin, S.; Brett, H.; Fawaz, A.; Berthier, R.; Zonouz, S. A multi-sensor energy theft detection framework for advanced metering infrastructures. *IEEE J. Select. Areas Commun.* **2013**, *31*, 1319–1330. [CrossRef]
9. Cárdenas; A.A.; Saurabh, A.; Schwartz, G.; Dong, R.; Sastry, S. A game theory model for electricity theft detection and privacy-aware control in AMI systems. In Proceedings of the 2012 50th Annual Allerton Conference on Communication, Control, and Computing (Allerton), Monticello, IL, USA, 1–5 October 2012; pp. 1830–1837.
10. Amin, S.; Schwartz, G.A.; Tembine, H. Incentives and security in electricity distribution networks. In Proceedings of the International Conference on Decision and Game Theory for Security, Virtual Conference, 25–27 October 2012; pp. 264–280.
11. Jokar, P.; Arianpoo, N.; Leung, V.C.M. Electricity theft detection in AMI using customers' consumption patterns. *IEEE Trans. Smart Grid* **2015**, *7*, 216–226. [CrossRef]
12. Gunturi Kumar, S.; Sarkar, D. Ensemble machine learning models for the detection of energy theft. *Electr. Power Syst. Res.* **2021**, *192*, 106904. [CrossRef]
13. Kong, X.; Zhao, X.; Liu, C.; Li, Q.; Dong, D.; Ye, L. Electricity theft detection in low-voltage stations based on similarity measure and DT-KSVM. *Int. J. Electr. Power Energy Syst.* **2021**, *125*, 106544. [CrossRef]
14. Buzau, M.M.; Tejedor-Aguilera, J.; Cruz-Romero, P.; Gómez-Expósito, A. Detection of non-technical losses using smart meter data and supervised learning. *IEEE Trans. Smart Grid* **2018**, *10*, 2661–2670. [CrossRef]
15. Qu, Z.; Li, H.; Wang, Y.; Zhang, J.; Abu-Siada, A.; Yao, Y. Detection of electricity theft behavior based on improved synthetic minority oversampling technique and random forest classifier. *Energies* **2020**, *13*, 2039. [CrossRef]
16. Lu, X.; Zhou, Y.; Wang, Z.; Yi, Y.; Feng, L.; Wang, F. Knowledge embedded semi-supervised deep learning for detecting non-technical losses in the smart grid. *Energies* **2019**, *12*, 3452. [CrossRef]
17. Ashraf Ullah, P.; Shoaib, M.; Muhammad, A.; Kabir, B.; Javaid, N. Synthetic Theft Attacks Implementation for Data Balancing and a Gated Recurrent Unit Based Electricity Theft Detection in Smart Grids. In Proceedings of the Conference on Complex, Intelligent, and Software Intensive Systems, Asan, Korea, 1–3 July 2021; Springer: Cham, Switzerland, 2021; pp. 395–405.
18. Hochreiter, S.; Schmidhuber, J. Long short-term memory. *Neural Comput.* **1997**, *9*, 1735–1780. [CrossRef] [PubMed]

19. Cho, K.B.; Merrienboer, V.; Bahdanau, D.; Bengio, Y. On the properties of neural machine translation: Encoder-decoder approaches. *arXiv* **2014**, arXiv:1409.1259.
20. Goodfellow, I.; Pouget-Abadie, J.; Mirza, M.; Xu, B.; Warde-Farley, D.; Ozair, S.; Courville, A.; Bengio, Y. Generative adversarial nets. *Adv. Neural Inform. Process. Syst.* **2014**, *27*.
21. Fabian, A.N.; Figueroa, G.; Chu, C. NTL detection in electric distribution systems using the maximal overlap discrete wavelet-packet transform and random undersampling boosting. *IEEE Trans. Power Syst.* **2018**, *33*, 7171–7180.
22. Hasan, M.; Toma, R.N.; Abdullah-Al, N.; Islam, M.M.; Kim, J. Electricity theft detection in smart grid systems: A CNN-LSTM based approach. *Energies* **2019**, *12*, 3310. [CrossRef]
23. Saeed Salman, M.; Mustafa, M.W.; Sheikh, U.U.; Jumani, T.A.; Mirjat, N.H. Ensemble bagged tree based classification for reducing non-technical losses in multan electric power company of Pakistan. *Electronics* **2019**, *8*, 860. [CrossRef]
24. Wang, X.; Yang, I.; Ahn, Su. Sample efficient home power anomaly detection in real time using semi-supervised learning. *IEEE Access* **2019**, *7*, 139712–139725. [CrossRef]
25. Liu, H.; Li, Z.; Li, Y. Noise reduction power stealing detection model based on self-balanced data set. *Energies* **2020**, *13*, 1763. [CrossRef]
26. Ibrahem, M.I.; Nabil, M.; Fouda, M.M.; Mahmoud, M.M.E.A.; Alasmary, W.; Alsolami, F. Efficient Privacy-Preserving Electricity Theft Detection with Dynamic Billing and Load Monitoring for AMI Networks. *IEEE Internet Things J.* **2020**, *8*, 1243–1258. [CrossRef]
27. Nabil, M.; Ismail, M.; Mahmoud, M.M.E.A.; Alasmary, W.; Serpedin, E. PPETD: Privacy-preserving electricity theft detection scheme with load monitoring and billing for AMI networks. *IEEE Access* **2019**, *7*, 96334–96348. [CrossRef]
28. Micheli, G.; Soda, E.; Vespucci, M.T.; Gobbi, M.; Bertani, A. Big data analytics: An aid to detection of non-technical losses in power utilities. *Comput. Manag. Sci.* **2019**, *16*, 329–343. [CrossRef]
29. Punmiya, R.; Choe, S. Energy theft detection using gradient boosting theft detector with feature engineering-based preprocessing. *IEEE Trans. Smart Grid* **2019**, *10*, 2326–2329. [CrossRef]
30. Razavi, R.; Gharipour, A.; Fleury, M.; Akpan, I.J. A practical feature-engineering framework for electricity theft detection in smart grids. *Appl. Energy* **2019**, *238*, 481–494. [CrossRef]
31. Ramos, C.; Rodrigues, D.; de Souza, A.N.; Papa, J.P. On the study of commercial losses in Brazil: A binary black hole algorithm for theft characterization. *IEEE Trans. Smart Grid* **2016**, *9*, 676–683. [CrossRef]
32. Ghasemi, A.A.; Gitizadeh, M. Detection of illegal consumers using pattern classification approach combined with Levenberg-Marquardt method in smart grid. *Int. J. Electr. Power Energy Syst.* **2018**, *99*, 363–375. [CrossRef]
33. LeCun, Y.; Bengio, Y. Convolutional networks for images, speech, and time series. In *The Handbook of Brain Theory and Neural Networks*; MIT Press: Cambridge, MA, USA, 1995; Volume 3361.
34. Li, S.; Han, Y.; Yao, X.; Yingchen, S.; Wang, J.; Zhao, Q. Electricity theft detection in power grids with deep learning and random forests. *J. Electr. Comput. Eng.* **2019**, *2019*, 4136874. [CrossRef]
35. Adil, M.; Javaid, N.; Qasim, U.; Ullah, I.; Shafiq, M.; Choi, J. LSTM and bat-based RUSBoost approach for electricity theft detection. *Appl. Sci.* **2020**, *10*, 4378. [CrossRef]
36. Rumelhart, D.E.; Hinton, G.E.; Williams, R.J. Learning representations by back-propagating errors. *Nature* **1986**, *323*, 533–536. [CrossRef]
37. Javaid, N. A PLSTM, AlexNet and ESNN Based Ensemble Learning Model for Detecting Electricity Theft in Smart Grids. *IEEE Access* **2021**, *9*, 162935–162950. [CrossRef]
38. Chung, J.; Gulcehre, C.; Cho, K.; Bengio, Y. Empirical evaluation of gated recurrent neural networks on sequence modeling. *arXiv* **2014**, arXiv:1412.3555.
39. Zhang, Y.G.; Tang, J.; He, Z.; Tan, J.; Li, C. A novel displacement prediction method using gated recurrent unit model with time series analysis in the Erdaohe landslide. *Nat. Hazards* **2021**, *105*, 783–813. [CrossRef]
40. Aniruddha, D.; Kumar, S.; Basu, M. A gated recurrent unit approach to bitcoin price prediction. *J. Risk Financ. Manag.* **2020**, *13*, 23.
41. Niu, Z.; Yu, Z.; Tang, W.; Wu, Q.; Reformat, M. Wind power forecasting using attention-based gated recurrent unit network. *Energy* **2020**, *196*, 117081. [CrossRef]
42. Luo, H.; Wang, M.; Wong, P.K.; Tang, J.; Cheng, J.C.P. Construction machine pose prediction considering historical motions and activity attributes using gated recurrent unit (GRU). *Automat. Construct.* **2021**, *121*, 103444. [CrossRef]
43. Cortes, C.; Vapnik, V. Support-vector networks. *Mach. Learn.* **1995**, *20*, 273–297. [CrossRef]
44. Available online: www.tutorialspoint.com (accessed on 5 June 2021).
45. Gul, H.; Javaid, N.; Ullah, I.; Qamar, A.M.; Afzal, M.K.; Joshi, G.P. Detection of non-technical losses using SOSTLink and bidirectional gated recurrent unit to secure smart meters. *Appl. Sci.* **2020**, *10*, 3151. [CrossRef]
46. Aslam, Z.; Javaid, N.; Ahmad, A.; Ahmed, A.; Gulfam, S.M. A Combined Deep Learning and Ensemble Learning Methodology to Avoid Electricity Theft in Smart Grids. *Energies* **2020**, *13*, 5599. [CrossRef]
47. Jindal, A.; Dua, A.; Kaur, K.; Singh, M.; Kumar, N.; Mishra, S. Decision tree and SVM-based data analytics for theft detection in smart grid. *IEEE Trans. Ind. Inform.* **2016**, *12*, 5–1016. [CrossRef]
48. Available online: www.machinelearningmastery.com (accessed on 17 April 2021).
49. Javaid, N.; Jan, N.; Javed, M.U. An adaptive synthesis to handle imbalanced big data with deep siamese network for electricity theft detection in smart grids. *J. Parallel Distrib. Comput.* **2021**, *153*, 44–52. [CrossRef]

energies

MDPI

Article

An Elastic Energy Management Algorithm in a Hierarchical Control System with Distributed Control Devices

Piotr Powroźnik [1,*], Paweł Szcześniak [2], Krzysztof Turchan [3], Miłosz Krysik [3], Igor Koropiecki [3] and Krzysztof Piotrowski [3]

1 Institute of Metrology, Electronics and Computer Science, University of Zielona Gora, 65-516 Zielona Gora, Poland
2 Institute of Automatic Control, Electronics and Electrical Engineering, University of Zielona Gora, 65-516 Zielona Gora, Poland; p.szczesniak@iee.uz.zgora.pl
3 IHP—Leibniz Institute for High Performance Microelectronics, 15236 Frankfurt (Oder), Germany; turchan@ihp-microelectronics.com (K.T.); krysik@ihp-microelectronics.com (M.K.); koropiecki@ihp-microelectronics.com (I.K.); piotrowski@ihp-microelectronics.com (K.P.)
* Correspondence: p.powroznik@imei.uz.zgora.pl; Tel.: +48-68-328-23-06

Abstract: In modern Electric Power Systems, emphasis is placed on the increasing share of electricity from renewable energy sources (PV, wind, hydro, etc.), at the expense of energy generated with the use of fossil fuels. This will lead to changes in energy supply. When there is excessive generation from RESs, there will be too much energy in the system, otherwise, there will be a shortage of energy. Therefore, smart devices should be introduced into the system, the operation of which can be initiated by the conditions of the power grid. This will allow the load profiles of the power grid to be changed and the electricity supply to be used more rationally. The article proposes an elastic energy management algorithm (EEM) in a hierarchical control system with distributed control devices for controlling domestic smart appliances (SA). In the simulation part, scenarios of the algorithm's operation were carried out for 1000 households with the use of the distribution of activities of individual SAs. In experimental studies, simplified results for three SA types and 100 devices for each type were presented. The obtained results confirm that, thanks to the use of SAs and the appropriate algorithm for their control, it is possible to change the load profile of the power grid. The efficacious operation of SAs will be possible thanks to the change of habits of electricity users, which is briefly described in the article.

Keywords: renewable energy sources; energy demand control; smart appliances; elastic energy management algorithm; GRASP algorithm; IoT

Citation: Powroźnik, P.; Szcześniak, P.; Turchan, K.; Krysik, M.; Koropiecki, I.; Piotrowski, K. An Elastic Energy Management Algorithm in a Hierarchical Control System with Distributed Control Devices. *Energies* **2022**, *15*, 4750. https://doi.org/10.3390/en15134750

Academic Editor: Miguel Jiménez Carrizosa

Received: 24 May 2022
Accepted: 24 June 2022
Published: 28 June 2022

1. Introduction

A number of countries have different approaches to problems related to sustainable development and environmental protection. The European Union (EU) countries have one of the most developed plans and strategies in this area. The European Commission has committed to achieving climate neutrality by 2050, as described in the documents of the "European Green Deal" [1]. Part of the European Green Deal is the "Fit for 55" package. Its aim is to put the EU firmly on the path to climate neutrality by 2050. The "Fit for 55" package envisages a revision of the Renewable Energy Directive (RED II) [2]. This change is expected to help the EU meet its new 55% greenhouse gas emissions target. The revised RED II directive sets a new EU target of at least 40% of the share of renewable energy in final energy consumption by 2030. Meeting these requirements will require the connection of a significant number of renewable energy sources (RES) to the distribution network. The RES energy generation profile is periodic and depends on weather conditions [3]. On the other hand, the reduction of greenhouse gases will result in the shutdown of a large number of conventional fossil fuel power plants processing brown coal, hard coal and petroleum [4]. Such a situation may lead to a very high generation of energy from renewable sources in favorable weather conditions, e.g., from Photovoltaic Panels (PV) on a sunny day. Such production will exceed the energy demand. In contrast, during unfavorable weather conditions, there will be a shortage of energy in the system, because RESs will not generate it. Additionally, solar energy from PVs will be available only during the day-time hours, and it will not be available in the morning, evening and night-time hours. The complementarity of PV and wind turbine energy generation may also not meet the energy demand. To meet the energy demand, either very expensive energy storage systems (ESS) [5] or fossil fuel generators with a high rate of inclusion should be used, e.g., gas microturbines. The geopolitical situation in Europe may mean that gas resources may not be sufficient to balance the electricity system with a large amount of RESs.

Figure 1 shows power measurements for one week, illustrating the load time courses for 110/15 kV substations, and a comparison with the generation profile of a PV installation with the power planned for installation in a given MV line. The overproduction of energy from the PV system is marked in green. This energy should be stored [6], used or sent to places where there is a demand [7]. As can be seen from the presented figure, the energy generated by PVs is unstable over time. During one week, there were days with very little, very high, and extremely variable energy outputs all day long, or with instantaneous energy bursts in production. Prosumer photovoltaic systems installed on buildings show a similar behavior. In the local balancing of the system, it is necessary to use/store as much energy as possible from both local prosumer low-voltage installations and large photovoltaic power plants connected to the MV.

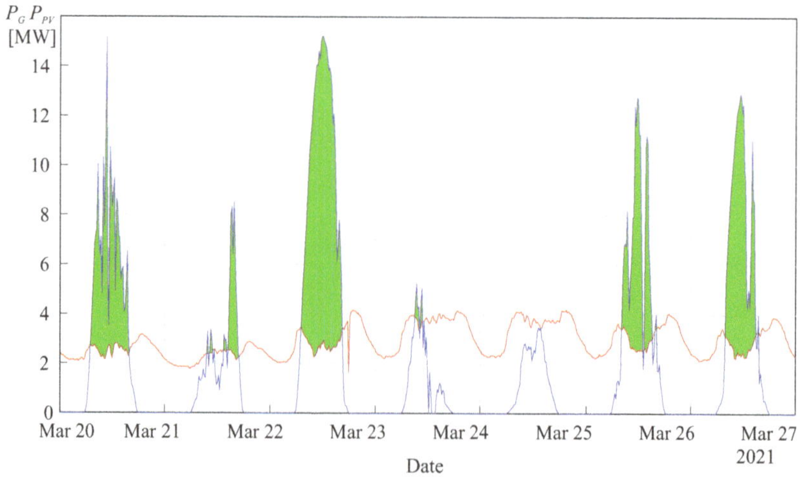

Figure 1. Sample measurement result in 110/15 kV substations; red: energy consumption from the transformer; blue: potential PV power generation, including sources to be connected in the future; the wave forms of PV generation were determined on the basis of measurements from a 1 MWp power plant; green field: energy that must be sent to the power grid to other loads or stored in the ESS.

The demand for electricity and the penetration of small-scale distributed generators are rapidly growing in LV distribution networks, thus, the existing electrical systems will need future adjustments. Inverters for distributed energy resources (DERs) have been released with standards and grid codes for interconnection with the distribution grid. To include increased DER penetration and optimize DER value in the grid, the new standard and grid codes anticipate that DERs will perform a range of power system support roles. The following are more widespread measures to detect violations of the upper voltage limit: voltage and reactive power control by PV inverters, demand response (DR) and storage.

One of the conceptual approaches to using energy overproduction is to manage energy demand in such a way that its consumption is as high as possible at times of high generation. Local self-consumption should be increased during a large surplus of energy production from RESs, and self-consumption should be reduced during an energy shortage in the system [8]. There are many concepts employed in controlling the demand for electricity by the use of a smart appliances (SA) described in the scientific literature. This article will present the concept of using and integrating smart home devices and consumer devices (e.g., electric vehicles [9], small energy storage) for energy balancing in a local low-voltage distribution system. An algorithm for elastic energy management will be proposed which takes into account the priorities of device operation set by the user. In order for such an energy management system to function properly, it should have information about the operating status of the system, i.e., the overproduction/shortage of energy in the system.

There are many solutions for the control of the demand for electricity in the scientific literature. Here, a brief overview of selected applications for use with home appliances will be provided. An interesting solution to the problem of excessive grid load was presented in [10]. The proposed Smart Home energy management system takes into account network constraints and user priorities. It is based on a heuristic technique for the device scheduling process. The limitation of this approach is the necessity to have information about the power available from the grid and about the power generated from RESs. As there is no single leading signal about the state of the power system, the system is quite complex. In another article [11], the predictive control model for optimal energy management in a district, involving buildings with vertically placed photovoltaic systems integrated into the building and energy storage systems, is discussed. The proposed control strategy aims to maximize the district's independence from the external power grid, which is achieved through the collaboration of interactive buildings. The proposed control scheme includes components such as forecast, optimizer and load in the form of the analyzed district. Another concept of energy management is presented in [12], which includes a comparative study of two home micro-grids, one of which is a vehicle-to-home network. Electric vehicle integration makes an economic contribution by reducing costs, supporting other MG components and reducing the load on the main network. The next publication [13] presents an overview of the most recent literature on residential household energy management, especially holistic solutions, and proposes new viewpoints on residential appliance scheduling in smart homes.

Due to the stochastic nature of RES generation, and in order to ensure the stable operation of the system, an increase in the number of control points and control devices must be achieved, which will significantly increase the burden on the system operator at the secondary control level. Therefore, a decentralized energy management and control system should be installed in the customer's system. Then, in such a system, information on regulatory needs will either be provided by the distribution system operator or it will be obtained from local measurement points. A hierarchical control system has been proposed, as shown in Figure 2 [14].

The simplest solution for a hierarchical control system is a structure with a local central unit located in an MV/LV transformer station. Information on production from large generating units (wind or PV) connected to the MV is provided to the central management unit, e.g., via Wi-Fi or other communication protocols. Additionally, information about the energy production from the RESs is sent to the central unit, if possible. This information can be taken directly from an intelligent energy meter, solar inverter or other device providing information about the generation level of prosumer sources. Information about the amount of energy in the grid can also be obtained from the voltage level of the consumer grid [8,15]. When there is a need to increase energy consumption or reduce it, the central management unit sends a signal to distributed control devices (DCD) located in a given building or at a location. All kinds of information exchange between individual units are realized using the Message Queue Telemetry Transport (MQTT) protocol. This protocol is based on published and subscribed topic patterns. Such a use of the MQTT protocol makes it possible to subscribe only to those data on topics that are essential for the operation of a given unit in the proposed hierarchical control system with DCDs. The role of the server with which clients will connect to publish topics for individual DCDs is performed by the MQTT broker. For the proposed hierarchical control system with DCDs, external and internal MQTT brokers are distinguished. Such an approach enables grouping of data from a given DCD. In the internal MQTT broker, topics may be grouped on energy consumption, e.g., on the level of one household and energy generation from RESs. The energy consumption can be the total value of the energy consumed by all devices, or it can also be grouped for individual devices of energy consumers. The greatest possibilities for providing additional data on configuration possibilities will be offered by SAs. For SAs, it is possible to read the current energy consumption and assess the possibility of modifying the behavior of a given SA. In this case, the modification will be to change the power level. The scope of possibilities for changing the power will result from the parameters set by the manufacturer of a given SA. The possibility of communication between the proposed hierarchical control system with the DCD and other hierarchical control systems is provided by the use of solutions based on the concept of the Internet of Things (IoT) with a cloud function. In this case, IoT Cloud solutions can be based on one of the following solutions: AWS IoT [16], Google Cloud IoT [17], IBM Cloud IoT [18], Microsoft Azure IoT [19], Oracle IoT Cloud Service [20] or others.

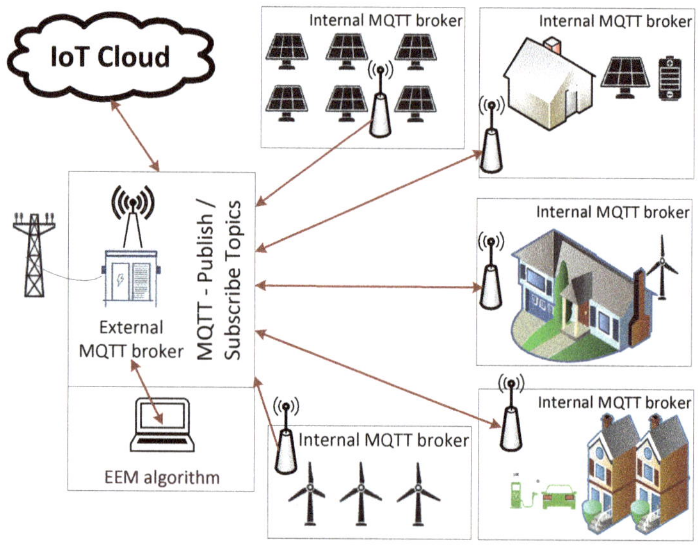

Figure 2. Proposed hierarchical control system.

To summarize the above, in the proposed hierarchical control, different system communication tasks can be distinguished. Some of these tasks play a measuring, controlling or measuring and controlling role. Measurement tasks involve collecting data on the current demand for and supply of energy. Data is provided from internal MQTT brokers. The control or measurement and control tasks concern the transfer of guidelines on the method of controlling a given intelligent device by the EEM algorithm. In addition, there are also tasks related to maintaining the correct functioning of the hierarchical control system. In this case, the tasks will concern the verification of the status of a given device. The tasks also include tasks related to ensuring the security level, such as system access control.

The DCD is also responsible for reacting to changes in the U_{DCD} voltage. After exceeding the range defined by the formula:

$$U_{DCD\_MIN} \leq U_{DCD} \quad \vee \quad U_{DCD\_MAX} \geq U_{DCD}, \tag{1}$$

actions are taken through the use of a dedicated algorithm. The proposed algorithm for the Elastic Energy Management (EEM) is implemented in DCD. The DCD communicates with individual smart appliances in order to achieve the set goal of reducing or increasing energy consumption [8]. For the proper operation of the EEM algorithm, all necessary data on energy consumption or generation will be provided from an external MQTT broker. The main aim of the article is to present the concept of an electricity management system in a smart home using SAs.

The innovative part of the research results presented in the article is the ensuring of the use of the EEM algorithm to respond to an overvoltage in the smart grid and to define user behavior and the means to model virtually any user device. Such innovation allows the creation of virtual device prototypes to evaluate their energy efficiency properties. Verification of the correctness of the assumptions was made based on the Probability Density Function (PDF) and the estimator PDF. Research studies have been conducted on datasets describing user behavior in an exemplary 1000-household study. The presented considerations take into account the IoT technology for data exchange between devices in the Smart Grid network.

The paper is organized as follows: Section 2 presents the Elastic Energy Management algorithm, Section 3 describes the simulation research conducted, Section 4 presents smart appliance functionality and user behavior, Section 5 includes assumptions and results for the conducted experimental results, and Section 6 includes the conclusions of this study.

2. The Elastic Energy Management Algorithm

In order to manage the demand and supply of energy in DCDs through EEM, it was necessary to define the parameters of the model. For this purpose, it was defined as:

$$(P_{NOM}, P, pr, ts) \tag{2}$$

The requirement for individual parameters of each SA is to have defined properties. The properties of a given SA are determined on the basis of the technical possibilities specified by the manufacturers. The P_{NOM} parameter defines the value of the nominal power with which the given SA works. P is used to define a list of other power values for which a given SA could adapt, depending on its working conditions. For example, a boiler could temporarily increase or decrease the power of the heaters. Considering the user's comfort and technical capabilities of SAs, the priorities pr were introduced. This feature makes it possible to determine which SAs can be changed first, and for which the modification should be made last. For this purpose, a three-step scale was introduced for pr:

- $EEM\_PRIORITY\_LOW$,
- $EEM\_PRIORITY\_MEDIUM$,
- $EEM\_PRIORITY\_HIGH$.

$pr = EEM\_PRIORITY\_LOW$ means that the SA with the greatest possibility of changing its power settings is given priority. The ts parameter is used to define the time delay property of the given SA startup. This functionality is already known, in particular in household appliances, such as washing machines. In this case, the user may determine a desired end time for the washing task in order to prioritize his other household tasks.

In EEM, after describing parameters (2) and taking into account dependencies (1), the need to use heuristic algorithms was established. The rationale behind this decision was the necessity to solve NP-hard problems [8]. The Greedy Randomized Adaptive Search Procedure (GRASP) heuristic algorithm allows the finding of the right solution (not necessarily the optimal one, but one acceptable due to the cost of calculations) and is suitable for solving difficult NP problems. For the purpose of the GRASP algorithm, a fitness function was developed:

$$F_{EEM} = \begin{cases} (F_{DCD} + F_{SA} - F_{RES})\left(1 - \frac{U_{SEL\_EEM_1} - U}{U}\right) & \text{for } EEM_1 \text{ variant,} \\ (F_{DCD} + F_{SA} - F_{RES})\left(1 - \frac{U_{SEL\_EEM_2} - U}{U}\right) & \text{for } EEM_2 \text{ variant,} \\ (F_{DCD} + F_{SA} - F_{RES}) & U_{DCD\_MIN} \leq U_{DCD} \quad \vee \quad U_{DCD\_MAX} \geq U_{DCD}. \end{cases} \quad (3)$$

where

$$F_{DCD} = C_{DCD} \sum_{i=1}^{N_{SA}} P_{SA\_SEL_i}, \quad (4)$$

$$F_{SA} = C_{DSO} \sum_{i=1}^{N_{SA}} \left(P_{NOM_i} - P_{SA\_SEL_i}\right), \quad (5)$$

$$F_{RES} = \sum_{j=1}^{N_{RES}} \left(C_{RES_j} - P_{RES\_SEL_j}\right), \quad (6)$$

$$U_{SEL\_EEM_1} = \frac{\sum_{i=1}^{N_{SA}} P_{SA\_SEL_i}}{\sum_{j=1}^{N_{RES}} P_{RES_j} + P_{HMAX}}, \quad (7)$$

$$U_{SEL\_EEM_2} = \frac{\sum_{i=1}^{N_{SA}} P_{SA\_SEL_i}}{\sum_{j=1}^{N_{RES}} P_{RES_j} + P_{HDCD}}. \quad (8)$$

C_{DSO} is the cost of purchasing 1 kWh from the Distribution System Operator (DSO); i and j are indices of SAs and RES, respectively. N_{RES} and N_{SA} are the numbers of RESs and SAs. P_{SA_SEL} is the power value after modification for the SA. C_{RES} is the cost of obtaining 1 kWh for a given RES. P_{RES} is the value of power obtained from a given RES. When $U = 1$, this means that there is a balanced value of power demand for its supply. A block diagram of the EEM algorithm is shown in Figure 3.

The complexity of the problem of selecting power settings required the development of an extensive F_{EEM} (3) fitness function. Among the set of all power selection options, one combination must be selected that meets criteria for the stated requirements, e.g., reduction of the peak demand phenomenon. The three variants of F_{EEM} are intended to respond to the EEM algorithm: user comfort (EEM_1), requirements set by the DCD (EEM_2) and the situation of exceeding the voltage U_{DCD} in the DCD. Equations (4)–(6) make it possible to determine the demand for energy supply in a given DCD. Equation (7), as a component of Equation (3), gives the opportunity to obtain results that, in terms of energy costs and user comfort, will be the most appropriate choice. In case (8), the greatest emphasis in selecting the power settings will be placed on the solutions that will be the best choice for a DCD. In this case, energy costs are a secondary criterion when the most important factor is to keep the energy system in balance, e.g., during environmental phenomena that would contribute to the damaging of the power grid infrastructure.

In F_{EEM}, three variants of the implementation of the EEM algorithm for DCDs were taken into account (3).

The first variant (EEM_1) concerns a situation where the task of the EEM algorithm is to ensure the comfort of users by, inter alia, controlling SAs so that the costs for the energy used are as low as possible. For this purpose, the value of $U_{SEL\_EEM_1}$ is determined from Equation (7). $U_{SEL\_EEM_1}$ is the current value of power demand for its supply, calculated by EEM_1, based on selected values of appliance power.

The second variant F_{EEM} (3) applies to (EEM_2) and takes into account the need to take action to respond to a crisis situation, e.g., a peak demand phenomenon [21] or damage to the power grid infrastructure. This variant takes into account $U_{SEL\_EEM_2}$ from Equation (8). P_{HMAX} is the maximum sum of power of all enabled consumers after the reduction process by EEM_1, determined individually by the energy consumer. $U_{SEL\_EEM_2}$ is for the EEM_2 variant which, due to a purpose of use other than EEM_1, requires the adjustment of Formula (7) of SA power reduction over a specified period of time (from t_{DCD_1} to t_{DCD_2}) to a value at least equal to P_{HDCD}.

The third variant F_{EEM} (3) takes into account situation (1) when the DCD voltage range is exceeded U_{DCD}. In this case, the EEM algorithm based on (2) will select P_{SEL} values for all SAs to ensure condition (9).

$$U_{DCD\_MIN} < U_{DCD} < U_{DCD\_MAX}. \tag{9}$$

The EEM algorithm performs its activities in an endless loop. This approach is required due to the dynamics of phenomena occurring over time in Electric Power Systems. Each change requires decisions to be made to ensure a balance between acting for the benefit of DCDs and the comfort of users.

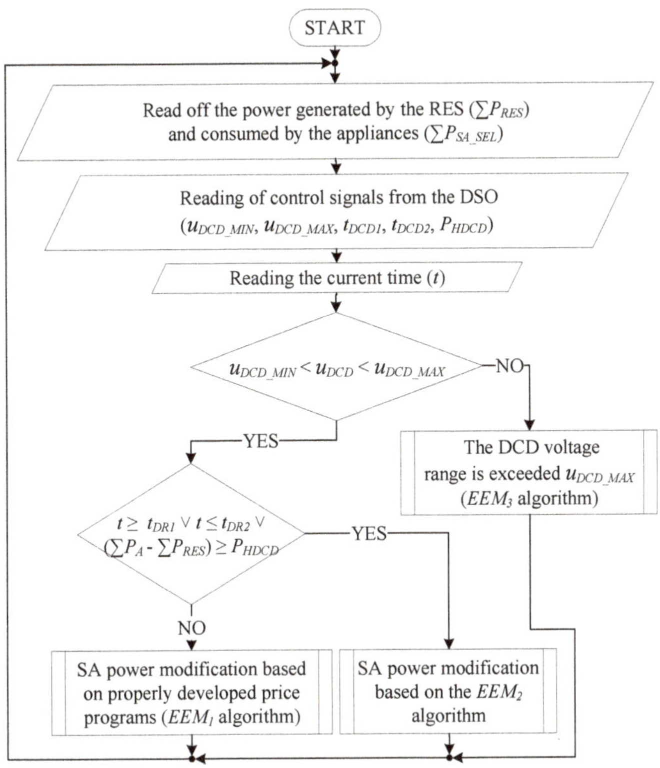

Figure 3. EEM flow chart.

3. Simulation Research

Simulation research for the EEM algorithm was performed in Java in the Eclipse IDE [22]. The implementation of the EEM algorithm in Java required the development of dedicated classes. Among others, the *EEM_Algorithm* and *EEM_AlgorithmData* classes were implemented for the EEM algorithm itself and for storing the required data. Additionally, the *EEM_Device_RES*, *EEM_Device_Source* and *EEM_Device_Appliance* classes, which are extensions to the *EEM_Device* class, were implemented. The *EEM_Priority* class introduces the ability to define one of the three available *pr* priorities.

The dataset for RESs that was used for simulation research is shown in Figure 4. The data presented in Figure 4 illustrate the situation of the largest generation of energy by a PV source at noon. The collection of such data was done to approximate a typical situation when energy from RESs is available during a selected period of time, so that its excess adversely affects the National Power Systems (NPS).

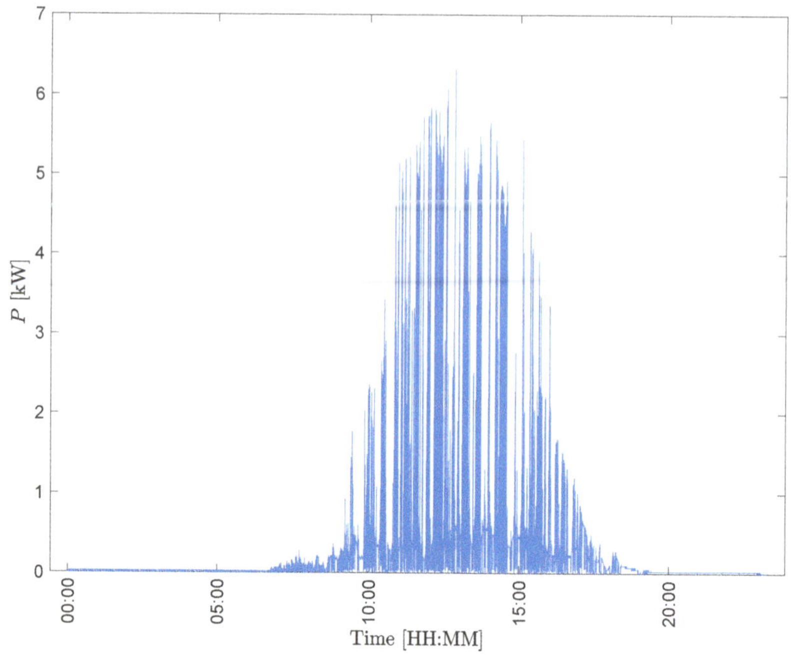

Figure 4. Sample data from a photovoltaic installation.

In earlier studies, research was conducted for one household [8,23]. The next part of the article presents the results of simulation research for an example of 1000 households. In each of the households, the possibility of using fifteen different SAs was assumed. The operating time profiles and the energy consumed by individual SAs at that time are presented in Figure 5.

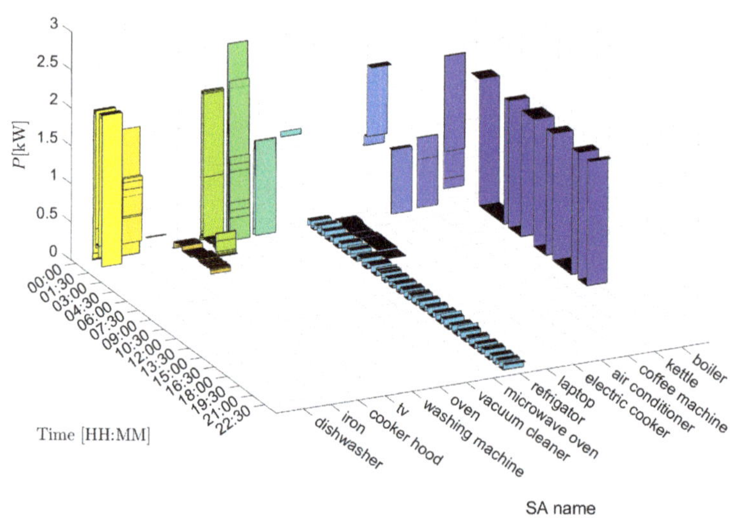

Figure 5. Profiles of SAs.

Figure 5 shows short profiles of SAs, from half an hour to all-day-long in duration. An example of a short profile is, e.g., the activity of ironing or boiling water in kettles. The operation of the refrigerator, which is switched on periodically throughout the day, can be considered a long profile. In the Figure 5 you can also see energy levels ranging from a few Watts to as much as 3 kW (oven).

The source of data sets defining individual profiles was the recording of measurement data using energy loggers. Data was recorded on exemplary devices that were intended to be general in nature. Particular characteristics were determined on the basis of values obtained from several days of operation of individual devices. The profiles were determined for the purpose of conducting simulation tests, which will be presented later in the article.

Table 1 lists the values of other parameters required for the correct operation of the EEM algorithm.

Table 1. Definition of other parameters for SAs.

SA	pr	$P(SA)$	-3σ	σ	$+3\sigma$
laptop	*EEM_PRIORITY_MEDIUM*	1	07:00	10:00	13:00
kettle	*EEM_PRIORITY_LOW*	0.9	07:30	08:00	08:30
			11:00	12:00	13:00
			17:45	18:00	18:30
washing machine	*EEM_PRIORITY_MEDIUM*	0.6	10:00	07:00	13:00
vacuum cleaner	*EEM_PRIORITY_LOW*	0.5	06:00	06:10	06:20
			14:15	14:30	14:50
tv	*EEM_PRIORITY_LOW*	0.8	07:00	08:00	09:00
			17:00	18:00	20:00
refrigerator	*EEM_PRIORITY_LOW*	1	00:10	00:30	01:00
dishwasher	*EEM_PRIORITY_LOW*	0.8	08:00	09:00	10:00
			18:00	19:00	20:00
iron	*EEM_PRIORITY_LOW*	0.8	08:00	09:00	10:00
			18:00	19:00	20:00
boiler	*EEM_PRIORITY_LOW*	0.9	06:30	07:00	07:30
oven	*EEM_PRIORITY_HIGH*	0.9	13:30	14:00	14:30
air conditioner	*EEM_PRIORITY_LOW*	0.9	09:30	10:00	11:00
			12:30	13:00	14:30
			15:30	16:00	17:30
microwave	*EEM_PRIORITY_LOW*	0.9	07:30	08:00	09:00
			15:30	16:00	17:30
			20:30	19:00	18:00
coffee maker	*EEM_PRIORITY_MEDIUM*	0.8	08:00	09:00	10:00
			18:00	19:00	20:00
cooker hood	*EEM_PRIORITY_LOW*	0.9	13:30	13:45	14:00
			14:00	14:15	14:30
			14:30	14:45	15:00
			15:00	15:15	15:30
electric cooker	*EEM_PRIORITY_HIGH*	0.9	12:30	13:00	13:30

The values of pr were adopted in such a way as to approximate the typical behavior of users and their needs. The $P(SA)$ parameter specifies the probability of a given SA occurrence and its launch within a specified time in the $\pm 3\sigma$ range. In order to differentiate individual households, a random selection was used with a given probability of the time when individual SAs are launched. For this purpose, a pseudo-random number generator with a Gaussian distribution, with mean $\mu = 0$ and standard deviation $\sigma = 1$, was used. The value of the time when the SA could be run most often is shown as σ. On the other hand, the statistical range from when to when a given SA could be run was marked with the following time intervals: from -3σ to $+3\sigma$. Due to the range of SA startup possibilities, it is assumed for all SAs that there will be no shifting during SA startup time ($ts = 0$).

Figure 6 shows a histogram for an example of 10,000 times when SA data could be run. In this case, it is assumed that σ will correspond to 1:45 p.m., when SAs are most likely to be run. The period for contingent activation of SAs was set to the following time range: $-3\sigma = 12{:}30$ and $+3\sigma = 15{:}00$.

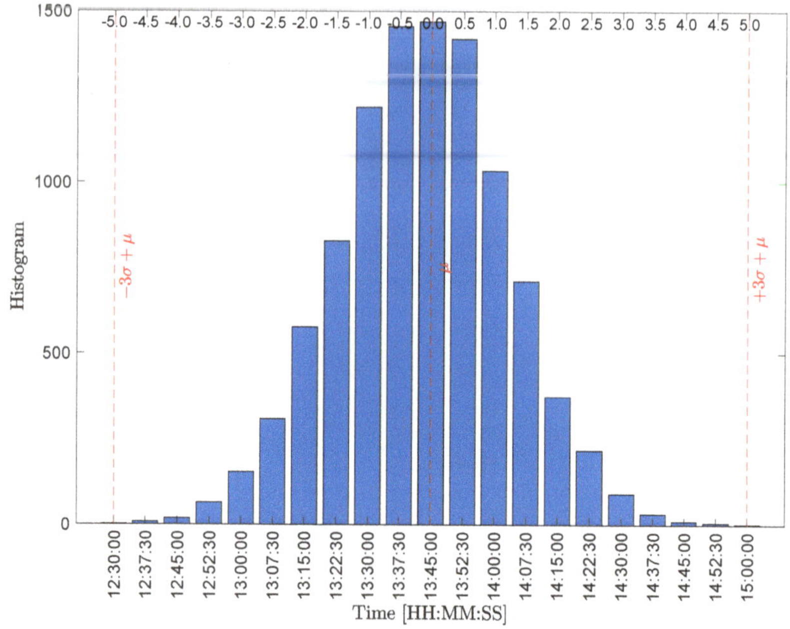

Figure 6. Histogram of the distribution of times for the interval from 12:30 to 15:00.

Using the 3σ rule [24–26], a distribution of time values was obtained with a probability of 99.7% falling within the $\pm 3\sigma$ time intervals. It was anticipated that SAs would be launched from 12:30 to 15:00.

In order to verify the correctness of the Java implementation of the activation algorithm of different SAs, the Probability Density Function (PDF) $f(x)$ was determined from Formula (10) and the estimator PDF $\hat{f}_k$ (11) [24–27]

$$f(x) = \frac{1}{\sqrt{2\pi}\sigma} \exp \frac{-(x-\mu)^2}{2\sigma^2}, \tag{10}$$

$$\hat{f}_k = S_k \frac{W}{N(H-L)}, \tag{11}$$

where S—the number of results in a given range, N—number of results, W—$\left\lfloor \sqrt[3]{N} \right\rfloor$, $k = 1 \ldots W$, $L = \min\limits_{1 \le n \le N} \{S_n\}$ and $H = \max\limits_{1 \le n \le N} \{S_n\}$.

In Figure 7, as a result of the statement $f(x)$ and $\hat{f}_k$, it can be confirmed that the randomly selected values of startup times of a given SA accumulate around 13:45 in the range $\pm 3\sigma$ (from 12:30 to 15:00).

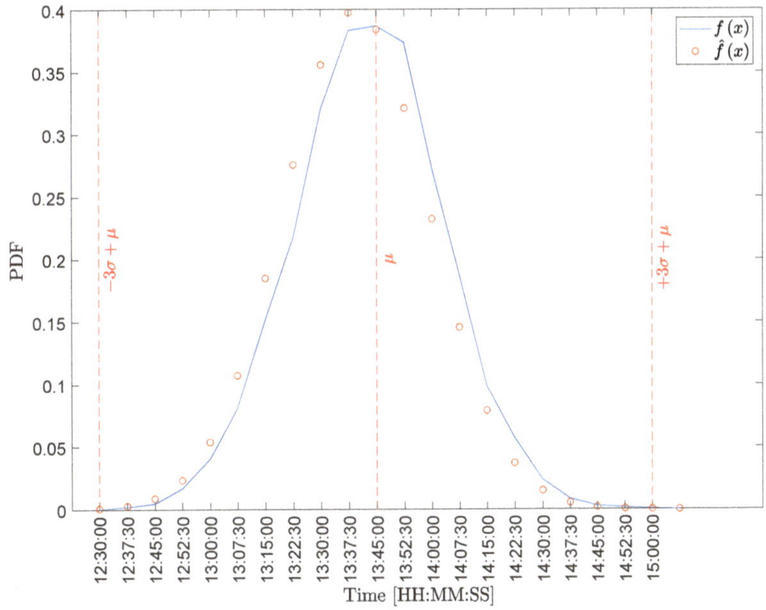

Figure 7. PDF $f(x)$ and estimator PDF $\hat{f}_k$ for the interval from 12:30 to 15:00.

The verification for starting individual SAs was repeated for a different time range. Again, it was assumed that σ would correspond to 13:45. The range $\pm 3\sigma$ was changed this time. The time range during which SAs could be run was reduced. The range is from $-3\sigma = 12{:}\ 30$ to $+3\sigma = 14{:}00$. Figure 8 shows a histogram of the times obtained when SAs could be run.

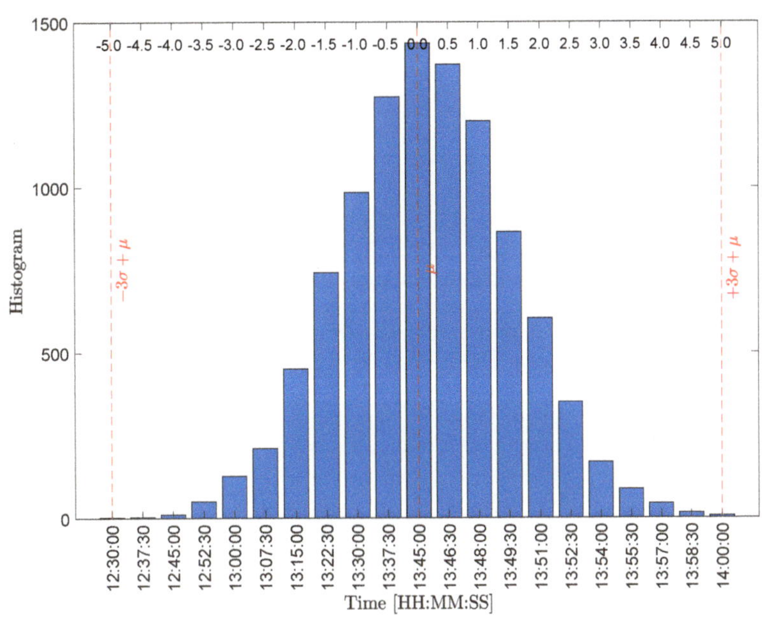

Figure 8. Histogram of the distribution of times for the interval from 12:30 to 14:00.

Based on the results shown in the histogram (Figure 8), it is unequivocally confirmed that the right-hand range was changed when SAs could be run. As before, the correctness of the obtained results was verified on the basis of the PDF analysis $f(x)$ (10) and the PDF estimator $\hat{f}_k$ (11). The results are presented in Figure 9.

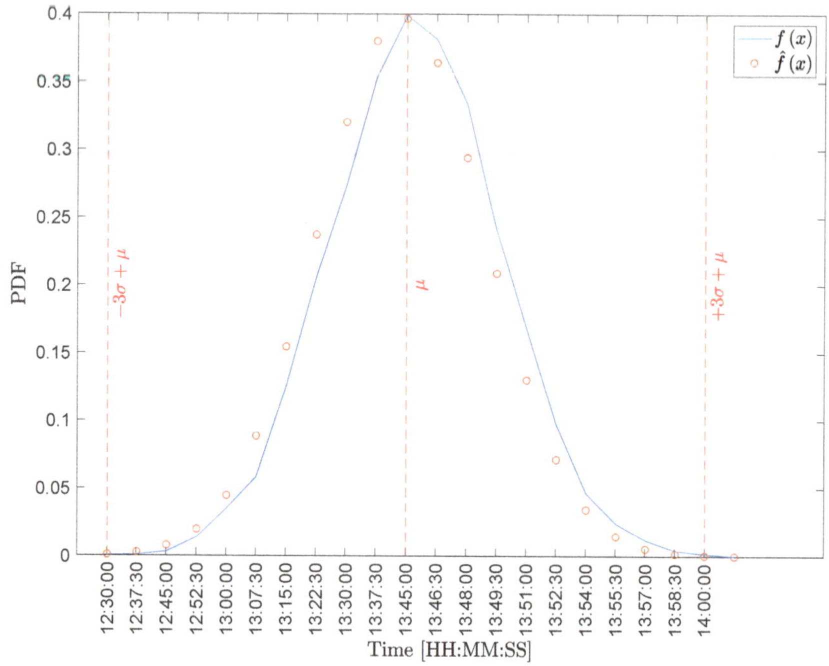

Figure 9. PDF $f(x)$ and estimator PDF $\hat{f}_k$ for the interval from 12:30 to 14:00.

By conducting an analysis of $f(x)$ and $\hat{f}_k$ from Figure 9, it can be stated that the narrowing of the ranges (on the right) did not adversely affect the randomized values of the activation times of a given SA. SAs are usually run at 13:45 in the range $\pm 3\sigma$ from $-3\sigma = 12{:}30$ until this time $+3\sigma = 14{:}00$.

The developed solution of differentiating the times when a given SA would be activated is important in reflecting the behavior in static households for which the EEM algorithm will be used. For example, Figure 10 shows a diagram of the schedule of air conditioners in 1000 households.

The differentiation of start-up times for air conditioners (Figure 10) gives an opportunity to approximate the behavior in typical statistical households. In this case, the data analysis approximates the pattern of behavior of users who, due to e.g., temperature, will statistically want to use air conditioning in similar time ranges. A complete summary of the SA usage pattern in the assumed simulation result is presented in Figure 11.

In the adopted schedule (Figure 11) assumptions were made to approximate the behavior in typical households. It is noticeable that household refrigerators are used all day long. There are also patterns resulting from the desire to use hot water from the boiler. Activities related to daily tasks, such as cooking, cleaning, etc., were also introduced. Based on the adopted schedule (Figure 11), simulation research was carried out. The test results are presented in Figure 12.

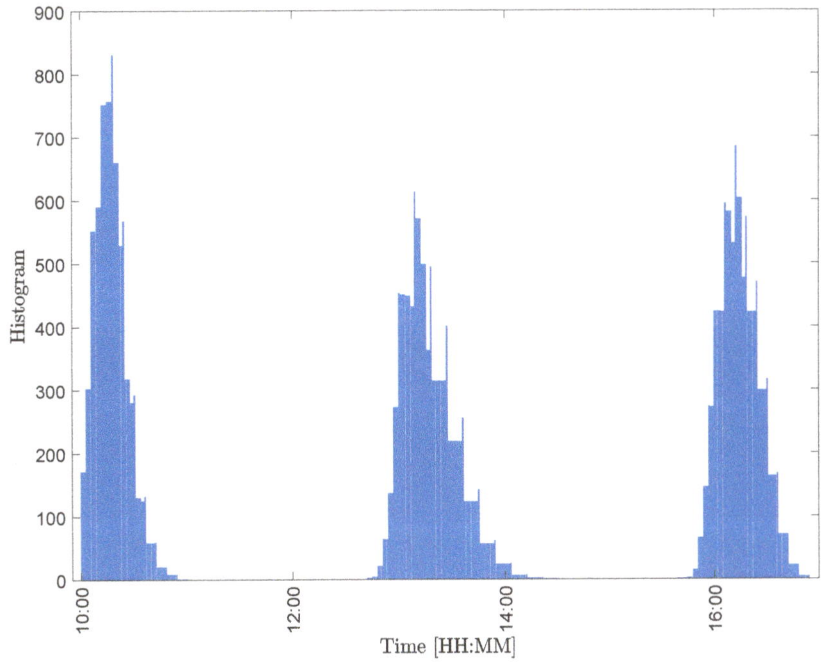

Figure 10. Air conditioner usage schedule.

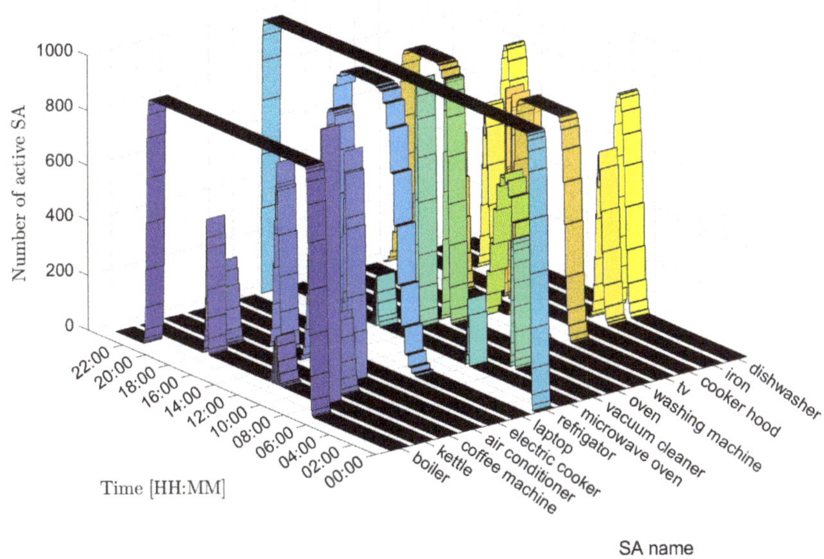

Figure 11. SA usage schedule.

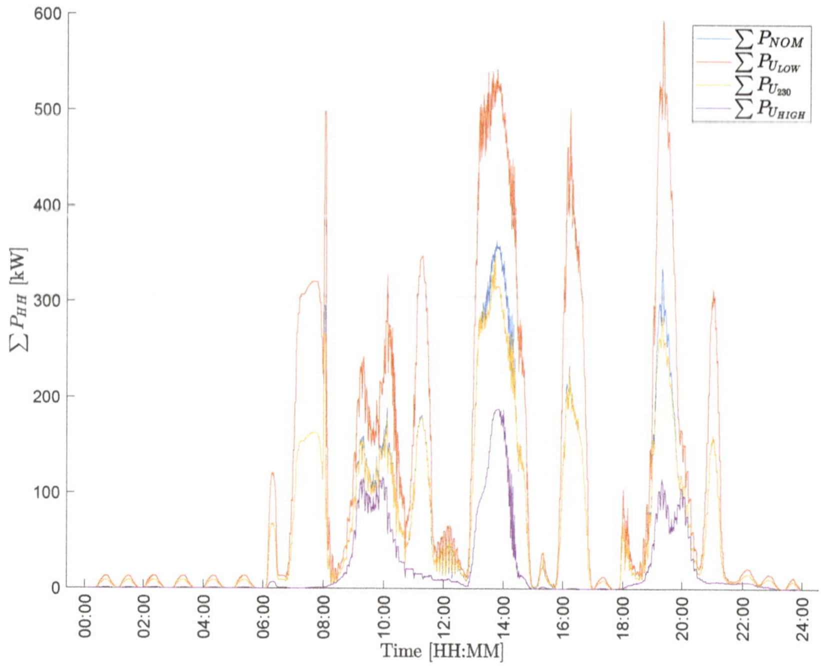

Figure 12. Energy consumption levels for different voltage ranges.

Figure 12 summarizes the total value of power consumed by 1000 households during one day ($\sum P_{HH}$) for four different variants. The first variant ($\sum P_{NOM}$) concerns the situation when the EEM algorithm is not used, so the power values of individual SAs were not modified. The next three variants concern the use of the EEM algorithm when the fitness function (3) is used. The values marked with $\sum P_{U_{LOW}}$ refer to the situation when the EEM algorithm would increase the load in the DCD network due to the excessively low voltage level U_{DCD}. Similarly, the EEM algorithm reduced the load in the network due to the excessively high value of U_{DCD}. In the case $\sum P_{U_{230}}$, the task of the EEM algorithm was to take into account the adjustment to the power generated from RESs (Figure 4). The use of the EEM algorithm resulted in a balance between acting for the benefit of DCDs and the comfort of users. When analyzing the data presented in the Figure 11, the value of U_{DCD} is adjusted based on the possibilities specified in *pr*.

The EEM algorithm modified the power settings of the SAs only when it was possible. The priorities and the possibility of modifying the power settings of individual SAs are taken into account. The biggest changes in power settings were made between 12:30 and 14:30. Additionally, in Figure 12, changes of power settings are visible from 18:30 to 20:15. Each modification of the power settings of SAs by the EEM algorithm has a direct impact on the total power consumed in the test households ($\sum P_{HH}$). Noticeable changes in energy consumption in households ($\sum P_{HH}$) are the result of activities that are performed statistically. In most cases, it will be activities related to cooking or cleaning. Using an air conditioner also affects energy consumption. These activities are mainly the most energy-intensive. The case marked as sum230 is not without significance for the user's comfort and expenses related to the purchase of energy. In this case, the availability of energy from RESs shapes how SAs are used. In such cases, devices such as boilers are activated. This is an example of storing the results of an electrical device.

4. SA Functionality and User Behavior

SA devices should be characterized by the possibility of a flexible switching-on moment activated by a given control signal, e.g., from a DCD or another signal that triggers their operation. In order to provide energy management services in the power grid, SAs should be able to set parameters such as: priority of action, time period during which it can be remotely activated and, if there is such a possibility, operating power. Devices that can be freely turned-on or turned-off are heating/cooling devices. In these cases, it may be possible to turn the device on/off at any point in time, as well as to set the operating power. These devices include fans, air conditioners, electric heaters, boilers, etc. This group of devices also includes all kinds of electric chargers, e.g., electric cars, other electric vehicles, and other battery-powered devices. ESSs should also be included in this group of SAs. They can be charged when there is energy production from RESs. When there is no RES energy production, it should be possible to regulate the charging time and current so as to minimize the impact on power system loading, of course taking into account the comfort of use. On the other hand, ESSs can be discharged at a time of increased demand for electricity by loads, when there is no energy from the power grid (power reduction on demand) [15].

Another group of SAs are those devices that operate continuously but can be deactivated for a short time without detriment to the user's comfort. These are devices such as a refrigerator or a food freezer. Their operation (on/off) depends on the temperature change set on the thermostat. Due to the large time constant of temperature changes, it is possible to temporarily turn off the devices without significant increases in temperature inside them, which may result in defrosting or spoiling the food [28]. Devices from this group will be used to control the load reduction rather than to increase the load on the power line.

The third group of SAs will be devices whose operation cannot be interrupted after they are turned on due to interference with and poor results of the function performed. This group includes appliances such as dishwashers, electric cookers, washing machines, electric dryers, etc. Disabling them may result in poor quality of washing/dishwashing/cooking etc. Often, interruption of the set operation scenario may result in higher electricity consumption after lowering the operating parameters such as temperature of washing/drying/cooking etc.

A separate group of SAs are devices that are used sporadically or depending on need. These are devices such as irons, hair dryers, electric kettles, coffee makers, etc. Their functionality should be set with a high priority for action, without the possibility of deactivating them. Nevertheless, devices of this type can give information about the state of the power system through some indicator or information on the operator panel. The functionality of individual SAs is summarized in Table 2.

In addition to the availability of smart home devices with the possibility of flexible control, in order to be able to effectively manage the electricity consumed in the home, the behavior and habits of electricity users should also be changed [29]. SA users should set their operation to when there is a need to increase the load on the power line, unless it is related to a drastic reduction in their operating comfort. Whenever it is possible, the settings of washing machines and dishwashers should be programmed for the period when the local production from renewable energy sources is at a high level. Similarly, you can change your eating habits for dinner. Often, these meals are prepared after people return home from work. Nevertheless, some of the meals can be remotely set to cook/bake. Then, after people come home from work, the dinner may be already cooked or partially prepared. These activities will require users to be particularly careful in setting up the devices for operation.

Table 2. Defining functionality for groups of devices.

Name of the SA	Possibility of Power Regulation	Any Time of Activation/Deactivation	Possibility of Operating Time Regulation	Ability to Feed Energy Back to the Grid	Possibility of Temporary Deactivation without Negative Effects for the Comfort of Use	Operation Cannot Be Interrupted after They Are Turned On	Devices with High Priority for Action with State of the Power System Indicator/Information
electric heaters	x	x	x				
boilers	x	x	x				
air conditioner	x	x	x				
fans	x	x	x				
electric cars	x	x	x	x			
electric scooters, bicycles, etc.	x	x	x	x			
battery-powered home devices (laptops, tablets, etc.)	x	x	x				
energy storage systems	x	x	x	x			
refrigerator					x		
food freezer					x		
dishwashers						x	
electric cookers						x	
washing machines						x	
electric dryers						x	
iron	x						
hair dryer	x						
electric kettle	x						x
coffee maker	x						x

Furthermore, some battery-powered devices can be connected and set to charge during energy generation from RESs. The group of such devices includes: home laptops, tablets, portable vacuum cleaners, etc. Electric vehicles should be charged at night, except for the evening peak load of the system. The exception will be when the user defines that he needs the availability of a charged electric vehicle at a designated time. Another change in the habits of electricity users is, for example, ironing clothes, which can be moved to a day when the user is not at work. Of course, this behavior makes sense if there will be energy generation from PV systems on that day. Ironing clothes before going to work increases the morning peak load on the power line.

It is not possible, in a short chapter, to indicate all the functionalities of SAs and changes in the behavior of electricity users. Nevertheless, the description indicates that the use of electricity in the future will involve the intelligent use of home appliances, especially when there is a shortage of power in the electricity system during peak demand and at night. High energy generation from PV systems during the day will imply the problem of excess energy in the system, which must either be stored (expensive solution) or consumed (change in energy use).

5. Experimental Results

Performing full-scale research on a network with available SAs is currently not feasible. Therefore, the article presents simplified results of experimental tests with simulation of SA load changes. The proposed EEM algorithm was implemented in the Programmable AC Load 3091LD [30], while the measurements were made using the HIOKI PW3337 Power Meter [31] (Figure 13). These changes were then scaled and applied to the grid load variability profile and the distributed generation profile from PV installations.

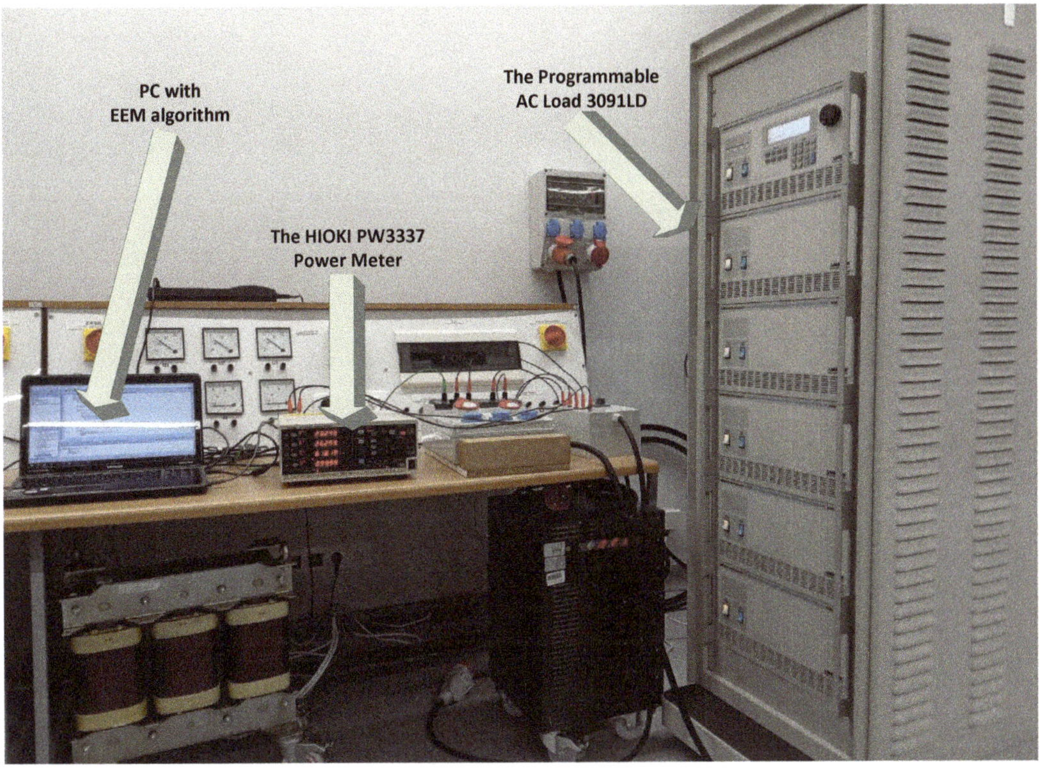

Figure 13. Photo of the laboratory experimental setup.

Experimental studies were conducted on a group of appliances limited to three SAs, which included a washing machine, an electric boiler and an electric cooker. With the use of the approach proposed in the article, a random load distribution was generated with the use of the indicated SAs.

Each new activation of the algorithm enables the generation of other load data resulting from the probability distribution. The research assumed that for each assay, 100 items of the indicated SM are attached, with an assumed distribution. The proposed experimental verification assumes that all 100 devices will always be in operation in the current study. An example of the variability of a SA switching on based on a washing machine is shown in Figure 14. The top of Figure 14 shows the load distribution for the operation of 1, 2, 5 and 10 washing machines, respectively, while at the bottom of Figure 14, the load distribution for 50, 75 and 100 washing machines is shown. In further studies, a load profile for 100 washing machines was adopted. In addition, similar load distributions were generated for 100 water-heating electric boilers as well as 100 electric cookers.

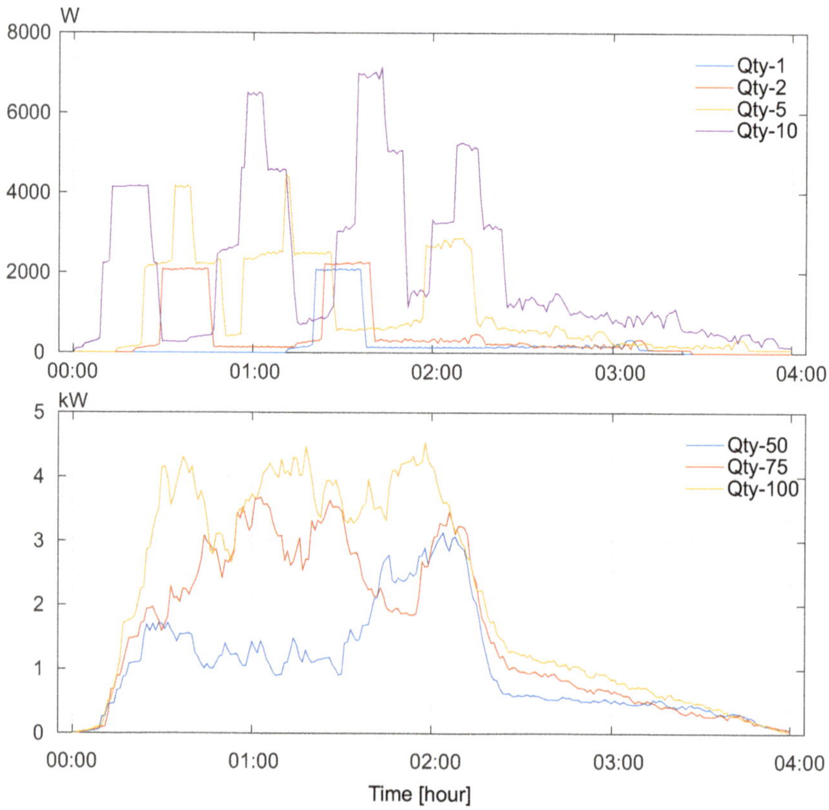

Figure 14. Load profile for different number of washing machines.

Figure 15 shows an example of the transformer station load profile and the generation of 500 kWp PV installations. Generation values are averaged over 15 min intervals (magenta line at the top of Figure 15). These are the values for a solar day with good PV energy generation. The typical load profile of the power grid is shown with the blue line at the top of Figure 15. The morning and evening grid load peaks are visible. In the hours from about 8:00 to about 15:00, a decrease in the load is visible. At the same time, during these hours, the largest generation profile for PV installations is observed.

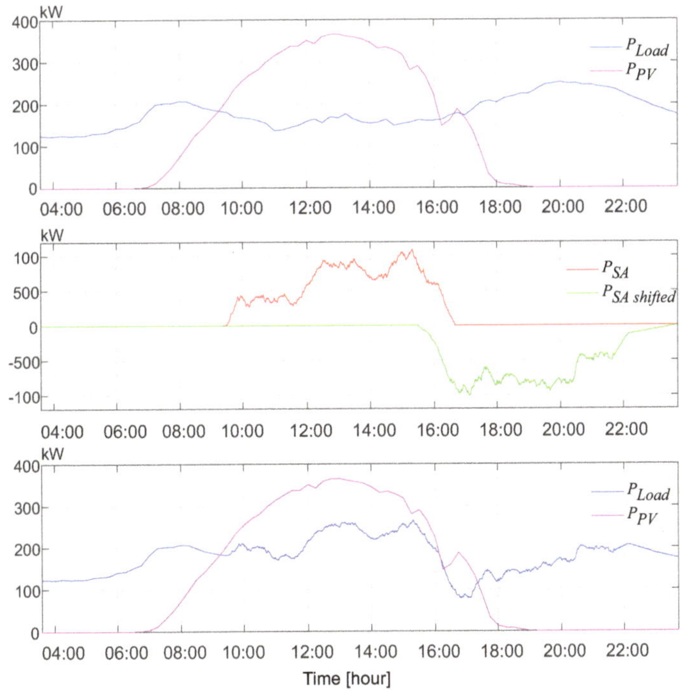

Figure 15. Experimental results: (**top**) load profile and generation profile of PV installations; (**middle**) the power consumed by the SA devices as a result of the proposed EEM algorithm, and an exemplary power profile shifted from the evening hours as a result of not switching on the SA; (**bottom**) load profile after applying the proposed EEM algorithm and PV installation generation profile.

The task of the algorithm was to shift the load for 100 SAs (washing machine, electric cooker, electric boiler) from evening hours to hours with the highest PV energy generation. The switching-on of the appropriate SAs took place depending on their suitability for the user. The initiation of switching on a selected group of devices and the period of their operation were defined in the tests. Throughout the period of operation of a given group of devices, they were switched on with a given probability distribution. The switching-on of other devices resulted in an increase in the instantaneous load power. Summing up the power consumed in a given period, the energy consumed by a given group of devices was calculated. Then, the power consumed by the given SA group was subtracted from the evening load profile. The power profile that was subtracted was identical to the SA load power profile. The timing of the evening hours during which the load power was shifted was also defined in the proposed test. The data from the performed test are summarized in Table 3. The washing machines were the first to start, as their operation does not have to be coordinated with the user. The wash cycle can start and end at any time of the day. Therefore, the operation of the washing machines was activated at the beginning of the day when there was a large generation of energy from the PV plant. The power consumed by the washing machines was shifted from the evening hours around 19:50 to 23:40. Other devices that were switched on are electric boilers for heating domestic hot water. Due to the good accumulation of heat energy, the heating of the water may be a little earlier than the return of people from work/school. They were turned on from around 11:30 to 14:30. The corresponding power needed to heat the water was shifted from the afternoon hours of 17:40 to 20:40. The last group of devices are electric cookers for preparing dinner. Usually, dinner is prepared after the residents return home from 15:00 to 17:00. If there is a possibility of cooking/baking some dishes, this activity can be performed earlier without the participation of people. The inclusion of smart electric cookers in our research was initiated from 13:40 to 16:40. The power needed to prepare dinner was shifted from the afternoon hours 15:30–18:30. The red color in the middle part of Figure 15 shows the power of devices connected by the proposed EEM algorithm, while the green color shows one of the possible reductions of the peak load. The lower diagram in Figure 15 shows the load profile taking into account the EEM algorithm. There was an increase in load during higher generation from PV and a reduction in power for the evening peak.

Table 3. Selected SA operating times and energy.

Device Name	SA Operating Period	Power Shifted from Hours	Consumed/Shifted Energy [W]
washing machines	9:30–13:20	19:50–23:40	84,790
electric boilers	11:30–14:30	17:40–20:40	190,130
electric cookers	13:40–16:40	15:30–18:30	116,930
			∑391,850

Figure 16 compares the load profile with and without the EEM algorithm. Additionally, at the bottom of Figure 16, the surplus of energy production from PV generation, which should be stored or sent to other loads, is presented. Without the proposed algorithm, the unused energy is exactly 1,039,892 Wh (around 1 MWh), and with the EEM algorithm, it is 648,042 Wh (around 0.65 MWh). The use of the proposed algorithm increased the consumption of electricity during the high output from PV installations and at the same time allowed for a reduction in the evening load on the line. Surplus energy produced from PV sources should be sent to other consumers or stored in ESSs. In the proposed algorithm and the analyzed test, a reduction of the volume of energy to be transmitted or stored by 391,850 Wh was obtained.

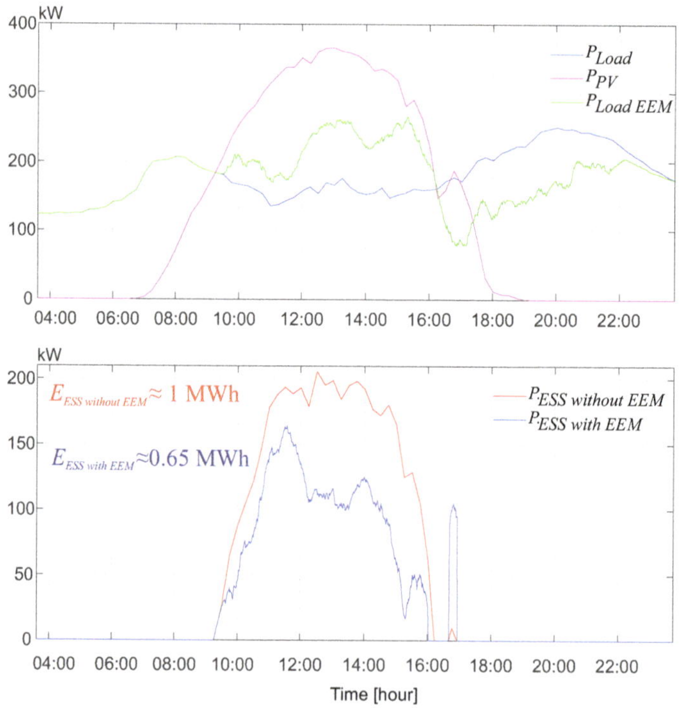

Figure 16. Experimental results: (**top**) load profile comparison with and without EEM algorithm; (**bottom**) comparison of unused energy from PV installation.

6. Conclusions

The use of RESs and temporary load changes (e.g., as a result of the peak demand phenomenon) in Smart Grid (SG) networks require the use of energy management algorithms. The lack of such algorithms would in the worst case result in unstable operation or damage to the electric power system. This article proposes the use of the EEM algorithm for energy management in a hierarchical control system with DCDs. We added the use of the MQTT protocol to our considerations. Such an approach would effectively enable data collection from DCDs. We proposed to use external and internal MQTT brokers to publish data. The data are published with messages that are organized into topics. The topics can contain plain text information. Structures in JavaScript Object Notation (JSON) format can be stored in the text. This format is independent of the specific programming language. This gives the possibility of scalability of the implemented system with new functionalities that would result from the addition of additional IoT devices.

Earlier works [8] considered the use of the EEM algorithm in a single household. Here, the application of the EEM algorithm in 1000 households during simulation studies was outlined. In order to verify the correct operation of the EEM algorithm, test data were developed from the example of a photovoltaic installation (Figure 4). The load profiles of SA devices, which were used in the energy management process, were also developed (Figure 5). Simulation research was carried out with additional assumptions about the priorities for the use of SAs pr. Differentiation in the use of individual SAs was also obtained on the basis of their parameter specificities and the probability of a given SA occurrence ($P(SA)$) launched within a specified time in the $\pm 3\sigma$ range. The correctness of a SA's usability being differentiated from others was confirmed by the determination of the PDF $f(x)$ (10) and the estimator PDF $\hat{f}_k$ (11).

On the basis of the simulation results, it was determined how SAs are used in the exemplary households (Figure 11). The simulation results are presented in Figure 12. In this case, the results of the EEM algorithm for four scenarios are compared. The individual variants reflect the situation where the EEM algorithm is not used, the EEM algorithm is used to increase/decrease the load in the DCD network, or it is used to produce a balance between acting for the benefit of DCDs and the comfort of users.

Experimental studies were carried out for the simplified case with the use of three SAs—a washing machine, an electric boiler and an electric cooker. The algorithm was implemented on an appropriate scale with an adjustable AC load. Then, the obtained results, after scaling, were added/subtracted to/from the previously measured load profile and compared with the generation profile of the PV installation. The obtained results show that by using the proposed EEM algorithm, with the use of SAs, it is possible to change the load profile of the power grid. The evening peak load on the line was reduced and the use of energy produced from the PV installation increased. About 0.35 MWh of energy was shifted using these three types of SA (Figure 16). This means that much less energy from RESs needs to be stored in very expensive ESSs, or that such a volume of energy should be sent to other loads. However, such activities will only be possible if the SAs have the functionality described briefly in Section 4. Additionally, such changes will be related to the change of habits of electricity users. They will have to consciously set the operation of their home appliances to operate at times of high generation coming from renewable energy, i.e., at times when there is a large supply of energy in the grid. They will also have to avoid using their electrical devices (unless there is specific need) in times of heavy load on the grid, when the energy supply in the grid is too small.

At the present stage of research, the biggest limitation for the EEM algorithm is the lack of mass use of SAs in all households. Additionally, data privacy may be an important issue. In this case, it may be about issues related to a lack of will to provide data on activities and tasks carried out by members of society. A solution to this problem may be price incentives encouraging the exchange and sharing of data from new smart devices (SAs).

In our further plans, we intend to design and implement selected types of SAs. Original versions of SAs would be created. Additionally, we plan to implement a prototype of a complete, universal SG network system with SAs. The scope of this work would be the analysis and verification of the EEM algorithm for the IoT concept in electronic device systems. These devices could automatically communicate and exchange data over the network without human intervention. We are also interested in studying the use of energy storage and their inclusion in the EEM algorithm.

Author Contributions: Conceptualization, P.P., P.S. and K.P.; methodology, P.P., P.S., K.P., K.T., M.K. and I.K.; software, P.P.; validation, P.P. and P.S.; formal analysis, P.P. and P.S.; investigation, P.P. and P.S.; resources, P.P. and P.S.; data curation, P.P. and P.S.; writing—original draft preparation, P.P., P.S. and K.P.; writing—review and editing, P.P., P.S. and K.P.; visualization, P.P., P.S., K.P., K.T., M.K. and I.K.; supervision, P.P.; project administration, P.P.; funding acquisition, K.P. All authors have read and agreed to the published version of the manuscript.

Funding: This work was partially funded by the European Union within the European Regional Development Fund (ERDF), in the context of the INTERREG V A BB-PL Programme. Part of this work was carried out as part of the INTERREG Project on "Smart Grid Platform for research on energy management", project number 85024423 as well as part of the H2020 ebalance plus project (grant number 864283).

Institutional Review Board Statement: Not applicable.

Informed Consent Statement: Not applicable.

Data Availability Statement: Data sharing not applicable.

Conflicts of Interest: The authors declare no conflict of interest.

Abbreviations

The following abbreviations are used in this manuscript:

DCD	Distributed Control Devices
DER	Distributed Energy Resources
DR	Demand Response
DSO	Distribution System Operator
EEM	Elastic Energy Management
ESS	Energy Storage Systems
GRASP	Greedy Randomized Adaptive Search Procedure
IoT	Internet of Things
JSON	JavaScript Object Notation
MQTT	Message Queue Telemetry Transport
NPS	National Power Systems
PDF	Probability Density Function
PV	Photovoltaic Panels
RED II	Renewable Energy Directive
RES	Renewable Energy Sources
SA	Smart Appliance
SG	Smart Grid
$\sum P_{HH}$	sum of power consumed in test households during one day
$\sum P_{U_{230}}$	sum of power consumed when the EEM algorithm was to take into account the adjustment to the power generated from RESs
$\sum P_{U_{HIGH}}$	sum of power consumed when the EEM algorithm would decrease the load in the DCD network
$\sum P_{U_{LOW}}$	sum of power consumed when the EEM algorithm would increase the load in the DCD network
$f(x)$	The Probability Density Function (PDF)
$\hat{f}_k$	The estimator Probability Density Function (PDF)
F_{DCD}	Component part of the fitness function (F_{EEM}) aimed at determining the influence of DCD on the choice of power settings
F_{EEM}	The fitness function
F_{RES}	Component part of the fitness function (F_{EEM}) aimed at determining the influence of RESs on the choice of power settings

F_{SA}	Component part of the fitness function (F_{EEM}) aimed at determining the influence of SAs on the choice of power settings
P	A list of other power values for which a given SA could change depending on its working conditions
$P(SA)$	The probability of a given SA occurrence and its launch within a specified time
P_{NOM}	The nominal power with which the given SA works
pr	SA priority
ts	The time delay property of the given SA startup
U_{DCD}	DCD voltage
$U_{SEL\_EEM_1}$	Component part of the fitness function (F_{EEM}) aimed at taking into account the user's comfort
$U_{SEL\_EEM_2}$	Component part of the fitness function (F_{EEM}) aimed at taking into account the DCD requirements

References

1. European Commission (EC). The European Green Deal, COM(2019) 640. 2019. Available online: https://ec.europa.eu/info/sites/info/files/european-green-deal-communication_en.pdf (accessed on 5 April 2022).
2. Directive (EU) 2018/2001 of the European Parliament and of the Council of 11 December 2018 on the Promotion of the Use of Energy from Renewable Sources (Text with EEA Relevance). Available online: http://data.europa.eu/eli/dir/2018/2001/oj (accessed on 5 April 2022).
3. Iweh, C.D.; Gyamfi, S.; Tanyi, E.; Effah-Donyina, E. Distributed Generation and Renewable Energy Integration into the Grid: Prerequisites, Push Factors, Practical Options, Issues and Merits. *Energies* **2021**, *14*, 5375. [CrossRef]
4. Naddeo, V.; Korshin, G. Water, energy and waste: The great European deal for the environment. *Sci. Total Environ.* **2021**, *764*, 142911. [CrossRef] [PubMed]
5. Masebinu, S.O.; Akinlabi, E.T.; Muzenda, E.; Aboyade, A.O. Techno-economic analysis of grid-tied energy storage. *Int. J. Environ. Sci. Technol.* **2018**, *15*, 231–242. [CrossRef]
6. Jannesar, M.R.; Sedighi, A.; Savaghebi, M.; Guerrero, J.M. Optimal placement, sizing, and daily charge/discharge of battery energy storage in low voltage distribution network with high photovoltaic penetration. *Appl. Energy* **2018**, *226*, 957–966. [CrossRef]
7. Jabir, H.J.; Teh, J.; Ishak, D.; Abunima, H. Impacts of Demand-Side Management on Electrical Power Systems: A Review. *Energies* **2018**, *11*, 1050. [CrossRef]
8. Powroźnik, P.; Szcześniak, P.; Piotrowski, K. Elastic Energy Management Algorithm Using IoT Technology for Devices with Smart Appliance Functionality for Applications in Smart-Grid. *Energies* **2022**, *15*, 109. [CrossRef]
9. Zhang, F.; Hu, X.; Langari, R.; Wang, L.; Cui, Y.; Pang, H. Adaptive energy management in automated hybrid electric vehicles with flexible torque request. *Energy* **2021**, *214*, 118873. [CrossRef]
10. Jindal, A.; Bhambhu, B.S.; Singh, M.; Kumar, N.; Naik, K. A Heuristic-Based Appliance Scheduling Scheme for Smart Homes. *IEEE Trans. Ind. Inform.* **2020**, *16*, 3242–3255. [CrossRef]
11. Fotopoulou, M.C.; Drosatos, P.; Petridis, S.; Rakopoulos, D.; Stergiopoulos, F.; Nikolopoulos, N. Model Predictive Control for the Energy Management in a District of Buildings Equipped with Building Integrated Photovoltaic Systems and Batteries. *Energies* **2021**, *14*, 3369. [CrossRef]
12. Ouramdane, O.; Elbouchikhi, E.; Amirat, Y.; Le Gall, F.; Sedgh Gooya, E. Home Energy Management Considering Renewable Resources, Energy Storage, and an Electric Vehicle as a Backup. *Energies* **2022**, *15*, 2830. [CrossRef]
13. Shewale, A.; Mokhade, A.; Funde, N.; Bokde, N.D. A Survey of Efficient Demand-Side Management Techniques for the Residential Appliance Scheduling Problem in Smart Homes. *Energies* **2022**, *15*, 2863. [CrossRef]
14. Stanelytė, D.; Radziukynas, V. Analysis of Voltage and Reactive Power Algorithms in Low Voltage Networks. *Energies* **2022**, *15*, 1843. [CrossRef]
15. Smoleński, R.; Szcześniak, P.; Drożdż, W.; Kasperski, L. Advanced metering infrastructure and energy storage for location and mitigation of power quality disturbances in the utility grid with high penetration of renewables. *Renew. Sustain. Energy Rev.* **2022**, *157*, 111988. [CrossRef]
16. AWS IoT Core Features. Available online: https://aws.amazon.com/iot/ (accessed on 2 May 2022).
17. Google Cloud IoT. Available online: https://cloud.google.com/solutions/iot (accessed on 2 May 2022).
18. IBM Cloud IoT. Available online: https://www.ibm.com/cloud/internet-of-things (accessed on 2 May 2022).
19. Microsoft Azure IoT. Available online: https://azure.microsoft.com/en-us/overview/iot/ (accessed on 2 May 2022).
20. Oracle IoT Cloud Service. Available online: https://www.oracle.com/cloud/ (accessed on 2 May 2022).
21. Delberis, A.L.; Andrés, M.C.G.; Érica, T.; Eidy, M.M.B. Peak demand contract for big consumers computed based on the combination of a statistical model and a mixed integer linear programming stochastic optimization model. *Electr. Power Syst. Res.* **2018**, *154*, 122–129. [CrossRef]
22. Eclipse IDE and Web IDEs. Available online: https://www.eclipse.org/ide/ (accessed on 5 April 2022).
23. Powroźnik, P.; Szulim, R.; Miczulski, W.; Piotrowski, K. Household Energy Management. *Appl. Sci.* **2021**, *11*, 1626. [CrossRef]

24. Ruzgas, T.; Lukauskas, M.; Čepkauskas, G. Nonparametric Multivariate Density Estimation: Case Study of Cauchy Mixture Model. *Mathematics* **2021**, *9*, 2717. [CrossRef]
25. Brajčić Kurbaša, N.; Gotovac, B.; Kozulić, V.; Gotovac, H. Numerical Algorithms for Estimating Probability Density Function Based on the Maximum Entropy Principle and Fup Basis Functions. *Entropy* **2021**, *23*, 1559. [CrossRef]
26. Ngatchou-Wandji, J.; Ltaifa, M.; Njamen Njomen, D.A.; Shen, J. Nonparametric Estimation of the Density Function of the Distribution of the Noise in CHARN Models. *Mathematics* **2022**, *10*, 624. [CrossRef]
27. Lal-Jadziak, J.; Sienkowski, S. Models of bias of mean square value digital estimator for selected deterministic and random signals. *Metrol. Meas. Syst.* **2008**, *15*, 55–67.
28. Werminski, S.; Jarnut, M.; Benysek, G.; Bojarski, J. Demand side management using DADR automation in the peak load reduction. *Renew. Sustain. Energy Rev.* **2017**, *67*, 998–1007. [CrossRef]
29. Zhai, S.; Wang, Z.; Yan, X.; He, G. Appliance Flexibility Analysis Considering User Behavior in Home Energy Management System Using Smart Plugs. *IEEE Trans. Ind. Electron.* **2019**, *66*, 1391–1401. [CrossRef]
30. AC Load 3091LD. Available online: https://www.powerandtest.com/power/electronic-loads/ac-electronic-load-3091ld (accessed on 5 April 2022).
31. Power Meter HIOKI PW3337. Available online: https://www.hioki.com/global/products/power-meters/3phase-ac-dc/id_5929 (accessed on 5 April 2022).

energies

MDPI

Review

Microgrid Energy Management and Methods for Managing Forecast Uncertainties

Shanmugarajah Vinothine [1,*], Lidula N. Widanagama Arachchige [1], Athula D. Rajapakse [2,*] and Roshani Kaluthanthrige [2]

1 Department of Electrical Engineering, University of Moratuwa, Moratuwa 10400, Sri Lanka
2 Department of Electrical and Computer Engineering, University of Manitoba, Winnipeg, MB R3T 2N2, Canada
* Correspondence: vinothines.21@uom.lk (S.V.); Athula.Rajapakse@umanitoba.ca (A.D.R.); Tel.: +94-77-901-8126 (S.V.)

Abstract: The rising demand for electricity, economic benefits, and environmental pressures related to the use of fossil fuels are driving electricity generation mostly from renewable energy sources. One of the main challenges in renewable energy generation is uncertainty involved in forecasting because of the intermittent nature of renewable sources. The demand also varies according to the time of day, the season, the location, the climate, and the availability of resources. Microgrids offer a potential solution for the integration of small-scale renewable energy sources and loads along with energy storage systems and other non-renewable sources. However, intermittent generation and varying demand need to be matched to provide stable power to consumers. Therefore, it is crucial to design an energy management system to effectively manage the energy sources and supply loads for reliable and efficient operation. This paper reviews different techniques proposed in the literature to achieve the objectives of a microgrid energy management system. The benefits of existing energy management systems and their challenges are also discussed. The challenges associated with uncertainties and methods to overcome them are critically reviewed.

Keywords: energy management; forecast uncertainties; microgrids; optimization; renewable energy integrations

Citation: Vinothine, S.; Widanagama Arachchige, L.N.; Rajapakse, A.D.; Kaluthanthrige, R. Microgrid Energy Management and Methods for Managing Forecast Uncertainties. *Energies* **2022**, *15*, 8525. https://doi.org/10.3390/en15228525

Academic Editor: Miguel Jiménez Carrizosa

Received: 6 October 2022
Accepted: 9 November 2022
Published: 15 November 2022

1. Introduction

In the last two decades, the electric power industry strived to increase the electricity generation from renewable energy sources (RES) due to the environmental issues associated with the use of fossil fuels and associated economic benefits. Being proven cost-effective technologies, currently solar PV and wind are the fastest deployed RESs in power generation [1]. However, solar PV power generation is impacted by changing weather conditions and passing cloud cover, while the amount of energy generated by wind varies with wind speed. The intermittent nature of renewable energy resources complicates power system operation and control. These uncertainties introduced to the generation of resources, in addition to the varying electric demand, make energy management more challenging.

The microgrid (MG) concept, schematically illustrated in Figure 1, has become a smart candidate for integrating RESs, as it can be operated as a single controllable system. A microgrid is usually comprised of energy resources, energy storages, and loads and operated within a clearly defined electrical boundary. The energy mix of a microgrid usually includes solar PV and wind as primary sources of renewable energy, and a few non-renewable resources, such as diesel generators, micro turbines, and gas turbines are also used as backup energy resources. Various energy storages, such as batteries, super capacitors, fuel cells, are considered to ensure the availability of power throughout the entire time horizon [2–4].

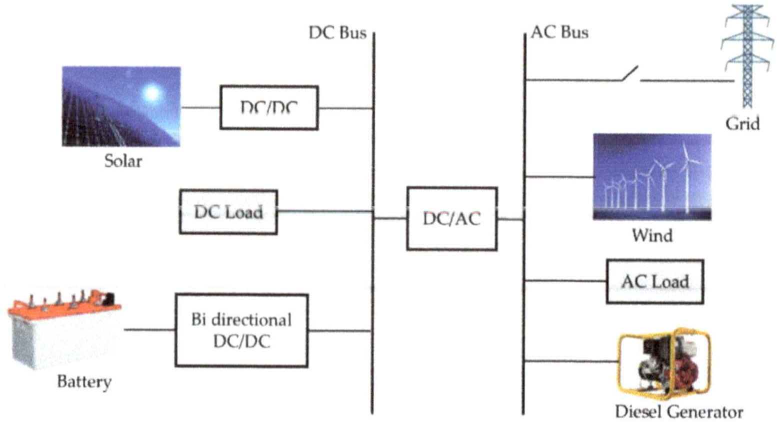

Figure 1. Generalized structure of a microgrid.

A microgrid can be either connected to or isolated from the grid and operate with full controllability. The output power from energy sources must, therefore, meet the requirements of local loads in the islanded mode. In the grid-connected mode, the microgrid shares the energy with the main grid (supply or absorb) via the point of common coupling (PCC). Microgrids can be classified based on voltage, such as AC microgrids, DC microgrids, and hybrid AC/DC microgrids. In AC microgrids, DC generating RES such as solar PV and wind are connected via DC/AC power inverters. The DC microgrid is similar to its AC counterpart, possessing a common DC bus. A hybrid microgrid is a combination of both AC and DC microgrids, offering the best solution for grid integration of RES. Various models and layouts are used to describe the microgrid operations in the literature [5].

A microgrid control system is responsible for ensuring desired voltages, currents, and frequency through proper management and control, including performing economic dispatch, balancing power supply and demand, demand side management, etc., under all modes of microgrid operation. An energy management system (EMS) is usually designed to optimize power generation to meet the demand at the minimum operational cost while maintaining the integrity of the system. Among the various definitions, the IEC 61970 standard has defined EMS as "a computer system comprising a software platform providing basic support services and a set of applications providing the functionality needed for the effective operation of electrical generation and transmission facilities so as to assure adequate security of energy supply at minimum cost" [6]. Microgrid energy management systems (MG EMS) also have the same aforementioned features to provide the required functions to ensure safe and efficient operation. An energy management problem is typically formulated as an optimization problem with the objective of minimizing the total cost of operation over a chosen time horizon (often over 24 h), subjected to operational constraints. The optimization is based on the forecasted load variation. When intermittent generation is involved, a resource forecast is also required to solve the optimization problem. The MG energy management is complicated by forecast uncertainties. The forecast uncertainty, which is the deviation of actual load and renewable generation from their respective forecast values, affects optimum scheduling and raises new challenges in microgrid systems with a high penetration of renewables. Therefore, uncertainty management needs to be incorporated into the energy management problems.

Several comprehensive reviews related to MG EMS can be found in the literature, and they address different aspects of energy management function. The review of microgrid EMS presented in [7] is organized based on four categories: (1) the optimization techniques employed, (2) the type of grid taken into consideration, (3) the mode of operation of the microgrid, and (4) the software used as a platform for solving the EMS problems. Two major categories of microgrid energy management strategies are discussed, including classical and intelligent methods for residential applications in [8]. A comparative and critical analysis of the literature on decision-making strategies and their solution methods for MG EMSs is presented in [6]. A comprehensive description of control and optimization methods to identify the most common and effective methods for MG EMS is highlighted in [9]. In [10], the review is conducted in terms of uncertainty modeling approaches, objective functions, constraints, optimization techniques, and simulation and experiment results for EMSs. However, the uncertainty issues are not comprehensively addressed. Recent techniques to model the uncertainties from renewable energy sources and loads in microgrids are reviewed in [11]. Methods of uncertainty management, parameter modelling, simulation tools, and test system in unit commitment in power systems are discussed in [12]. Methods for uncertainty modelling in power systems, comparison between these methods, strengths, and weaknesses are studied in [13]. A standard classification of uncertainty handling methods is proposed in [14], where the models are compared, and their strengths and weaknesses are investigated.

Proper modelling and managing forecast uncertainties are an important aspect of an efficient and effective energy management system for a microgrid with high penetration of renewables. The main objective of this paper is to present a comparative review of effective energy management methods used in microgrids along with forecast uncertainty management. This study identifies the techniques used for managing forecast uncertainties and modeling those uncertainties in microgrid systems. The paper is organized as follows: the concept of a microgrid energy management system is discussed in Section 2. Energy management problem formulation and solution approaches are discussed in Sections 3 and 4, respectively. Section 5 addresses the potential challenges caused by the uncertainties from forecasted data along with managing methods in MG EMS. Section 6 discusses the application of artificial intelligence (AI) and machine learning (ML) in MG EMS. The paper concludes with future research opportunities related to microgrid energy management under source and load uncertainties.

2. Microgrid Energy Management System (MG EMS): The Concept

A microgrid energy management system (MG EMS) performs a variety of functions for the efficient and effective operation of the system. Energy management is an optimization problem with the target of properly scheduling the short-term operation of production by generators, storage, as well as controllable loads, to cover the system demand and minimize the generation costs. The EMS generates a schedule of unit commitment and the optimized output of each source considering the results of the optimization. Figure 2 illustrates the overall outline of the MG EMS.

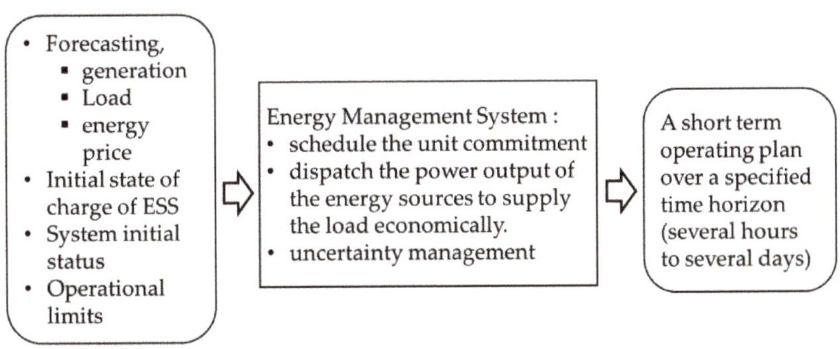

Figure 2. Energy management system outline.

Control Systems Used in EMS

The control system associated with MG EMS can be implemented using centralized, decentralized, and hierarchical control methods [7,15,16]. In centralized control-based EMS, a single central controller that receives all the information, such as RES energy generation, load profile, market price, weather conditions, etc., is used. Based on the inputs, a central controller decides the optimum microgrid energy schedule and then sends these decisions to all local controllers. The basic structure of the centralized control is shown in Figure 3. However, the failure of the central control could cause the entire system to fail. Unlike centralized control-based EMS, in decentralized control shown in Figure 4, a few local connections are needed, and only local measurements are used to make control decisions.

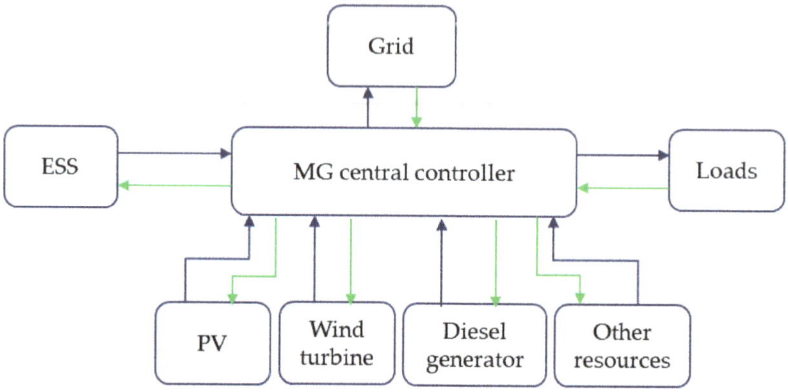

Figure 3. Centralized control structure.

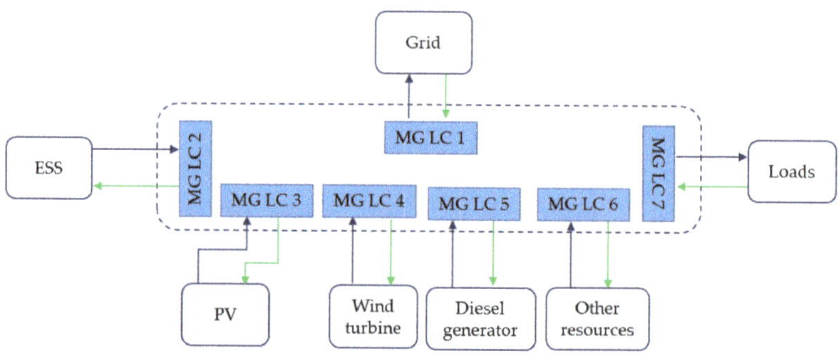

Figure 4. Decentralized control structure.

Hierarchical control approaches are used to provide a compromise between totally centralized and decentralized control structures, and they includes primary, secondary, and tertiary controls. The primary control provides local voltage and current control, as well as power sharing control. It generally follows the instructions of higher-level controllers. The secondary control is responsible for the power management of the system. It is also used for microgrid synchronization to the main grid when switching from islanded to grid connected mode. Tertiary control is used to control the power flow. It can also be used for other objectives, such as islanding detection. The hierarchical control approach is the most widely used conventional method, and its objective is to enhance the efficiency and effectiveness of microgrid operation [17]. However, hierarchical control is challenging with the consideration of the intermittency of RES. Recent studies have extensively focused on hierarchical control approaches to improve the energy management aspects of microgrid systems. A typical hierarchical control scheme is illustrated in Figure 5.

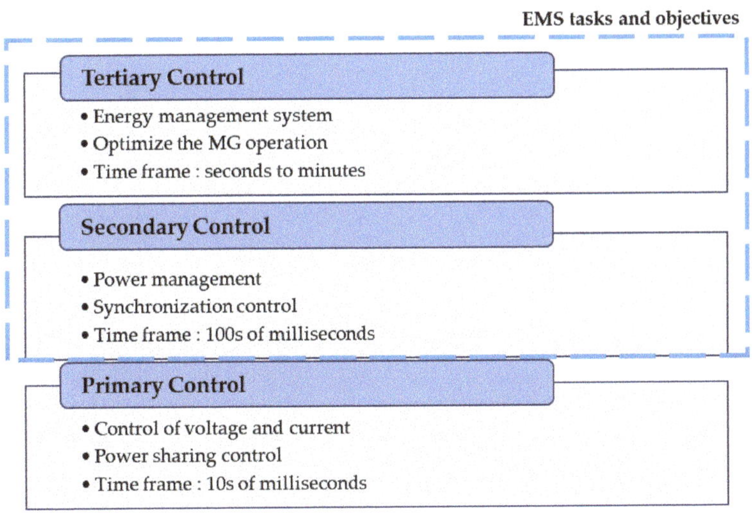

Figure 5. Hierarchical control.

Table 1 summarizes the features, advantages, and disadvantages of implementing EMS based on different control aspects.

Table 1. Comparison of control methods used in EMS.

	Features	Advantages	Disadvantages
Centralized control	Centralized control has complete knowledge of the entire system and is in charge of ensuring its optimal operation.	It provides wide control over the entire system. Established control approach. Simple architecture. Easy to implement and maintain. It ensures optimal decision.	It requires a high performance computing unit and communication network. The failure of central control could cause the entire system to fail. Computational complexity is high. Low flexibility.
Decentralized control	Functions provided by centralized control are realized in a decentralized way. Local decisions contribute to achieving the goal.	It does not require a high performance computing unit and a high-level connectivity. Easy realization of plug-and-play functionality.	It requires an effective method to ensure corporation among local controllers. Low performance compared to centralized control due to low response time and incomplete information about the total microgrid system installation. High implementation complexity.
Hierarchical control Based methods	Each level provides supervisory control over lower-level systems. Three layers: tertiary, secondary, and primary control layers. The bandwidths of different control levels are separated.	Combining the centralized and decentralized controllers. Higher levels attempt to optimize the microgrid operation. Local controls regulate the voltage and current. It simplifies modelling and analysis of microgrid systems.	Proper coordination of all three layers is required.

3. Microgrid Energy Management: Problem Formulation

Microgrid energy management is used to either minimize or maximize an objective or set of objectives while ensuring the constraints of individual units and the system as a whole. These objectives are quantitative in nature and usually include cost reduction, emission reduction, increased renewable energy integration, etc. The associated constraints include power balance, individual unit ratings, charge and discharge rates of ESS, maximum and minimum limits of the state of charge (SOC) of ESS, power import and export limits, and other technical constraints of the microgrid. Most of the existing literature focuses on microgrid cost minimization in a single-objective format. The considered cost factors are related to fuel, start-up, shut-down, maintenance, degradation, utility purchases, etc. When several objectives are optimized, the optimization framework is formulated in a multi-objective framework. In such cases, each objective is assigned a weighting factor. These weighting factors are usually assigned based on the significance of individual objectives in relation to the final objective function.

Various solving techniques, such as mixed integer linear and non-linear programming (MILP and MINLP) methods, heuristic optimization methods, etc. are used to solve the optimization problem, sometimes together with rule-based and fuzzy logic control methods to simplify the problem. These optimization strategies use various optimization time windows (horizon) on different time scales. A suitable selection is used to improve the energy management system. Recently, the rolling horizon is considered to reduce the impact of uncertainties from the renewable energy output and load forecasting.

The design of an EMS for a microgrid includes the task of the mathematical formulation of objective functions and constraints, selection of the optimization time horizon and the time step, as well as choosing an optimization technique to solve the problem.

The typical mathematical representation of the EMS problem is shown below:

Objective function:

Minimize the total cost of the microgrid operation;

- Operational cost = fuel cost + maintenance cost + startup cost of the thermal unit + shutdown cost of the thermal unit + cost of buying and selling power to the main grid + load shedding penalty cost + losses cost
- Environmental cost = carbon emission + penalties for emissions
- Energy storage cost = charging cost + discharging cost + degradation cost
- Constraints:
- Power balance: load demand at each time must be equal to the summation of power from microgrid resources and receiving/sending power from the main grid.
- Emission constraints: emissions caused by each fossil-fueled thermal generators cannot exceed the maximum limits at each time.
- Capacity limits: each RESs, ESS, and interconnection has a maximum and minimum capacity during the operating mode.
- Limit of ESS: charging and discharging power rates for batteries during operation mode and the operating SOC range must be limited as it may affect battery life time.
- Operating reserve: extra storage and generation capacity
- Generator start/stop limits: the number of generator starts/stops cannot exceed a certain number.
- Ramp rate power limit: the maximum power fluctuation of each unit is defined.

System variables:

- Load profile: the demand forecast varies according to time, geographical location, season, weather, and other factors.
- PV and wind sources: the wind and PV power availability depends on wind speed forecasts and solar irradiation forecasts, respectively. Seasonal and local weather impacts these forecasts, and there is always some uncertainty associated with the forecasts.
- Electricity price: it is related to the price of the buying/selling power to the main grid. Prices may be time-sensitive.

The energy management problem in a microgrid becomes a mono objective when a single cost function is presented. The problem becomes multiple objectives when it simultaneously presents a solution to the competing technical, economic, and environmental problems. The weighting coefficients of each function must be properly defined when multiple objectives, such as operational cost minimization, emission reduction, and other objectives, are taken into account for the optimization problem. Effectively setting the weighting factors of the objective function is still being researched.

In addition to the typical objectives and constraints, there are other elements that needs to be incorporated into the MG EMS. Some such aspects include real-time or time-varying electricity tariffs and demand response, which add further benefits for both energy providers and consumers. From the consumer perspective, consumer comfort and a profitable electricity bill are important considerations. From the energy providers' perspective, efficient load profile reshaping is essential. Techniques such as peak clipping, valley filling, and load shifting can be employed to successfully execute the reshaping of the load profile while considering factors such as cost, dependability, control strategies, targeted customers, and supporting infrastructure [18]. The impact of electric vehicles (EVs) is an emerging factor because the use of EVs is expected to significantly increase in the next decade, causing a major increase in demand and demand pattern. Energy storage available in EV batteries can be used in the MG EMS with proper infrastructure for EV charging. Consumer comfort maximization can be defined as one of the objective functions in the formulation of the MG energy management problem, which includes demand response and EV energy storage. However, it makes optimization tasks computationally complex. The proposed method in [19] ensures customer satisfaction by optimal allocation of demand in a distribution feeder using autonomous decision-making entities.

4. Microgrid Energy Management: Solution Approaches

The selection of EMS methods depends on the microgrid system and the requirements. The solution methods for energy management problems can be classified in various ways. In this paper, those EMS solution methods are classified as shown in Figure 6.

Approaches for solving basic energy management problems

- Mixed Integer Linear and Non Linear Programming (MILP & MINLP)
- Heuristic optimization methods
  - Genetic Algorithm
  - Particle Swarm Optimization
- Rule based control
- Fuzzy logic control

Approaches for Uncertainty Management

- Stochastic optimization
- Robust optimization
- Chance Constrained Programming (CCP)
- Model Predictive Control (MPC)

Figure 6. Classification of microgrid EMS methods.

4.1. Mixed Integer Linear and Non-Linear Programming Methods

Mixed integer programming methods deal with optimization problems where variables can be discrete and continuous. Therefore, the methods fit perfectly for applications in microgrid EMS. In mixed integer linear programming (MILP) based EMS, mathematical models of microgrid components are developed using MILP to optimize the cost function. The MILP model considers wind speed and irradiation, loads, and cost parameters of the components. For MILP methods, the objective function and constraints are linear, but for mixed integer non-linear programming (MINLP) methods, they are non-linear. Typically, MINLP models make use of approximations to obtain a linear model. In MINLP models, continuous variables are the power imported/exported at PCC, the power generation from available generators, and the power injection of the ESS. When considering the power flow equation of microgrids, they introduce non-linearity and complexity to the energy management problem and increase the computational burden.

In [20], considering a distribution network with radial structure, power flow of a line (referred to as $S_{ij}(t)$) is expressed using Equation (1), which shows non-linearity.

$$S_{ij}(t) = V_i(t)\ I^*_{ij}(t) \tag{1}$$

$$[V_i(t)]^2 \times [I_{ij}(t)]^2 = P_{ij}{}^2 + Q_{ij}{}^2 \tag{2}$$

Thus, voltage is considered equal to a constant value as voltage changes are very small. With the said assumption, a linear approximation of the current in power flow is calculated using the piecewise linearization method. In a similar manner, nonlinear constraints and equations are converted into linear form to simplify.

Mixed integer linear and non-linear optimizers are often used because of their high efficiency and modelling flexibility. However, complex problems with a large number of variables take a long time to calculate. The MILP and MINLP-based EMS are applied in a variety of ways in the reported literature [21,22].

4.2. Heuristic Optimization Methods

There are many heuristic optimization methods that can be applied to the MG EMS optimization problem. Among them, the genetic algorithm (GA) and particle swarm optimization (PSO) are used frequently to solve energy management optimization problems. Similar techniques, such as the ant colony algorithm [23], Lagrange algorithm [24], crow search algorithm [25], and simulated annealing [26] are also utilized for microgrid energy management.

a. Genetic Algorithm

The genetic algorithm (GA) is a frequently used heuristic optimization method to solve optimization problems, and it has the capacity to optimize multi-dimensional problems. The genetic algorithm is developed through various stages, as shown in Figure 7. The GA repeatedly modifies a population of individual solutions. Individuals can be defined as arrays, trees, or lists of variable values to be optimized. Solutions are typically represented as strings of zeros and ones in binary, but other representations are also available. In a binary implementation of genetic algorithms, GA starts from a population of randomly created individuals. The objective function of every individual in a population will be evaluated and ranked. Selection determines which ones are to be selected from the current population and allowed to reproduce. There are various approaches to implementing selection in GA, such as roulette selection, tournament selection, stochastic universal selection, and Boltzmann selection. Frequently used genetic operators are crossover and mutation. Crossover is a recombination operator by swapping the values between two strings to create new solutions from the existing population. There are various crossover methods, such as uniform crossover, arithmetic crossover, permutation encoding crossover, value encoding crossover, tree encoding crossover etc., being applied based on the application [27]. A new population is full of individuals after selection and crossover. Some are formed by crossover, while others are directly copied. Mutation is the change of a small amount or the replacement of a value with a new one in order to ensure genetic variability among the population. The probability of a mutation is typically 1 to 2%.

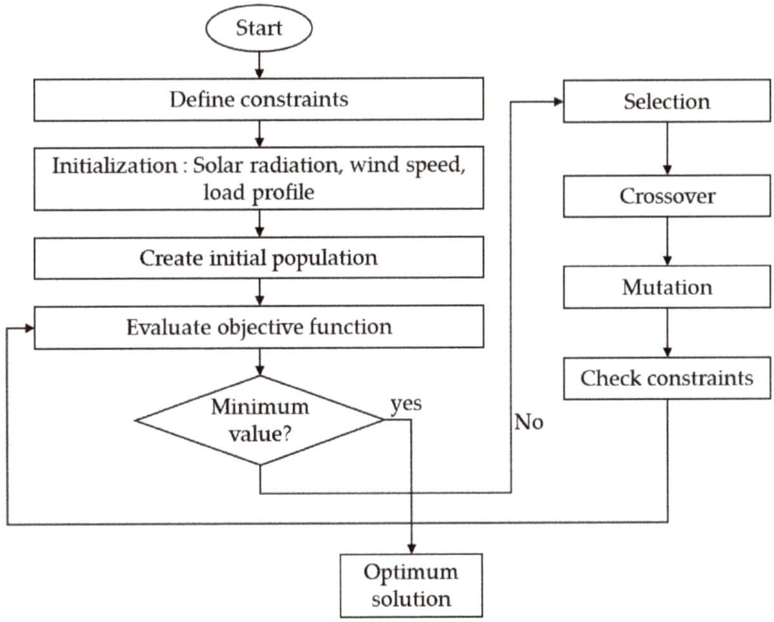

Figure 7. Basic GA algorithm for MG EMS.

In order to solve the optimization problem for MG EMS, GA considers the system variables, constraints, parameters, and the objective function, as shown in Figure 7. The GA is developed to schedule the generators, battery storage systems, and controllable loads. The total running cost, microgrid emission, and other objectives are optimized while satisfying all equality and inequality requirements using a suitable ON or OFF state. The GA optimizes with continuous or discrete variables, and it can also optimize variables in an extremely complex problem. It can handle a large number of variables. However, the challenges associated with GA are the long calculation time involved and the possibility of ending up with one or more solutions. Each run could produce a different result. In the case of EMS in microgrids, it is challenging to find the optimal solution through GA.

b. Particle Swarm Optimization

Particle Swarm Optimization (PSO) is a population-based heuristic optimization technique and can be deployed in a wide range of microgrid-related applications. Particles use both the personal best solution and the best solution found by the swarm to collectively move towards the optimum. Figure 8 illustrates the typical PSO algorithm used for microgrid energy management applications.

The system starts with a population of random solutions and then updates generation to search for an optimum solution. It can handle a wide range of problems while achieving a set of goals, such as minimizing energy costs and reducing emissions. It is feasible to update the PSO objective function at a smaller interval. In PSO methods, it is important to define parameters such as population size, maximum number of iterations, etc. The initial particles are selected randomly between the minimum and maximum values of the variables. Each particle is evaluated using the objective function to find the best solution. The performance of PSO depends on the selection of suitable PSO parameters and stopping criteria. The main advantage of PSO-based EMS is its fast convergence time, which is essential for real-time energy management applications. The PSO is also used by many researchers to solve microgrid optimal sizing problems [28,29].

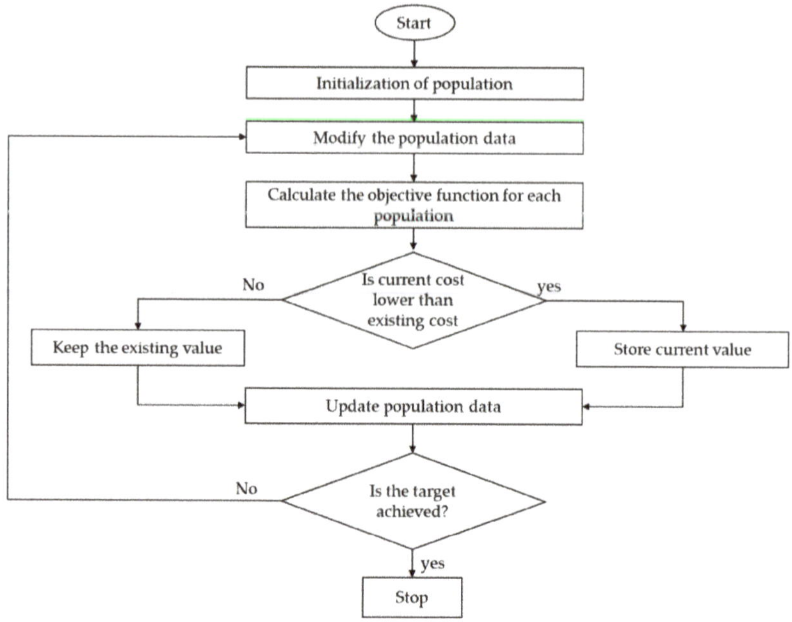

Figure 8. Typical PSO algorithm.

4.3. Rule-Based Methods

With the rule-based methods, the inter-dependability of the resources of the microgrid are defined using predefined logical rules. If-then-else rules are applied for assigning values to binary decision variables used in the optimization procedure of the microgrid operation. Load, PV generation, and wind generation values are not affected by values from the present or the past. However, changes in battery storage have an impact on future schedules. Rule-based algorithms can also be used to develop a method for BESS scheduling to determine the optimal charge/discharge of the battery. Such methods use the energy generated by RESs and the battery SOC level to determine whether the battery energy storage system (BESS) should be charged or discharged. Rules also ensure that SOC is kept under allowable levels while performing perfect dispatch. The analysis takes into account a variety of operational modes. For example, in the case of islanded microgrids, if the output power from renewable resources (solar and wind) is higher than the power demand, the excess power will be used to charge the battery storage up to its maximum capacity. If the generation is higher than the demand and the battery is at its maximum capacity, the remaining power will be discarded to a dump load. Similarly, rules are developed for the optimum operation of microgrid systems. The rule-based method is easy to execute on various storage types once the essential rules have been established [30,31]. The rule-based method allows for a significant reduction in computational complexity.

4.4. Fuzzy Logic Control Methods

Figure 9 shows the block diagram of a fuzzy logic system. The fuzzy logic controller (FLC) contains mainly three parts: the fuzzifier, the fuzzy interface, and the defuzzifier. The fuzzifer converts the input into a linguistic variable, and this process is referred to as fuzzification. The defuzzifier returns the output by converting from linguistic variable, and the process is called defuzzification. The interface system is a rule-based system. Fuzzy controllers do not require complex mathematical modelling.

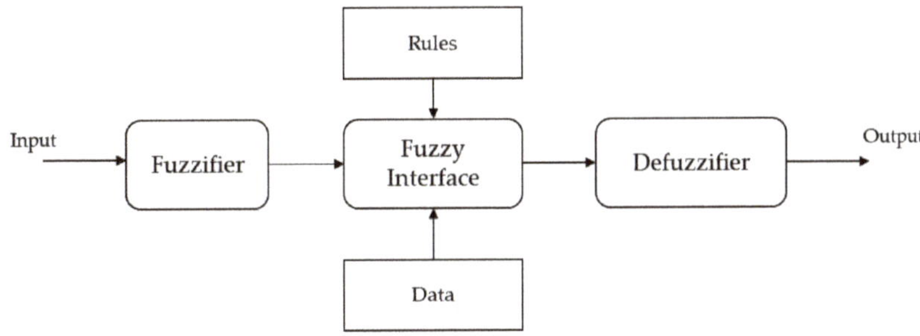

Figure 9. Fuzzy logic system.

Fuzzy logic-based energy management is used for various objectives. Fuzzy logic controller generates the output whether the battery should be charged or discharged, as well as the charging or discharging rates. The fuzzy logic control is designed to meet energy demands and maintain the SOC of the energy storage level within predefined margins while attempting to optimize the utilization cost and lifetime of the ESS [32]. The fuzzy controller is also used to find the status of the power received or supplied from the grid. An optimal fuzzy-based EMS for a residential grid-connected microgrid is used to minimize grid fluctuation and to preserve the storage system lifetime in [33].

In microgrid operations, hybridization of fuzzy logic with other methods such as GA and PSO are used. The problem of energy flow management system optimization in a microgrid is investigated using a fuzzy-GA paradigm in [34]. A hybrid approach combining a fuzzy approach and a neural network is used in EMS presented in [35].

The advantages and disadvantages of the energy management solution approaches discussed above are summarized in Table 2.

Table 2. Advantages and disadvantages of optimization techniques.

Techniques	Advantages	Disadvantages
Mixed integer linear and non-linear programming methods	Availability of efficient software packages. Most flexible modelling. Optimal solution.	Computational complexity.
Generic algorithm	Possibility to use complex formulation. It can handle many objectives and constraints. Widely used in many fields.	GA is unable to ensure mathematical optimality in its output.
Particle swarm optimization (PSO)	It has fast convergence time. Commonly used in the sizing of distributed generators. It can handle a wide range of problems while achieving a set of goals.	PSO is unable to ensure mathematical optimality in its output.
Rule-based methods	The approach allows for a significant reduction in computational complexity. The method is easy to execute on various storage types once the essential rules have been established.	Solution could be a sub-optimal solution.
Fuzzy logic control methods	Gain more flexibility. It can be easily incorporated with other methods.	Solution could be a sub-optimal solution. High-quality processing unit is required.

5. Uncertainties in Microgrid Energy Management

Power generation from the RESs offers an intermittent and uncertain power supply. Solar and wind are the most popular and widely used resources among all the renewable energy resources used in microgrid applications. However, the intermittent nature of solar and wind energy is always a challenge. Solar energy is only available during the day, and it also varies with other factors such as cloud movements and shadow. Wind patterns change according to the weather. Consumer loads connected to the grid are also continuously varying, and these variations can become more complex with the introduction of demand response and EV charging. The high intermittency leads to an uncertain operational environment for microgrids. Therefore, one of the main challenges is to handle the uncertainty of renewable energy generation and power demand.

In this regard, it is important to properly model the uncertainties in the parameters and components. Researchers consider a variety of sources of uncertainty, such as wind power, load demand, electricity prices, PV generation, EV demand, etc. [36]. In MG EMS, the uncertainty from renewable energy sources and load demand are important factors. To address uncertainty management, modelling the uncertainty of renewable sources and load becomes the consequential issue. Accurate modelling has a high effect on the operational cost of a microgrid. Modelling uncertainty is always a challenge; hence, several approaches are employed to model these uncertainties with respect to their applications. This section provides an overview of all recent uncertainty modeling approaches used by an EMS.

a. Monte Carlo Simulation (MCS)

The MCS is used to calculate the probabilities of various outcomes in a process that is difficult to forecast because it contains random variables. This method can accurately handle the uncertainty variable. For each input parameter, a sample is generated using its probability density function (PDF), and the sample generation process is repeated for many iterations. Therefore, the method is computationally complex. Most of the studies are focused on developing uncertainty models for PV, wind power, and load demand [12].

b. Worst Case Scenario Method

Even though it is not a new concept, the worst-case scenario approach is frequently used in recent studies. The worst-case scenario approach restricts the range of the random variables to a set of predetermined uncertainty with defined upper and lower boundaries. Prediction intervals (PIs) are calculated to evaluate the measure of prediction uncertainty. Upper and lower limits are used to define PIs [11].

c. Point Estimate Method (PEM)

The PEM is one of the approximate methods with a low computation burden. The method focuses the statistical data of a random variable on a specific number (K) of points in order to create a connection between input and output variables. Solar radiation and wind speed are treated as two random variables, and the function is developed using power flow equations in [37]. In [38], PEM is used to determine power exchanges between MGs and evaluates the optimal solutions in terms of accuracy and computational effort.

d. Fuzzy Method

Each uncertain parameter can be assigned a degree of membership based on fuzzy theory by using membership functions. After a suitable fuzzy membership function is applied to each parameter, the defuzzification will be carried out. The fuzzy method is used to model the uncertainty in forecasting day-head demand in [39]. Although uncertainty is handled in fuzzy systems, the issue of randomness is not properly accounted for. Approaches, such as probabilistic fuzzy systems, have been introduced for overcoming this issue [18].

e. Autoregressive Moving Average

It is another model used in recent days to model uncertainties from load demand and wind power. The autoregressive moving average model is a combination of auto regression and moving average. This method can be used to forecast future estimates of a variable if historical data of the variable with uncertainty is presented by a time series, such as load demand, wind, etc. A significant amount of historical data, as well as data mining and analysis, are required for developing proper autoregressive models, and the predictions with these models are valid only over a short horizon [11].

Other methods, such as kernel density estimation, hyper-heuristics, and two stage scheduling strategy, are also used to model these uncertainties. Each model has its own advantages and disadvantages that determine its application.

There are various approaches that could be used to deal with different sources of uncertainty. Generally, optimization under uncertainties can be broadly categorized as stochastic programming [40,41], robust optimization [42,43], and other methods, such as model predictive control and chance constrained programming. These methods are implemented as either a single-layered or multi-layered framework.

5.1. Stochastic Optimization

In stochastic programming methods, uncertain parameters are described using probabilistic distributions. Stochastic optimization, which is based on statistical data, is widely used for energy scheduling under uncertainty due to its effective performance in the case of uncertainties [44]. Stochastic programming can be classified into four main methods, including the three-point estimating technique, the Monte Carlo simulation-based method (MCS), the scenario-based modelling approach, and the approximate analytical method. The scenario-based modelling approach is utilized to handle the uncertainties from wind in [45–47].

However, there are significant limitations to stochastic optimization in some situations, such as the large presence of uncertain data, dependence between uncertain parameters, and a lack of historical data. The computational complexity of the model also increases along with its scale. Stochastic programming methods may lead to an infeasible solution due to the constrained violation [12].

5.2. Robust Optimization (RO)

Robust optimization (RO) is an interval-based approach, and RO methods do not require prior knowledge of the probability distribution of the uncertain parameter. The RO method addresses data uncertainty by considering a single worst-case scenario over an uncertainty set, and it can improve the performance of MG EMS even under the lack of full information on the nature of uncertainty. The typical extreme cases in real-world applications can be included in the uncertainty set in the RO methods. Robust optimization has significantly reduced the computational complexity compared with stochastic approaches [48]. The RO method has received attention for being able to handle the uncertainty optimization problem. However, complexities in deriving the uncertainty set can lead to a computationally intractable solution. Additionally, focusing on the worst-case scenario can lead to a more conservative resource-handling option degrading the benefits achievable through an optimized solution.

A two-stage robust optimization approach is proposed for the islanded microgrid system to reduce the uncertainty impacts from energy sources and loads in [49]. Uncertain parameters were used to classify decision variables for two stages: (1) the day ahead stage decision (pre-scheduling) and (2) the real-time stage decision (rescheduling). This approach is moderately effective in reducing the impact of uncertain factors.

5.3. Chance Constrained Programming (CCP)

The chance constrained programming (CCP) approach is also used to solve energy management problems under uncertainties. It is a mathematical program containing chance constraints which only needs to be satisfied with a probability. The requirement for power balance in the microgrid is formulated as chance constraints. Approximated methods have been commonly employed to solve the CCP. In general, CCP is employed only for special cases due to being computationally challenging, conservative, or incapable of guaranteeing the satisfaction of chance constraints [50].

5.4. Model Predictive Control

The traditional deterministic frameworks have a feedback mechanism to adjust the initial dispatch solution to compensate for variations in the uncertain decision variables. Among them, model predictive control (MPC) is gaining considerable attention in microgrid systems as a promising control scheme with several advantages, such as the possibility of incorporating optimization techniques and the ability to integrate the constraints and disturbances in forecasted control decisions. The MPC is a discrete time control scheme in which each time step solves an open-loop optimal control problem for a chosen control horizon. Figure 10 illustrates the block diagram of a typical MPC structure.

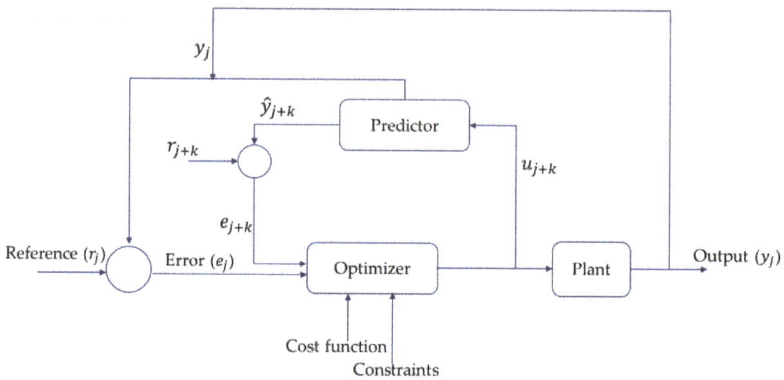

Figure 10. Basic structure of MPC.

Model predictive control is usually implemented by developing a model with relevant and controllable variables and then minimizing a cost function between reference values and candidate values of controlled variables. The minimal difference to actuate the next period is chosen. The general MPC-based optimization process is shown in Figure 11.

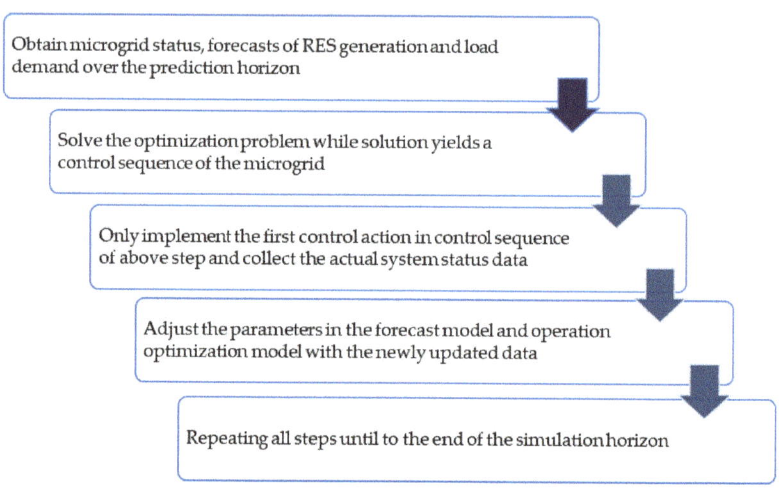

Figure 11. MPC-based optimization process.

The MPC framework models for microgrids have advantages, as they use the dynamic behavior of microgrid components, particularly the battery charge and discharge models [51]. Typical control methods are no longer effective against fluctuations, but MPC has received more attention due to its flexibility to include constraints and non-linearity [52]. Another advantage of MPC over other control techniques is its simple expansion to the multivariable case. The MPC method can provide a receding prediction horizon with a feedback mechanism that also effectively reduces the impacts of uncertainties [53]. Model predictive control is typically used for MG EMS with a prediction horizon of 24 h, considering the daily periodicity of generation and load profiles. The performance of MPC is affected by the selection of the prediction horizon. One of the main drawbacks of MPC is that computational complexity increases with a larger prediction horizon [54]. With significant development in recent years, MPC with its variations is used at different levels.

A model predictive control-based energy management is proposed in [55] to provide an economically optimal operation, and the system is represented by a linear invariant discrete time model. In this study, the minimization of the cost of energy consumed from the main grid is considered to be an optimization objective. When renewable generation is insufficient to supply demand, the battery will start supplying the load. If this energy also becomes insufficient, EMS has to acquire energy from the main grid. Another economic MPC is proposed in order to achieve optimal economic performance in terms of the operational cost of a microgrid in [56].

A MPC rolling horizon approach for the optimal operation of a microgrid for residential network is proposed in [57]. The MPC problem is formulated using a MILP model used to optimize the total cost of the microgrid. The objective of the problem is to minimize the total cost for the DER system over the prediction horizon. Heat and electricity storage terms are additionally included in the objective function. The overall planning and performance could have been improved if the uncertainty on solar irradiance and energy demands was considered in the modelling of the network.

The conventional MPC approach, with techniques such as robust optimization and stochastic optimization methods, directly incorporate uncertainty in the optimization models to achieve effective and efficient operation of microgrids.

Stochastic MPC methods are used to take uncertainty due to the stochastic behavior of renewable energy generation and demand into consideration. A two-stage stochastic programming-based MPC strategy for microgrid energy management under uncertainties is proposed in [45]. Stochastic programming is used for the inclusion of uncertainties in the optimization model. The MPC can effectively compensate for the disturbances in load profile and power delivered from renewable sources that are connected to the AC microgrid through a feedback mechanism. The cost of the microgrid is commonly formulated using start-up/shut down costs, the fuel cost of diesel generators, the degradation cost of a battery, and the power purchase cost from the utility grid. Here, uncertainties of solar PV, wind, and load are also considered in the formulation of microgrid cost. In the first stage, decisions are made before the actual realization of available uncertainty. In the second stage, correction decisions are made after a particular realization.

Sliding mode control (SMC) is a reliable control technique that can address uncertainties. The combination of MPC and SMC can be applied for the effective and efficient operation of microgrids. A suboptimal second-order SMC is designed as a low-level controller to track the power references generated by a high-level MPC component for EMS [58]. Nonlinear MPC is implemented using nonlinear models that include non-linear constraints on the state and control variables, as well as the cost functions. The nonlinear MPC type control is implemented for energy management of the batteries and load shedding purposes in [59].

The reviewed literature shows successful applications of using MPC for handling uncertainties related to microgrid energy management.

Researchers in this field have been motivated by recent developments in machine learning techniques to work on more realistic predictive strategies. The MPC is formulated based on the predictions generated by the predictive model and possible desired targets. The MPC performance is directly impacted by the modelling quality and accuracy of the predictive model. This is one of the challenges when using the MPC scheme. When the MPC strategy is used, the prediction method must be considered. Various methods, such as physics-based models, the Markov chain, neural networks, the Kalman filter, etc., can be used for prediction. For example, in terms of Markov chain prediction, the future value depends on the current value and the transition probabilistic matrix, which is calculated based on a historical statistical data set. In real-time operation, actual measurements are continuously updated. To predict load demands and renewable outputs, online learning Markov chain prediction can be used. However, quite often, a physics-based model is needed for repetitively predicting the MG system behavior. It constitutes high computational complexity, depending on the type and complexity of the models used.

In MPC-based energy management problems, the design, arrangement, and allocation of all the control objectives, along with the proper constraints, can reduce the computational burden. Furthermore, the MPC-based energy management problem formulation can be simplified by using methods such as fuzzy logic control, rule-based methods, PSO, etc.

The MPC optimization problems are resolved by moving the time horizon window ahead. Once new predictions are available, the optimization problems can then be recalculated and resolved. The time horizon is often determined according to multiple factors. One of the most challenging aspects of MPC is that it needs to deal with short sampling times for the power electronic technology, and it also needs long-time horizons to improve the robustness and accuracy of the model. Combining short and long-time horizons at different stages is an approach used to obtain better system performance. Receding time horizon methods are also found in recent research because they aid in dealing with uncertainty and achieving effective performance [60].

Scalability is crucial to address the changing requirements when moving from industrial plants to very large-scale systems with operational and managerial independence. Distributed model predictive control (DMPC) [61] attempts to address this issue through the development of multi-level, multi-scale models. The goal is to avoid redesigning the overall energy management system (EMS) when the additional sources and storage are connected. In DMPC, the power of each energy source is optimized using an individual subsystem-based MPC to make the plug-and-play property easier to achieve and to reduce the computational load [60].

DMPC could be a suitable strategy for large-scale systems to solve the resulting economic optimization problem. Many researchers are focusing on using stochastic and distributed model predictive control techniques to optimize large-scale microgrid systems [59]. Another study, in [62], proposes fully distributed water-filling distributed algorithms that scale to large-scale situations. A decentralized charging control is formulated for large populations of plug-in EVs [63]. Approaches used to handle uncertainties in microgrid system are summarized in Table 3.

Table 3. Approaches used to handle uncertainties in microgrid system.

Proposed Approach	Modelling Uncertainty	Uncertainty Handling	Scalability Handling Possibility
Optimal probabilistic energy management in a typical microgrid based on robust optimization and point estimate method [37]	Uncertainties of wind, solar, and load are used.	PEM and RO are used. The data determined from PEM are used in the PSO-based energy management algorithm. RO generates and transfers the load demand scenarios to the PSO algorithm.	The proposed algorithm is used in order to perform an optimal operation on a low voltage (LV) MG, including renewables and conventional DGs, as well as a battery bank.
Two-stage stochastic programming based MPC strategy for microgrid energy management under uncertainties [45]	Load, PV, and wind uncertainties are used.	Two stage scheduling strategy is used. The first stage makes a decision before the actual reality of the uncertainty becomes available, and the second stage makes a correction decision to compensate for infeasibilities from the first stage.	The proposed method combines the advantages of both two-stage SP and MPC.
A two-stage robust optimization method based on the expected scenario for islanded microgrid energy management [50]	Uncertainties of wind, solar, and load are considered.	Two stage scheduling strategy is used. Prescheduling stage and rescheduling stage are applied to reduce the impact of uncertain factors.	To manage various constraints during the optimization process and ensure the feasibility of individuals in the evolving population, a constraint-handling technique is developed.

Table 3. *Cont.*

Proposed Approach	Modelling Uncertainty	Uncertainty Handling	Scalability Handling Possibility
Optimal operation of a smart residential microgrid based on model predictive control by considering uncertainties and storage impacts [59]	Uncertainties from solar, wind, load, and electricity price are used.	The MILP problem is incorporated into a MPC framework for compensating the potential disturbances.	Stochastic and distributed model predictive control techniques can be used to optimize large-scale microgrid systems.
Distributed MPC for grid-connected microgrid power management [60]	Uncertainty related to the availability wind and load are considered.	The MPC-based EMS is implemented under a distributed framework. Receding horizon methods are used to mitigate uncertainties.	The optimization problem is decomposed into several small-scale nonlinear continuous optimization problems and several integer programming problems.
A two-layer stochastic MPC scheme for microgrids [64]	Uncertainties from wind and PV are considered.	Shrinking-horizon MPC is implemented. A stochastic MPC runs at a higher frequency at the lower layer to compensate for uncertainties and maintain the energy exchange as close as to the desired value over each sampling period.	Stochastic MPC is used with high-level off-line economic optimization.
Stochastic programming and market equilibrium analysis of microgrid energy management systems [65]	Load, PV, and wind uncertainties are used.	Two-stage stochastic programming model is used. In the first stage, the decision for investment in microgrid devices is deter-mined, and energy management strategies are determined in the second stage.	A general algebraic modeling system is designed for solving large-scale, complex optimization problems.
Energy management system for hybrid PV-wind-battery microgrid using convex programming, model predictive and rolling horizon predictive control with experimental validation [66]	Uncertainties from solar, wind, load and electricity price are used.	A rolling horizon predictive controller with a MPC at the lower control layer with a one-minute sampling time reduces the impact of prediction and model uncertainties.	A rolling-horizon predictive controller does not require a complex optimization process.
Analysis of robust optimization for decentralized microgrid energy management under uncertainty [67]	Uncertainty related to the availability wind and load are considered. Prediction intervals are used.	The impact of different levels of uncertainty is evaluated.	Agent-based modelling (ABM) is used to describe the system, with each stakeholder modeled by an individual agent.
Robust optimization for dynamic economic dispatch under wind power uncertainty with different levels of uncertainty budget [68]	Wind uncertainties is used.	A robust optimization method with an adjustable uncertainty budget with different levels is proposed.	Constraint handling technique is also proposed to handle various constraints and ensure the feasibility of individuals in the evolutionary population.
Robust optimization of microgrid based on renewable distributed power generation and load demand uncertainty [69]	Uncertainties of wind, solar, and load are considered.	A two stage scheduling strategy is used. Robust adjustment parameters are optimized to make the microgrid have a reasonable robustness.	The robustness of grid operation is guaranteed by the proposed solution, which is more in line with technical realities and has better practical value.

6. Application of Artificial Intelligence and Machine Learning

The use of machine learning and data-driven techniques for MG energy management is becoming increasingly popular due to the recent development of machine learning (ML) and artificial intelligence (AI), as well as the availability of advanced processing in modern control systems. For example, ML has been introduced as a methodology either for energy management in microgrids or for forecasting weather conditions and loads. A hybrid approach of a nonlinear MPC controller integrating machine learning models is presented in [70]. A two-layer ensemble machine learning technique is used to construct a data-driven multi-model wind forecasting system [71]. Utilizing the statistically different characteristics of each machine learning algorithm is the focus of this two-layer model. Additionally, many of the heuristic optimization techniques used in MG EMSs are considered under the umbrella of AI. The opportunities for ML extend far beyond forecasting, model improvement, and adaption.

7. Conclusions

The literature review highlighted energy management methods, modelling uncertainties, and forecast uncertainty management in microgrids. The microgrid energy management systems are developed considering factors such as the operating mode, control system, intermittent nature of renewable sources, economic and environmental aspects, and other factors. Different approaches can be used to design microgrid energy management. It is necessary to choose proper methods based on the required application. This paper reviewed recent energy management strategies for microgrids. Several EMS methods were discussed, including MILP and MINLP methods, heuristic optimization methods, rule-based methods, fuzzy logic control methods, MPC methods, and others based on numerous studies. The suitable EMS method is determined by the microgrid system and its requirements. The MILP methods deal with optimization problems when variables may be either discrete or continuous. However, it takes a long time to calculate complex problems with a large number of variables. The GA and PSO are frequently used to solve heuristic optimization problems due to convenience in application to multidimensional problems. Finding the optimal solution through GA is difficult. Fast convergence time, which is necessary for real-time energy management applications, is the main advantage of PSO-based EMS. Many researchers also use the PSO to address issues with microgrid optimum sizing. Rule-based methods allow for a significant reduction in computational complexity compared to other methods. Fuzzy logic control methods gain more flexibility, and they can be easily incorporated with other methods.

It has been identified that the uncertainty present in microgrid systems must be considered for proper energy management. Recently, to model the uncertainty, the Monte Carlo simulation, worst case scenario method, point estimate method, fuzzy method, and auto regressive method were used. Few uncertainty modelling techniques depend on the latest data recorded for future projection. Some uses present forecast errors. There is still a lot of scope for the development of new techniques. Many approaches are used in the reviewed research to overcome the uncertainty problem. Generally, uncertainty optimization can be broadly categorized as stochastic programming, robust optimization, and other methods that include MPC. Among those methods, MPC methods have gained more attention, especially in the application of MG EMS because of its ability to manage future behavior as well as the feedback mechanism. The feedback mechanism introduced through the MPC partially compensates for the uncertainty associated with the microgrid system. To adopt future power systems, MPC-based research and development must be established. The MPC has been improved with multi levels and different time horizons, and it has been incorporated into other methods to mitigate uncertainties and to improve performance.

Uncertainty may also result from stochastic electricity price fluctuations and demand response. The MG EMS can be formulated to take into account such other sources of uncertainty; however, when more factors are considered in the formulation of the problem, it becomes more complex. Possible corrective measures to decrease this complexity require more investigations.

The use of machine learning and data-driven techniques for MPC is becoming increasingly popular due to the recent development of machine learning (ML) and artificial intelligence (AI), as well as the availability of advanced processing in modern control systems. There is great potential for applying machine learning (ML) to MG energy management for various tasks to improve solutions and computational efficiency.

This paper highlights the research areas in energy management, considering forecast uncertainty from renewable sources and load in microgrid environment, for further investigation.

Author Contributions: S.V.: writing—investigation, original draft preparation and editing, L.N.W.A., A.D.R. and R.K.: writing—review and editing, L.N.W.A. and A.D.R.: supervision. All authors have read and agreed to the published version of the manuscript.

Funding: This research received no external funding.

Conflicts of Interest: The authors declare no conflict of interest.

Abbreviations

AC	Alternative Current
AI	Artificial Intelligence
BESS	Battery Energy Storage System
CCP	Change Constrained Programming
DC	Direct Current
DER	Distributed Energy Resources
EMS	Energy Management System
ESS	Energy Storage System
EV	Electric Vehicles
FLC	Fuzzy Logic Controller
GA	Generic Algorithm
IEC	International Electrotechnical Commission
MCS	Monte Carlo Simulation
MG	Microgrid
MG LC	Microgrid Local Controller
MG EMS	Microgrid Energy Management System
MILP	Mixed Integer Linear Programming
MINLP	Mixed Integer Non-Linear Programming
ML	Machine Learning
MPC	Model Predictive Control
PCC	Point of Common Coupling
PDF	Probability Density function
PEM	Point Estimate Method
PI	Prediction Interval
PSO	Particle Swarm Optimization
PV	Photo Voltaic
RES	Renewable Energy Sources
RO	Robust Optimization
SOC	State of Charge
SOWGP	Sparse Online Warped Gaussian Process

References

1. Su, W.; Yuan, Z.; Chow, M.-Y. Microgrid Planning and Operation: Solar Energy and Wind Energy. In Proceedings of the IEEE PES General Meeting, Minneapolis, MN, USA, 25–29 July 2010. [CrossRef]
2. Elsied, M.; Oukaour, A.; Gualous, H.; Hassan, R.; Amin, A. An Advanced Energy Management of Microgrid System Based on Genetic Algorithm. In Proceedings of the IEEE 23rd International Symposium on Industrial Electronics (ISIE), Istanbul, Turkey, 1–4 June 2014. [CrossRef]
3. Radosavljević, J.; Jevtić, M.; Klimenta, D. Energy and Operation Management of a Microgrid Using Particle Swarm Optimization. *Eng. Optim.* **2015**, *48*, 811–830. [CrossRef]
4. Chen, Y.-K.; Wu, Y.-C.; Song, C.-C.; Chen, Y.-S. Design and Implementation of Energy Management System With Fuzzy Control for DC Microgrid Systems. *IEEE Trans. Power Electron.* **2012**, *28*, 1563–1570. [CrossRef]
5. Mahmoud, M.S.; Rahman, M.S.U.; Sunni, F.M.A.L. Review of Microgrid Architectures—A System of Systems Perspective. *IET Renew. Power Gener.* **2015**, *9*, 1064–1078. [CrossRef]
6. Zia, M.F.; Elbouchikhi, E.; Benbouzid, M. Microgrids Energy Management Systems: A Critical Review on Methods, Solutions, and Prospects. *Appl. Energy* **2018**, *222*, 1033–1055. [CrossRef]
7. Arunkumar, A.P.; Kuppusamy, S.; Muthusamy, S.; Pandiyan, S.; Panchal, H.; Nagaiyan, P. An Extensive Review on Energy Management System for Microgrids. *Energy Sources Part A Recovery Util. Environ. Eff.* **2022**, *44*, 4203–4228. [CrossRef]
8. Ali, S.; Zheng, Z.; Aillerie, M.; Sawicki, J.-P.; Péra, M.-C.; Hissel, D. A Review of DC Microgrid Energy Management Systems Dedicated to Residential Applications. *Energies* **2021**, *14*, 4308. [CrossRef]
9. Elmouatamid, A.; Ouladsine, R.; Bakhouya, M.; Kamoun, N.E.; Khaidar, M.; Zine-Dine, K. Review of Control and Energy Management Approaches in Micro-Grid Systems. *Energies* **2020**, *14*, 168. [CrossRef]
10. Shayeghi, H.; Shahryari, E.; Moradzadeh, M.; Siano, P. A Survey on Microgrid Energy Management Considering Flexible Energy Sources. *Energies* **2019**, *12*, 2156. [CrossRef]
11. Kumar, K.P.; Saravanan, B. Recent Techniques to Model Uncertainties in Power Generation from Renewable Energy Sources and Loads in Microgrids—A Review. *Renew. Sustain. Energy Rev.* **2017**, *71*, 348–358. [CrossRef]
12. Hong, Y.-Y.; Apolinario, G.F.D. Uncertainty in Unit Commitment in Power Systems: A Review of Models, Methods, and Applications. *Energies* **2021**, *14*, 6658. [CrossRef]

13. Aien, M.; Hajebrahimi, A.; Fotuhi-Firuzabad, M. A Comprehensive Review on Uncertainty Modeling Techniques in Power System Studies. *Renew. Sustain. Energy Rev.* **2016**, *57*, 1077–1089. [CrossRef]
14. Soroudi, A.; Amraee, T. Decision Making under Uncertainty in Energy Systems: State of the Art. *Renew. Sustain. Energy Rev.* **2013**, *28*, 376–384. [CrossRef]
15. Meng, L.; Sanseverino, E.R.; Luna, A.; Dragicevic, T.; Vasquez, J.C.; Guerrero, J.M. Microgrid Supervisory Controllers and Energy Management Systems: A Literature Review. *Renew. Sustain. Energy Rev.* **2016**, *60*, 1263–1273. [CrossRef]
16. Lefort, A.; Bourdais, R.; Ansanay-Alex, G.; Guéguen, H. Hierarchical Control Method Applied to Energy Management of a Residential House. *Energy Build.* **2013**, *64*, 53–61. [CrossRef]
17. Guerrero, J.M.; Vasquez, J.C.; Matas, J.; Vicuna, L.G.; de Castilla, M. Hierarchical Control of Droop-Controlled AC and DC Microgrids—A General Approach Toward Standardization. *IEEE Trans. Ind. Electron.* **2010**, *58*, 158–172. [CrossRef]
18. Kaluthanthrige, R.; Rajapakse, A. Demand Response Integrated Day-Ahead Energy Management Strategy for Remote off-Grid Hybrid Renewable Energy Systems. *Int. J. Electr. Power Energy Syst.* **2021**, *129*, 106731. [CrossRef]
19. Haider, Z.M.; Mehmood, K.K.; Khan, S.U.; Khan, M.O.; Wadood, A.; Rhee, S.-B. Optimal Management of a Distribution Feeder During Contingency and Overload Conditions by Harnessing the Flexibility of Smart Loads. *IEEE Access* **2021**, *9*, 40124–40139. [CrossRef]
20. Zaree, N.; Vahidinasab, V. An MILP Formulation for Centralized Energy Management Strategy of Microgrids. In Proceedings of the Smart Grids Conference (SGC), Kerman, Iran, 21–22 December 2016. [CrossRef]
21. Amrollahi, M.H.; Bathaee, S.M.T. Techno-Economic Optimization of Hybrid Photovoltaic/Wind Generation Together with Energy Storage System in a Stand-Alone Micro-Grid Subjected to Demand Response. *Appl. Energy* **2017**, *202*, 66–77. [CrossRef]
22. Tenfen, D.; Finardi, E.C. A Mixed Integer Linear Programming Model for the Energy Management Problem of Microgrids. *Electr. Power Syst. Res.* **2015**, *122*, 19–28. [CrossRef]
23. Trivedi, I.N.; Thesiya, D.K.; Esmat, A.; Jangir, P. A Multiple Environment Dispatch Problem Solution Using Ant Colony Optimization for Microgrids. In Proceedings of the International Conference on Power and Advanced Control Engineering (ICPACE), Bengaluru, India, 12–14 August 2015. [CrossRef]
24. Tran, H.G.; Thao, N.G.M.; TonThat, L. Energy Management and Optimization Method Based on Lagrange Multiplier for Microgrid with Considerations of Electricity Price and Vehicle. In Proceedings of the 10th Global Conference on Consumer Electronics (GCCE), Kyoto, Japan, 12–15 October 2021. [CrossRef]
25. Papari, B.; Edrington, C.S.; Vu, T.V.; Diaz-Franco, F. A Heuristic Method for Optimal Energy Management of DC Microgrid. In Proceedings of the IEEE Second International Conference on DC Microgrids (ICDCM), Nuremburg, Germany, 27–29 June 2017. [CrossRef]
26. Angelim, J.H.; Affonso, C.M. Energy Management on University Campus with Photovoltaic Generation and BESS Using Simulated Annealing. In Proceedings of the IEEE Texas Power and Energy Conference (TPEC), College Station, TX, USA, 8–9 February 2018. [CrossRef]
27. Provata, E. Development of Optimization Algorithms for a Smart Grid Community. Master's Thesis, Technical University of Crete, Chania, Greece, 2014.
28. Grisales-Noreña, L.F.; Montoya, D.G.; Ramos-Paja, C.A. Optimal Sizing and Location of Distributed Generators Based on PBIL and PSO Techniques. *Energies* **2018**, *11*, 1018. [CrossRef]
29. Okhuegbe, S.N.; Mwaniki, C.; Akorede, M.F. Optimal Sizing of Hybrid Energy Systems in a Microgrid: A Review. In Proceedings of the Sustainable Research and Innovation Conference, Nairobi, Kenya, 5–6 October 2022.
30. Bukar, A.L.; Tan, C.W.; Yiew, L.K.; Ayop, R.; Tan, W.-S. A Rule-Based Energy Management Scheme for Long-Term Optimal Capacity Planning of Grid-Independent Microgrid Optimized by Multi-Objective Grasshopper Optimization Algorithm. *Energy Convers. Manag.* **2020**, *221*, 113161. [CrossRef] [PubMed]
31. Moghimi, M.; Leskarac, D.; Bennett, C.; Lu, J.; Stegen, S. Rule-Based Energy Management System in an Experimental Microgrid with the Presence of Time of Use Tariffs. In *MATEC Web of Conferences*; EDP Sciences: Les Ulis, France, 2016; Volume 70. [CrossRef]
32. García, P.; Torreglosa, J.P.; Fernández, L.M.; Jurado, F. Optimal Energy Management System for Stand-Alone Wind Turbine/Photovoltaic/Hydrogen/Battery Hybrid System with Supervisory Control Based on Fuzzy Logic. *Int. J. Hydrog. Energy* **2013**, *38*, 14146–14158. [CrossRef]
33. Aviles, D.A.; Pascual, J.; Marroyo, L.; Sanchis, P.; Guinjoan, F.; Marietta, M.P. Optimal Fuzzy Logic EMS Design for Residential Grid-Connected Microgrid with Hybrid Renewable Generation and Storage. In Proceedings of the IEEE 24th International Symposium on Industrial Electronics (ISIE), Rio de Janeiro, Brazil, 3–5 June 2015. [CrossRef]
34. Santis, E.D.; Rizzi, A.; Sadeghian, A. Hierarchical Genetic Optimization of a Fuzzy Logic System for Energy Flows Management in Microgrids. *Appl. Soft Comput.* **2017**, *60*, 135–149. [CrossRef]
35. Battula, A.R.; Vuddanti, S.; Salkuti, S.R. Review of Energy Management System Approaches in Microgrids. *Energies* **2021**, *14*, 5459. [CrossRef]
36. Jirdehi, M.A.; Tabar, V.S.; Ghassemzadeh, S.; Tohidi, S. Different Aspects of Microgrid Management: A Comprehensive Review. *J. Energy Storage* **2020**, *30*, 101457. [CrossRef]
37. Alavi, S.A.; Ahmadian, A.; Aliakbar-Golkar, M. Optimal Probabilistic Energy Management in a Typical Micro-Grid Based-on Robust Optimization and Point Estimate Method. *Energy Convers. Manag.* **2015**, *95*, 314–325. [CrossRef]

38. Xiao, F.; Ai, Q. New Modeling Framework Considering Economy, Uncertainty, and Security for Estimating the Dynamic Interchange Capability of Multi-Microgrids. *Electr. Power Syst. Res.* **2017**, *152*, 237–248. [CrossRef]
39. Soares, J.; Ghazvini, M.A.; Vale, Z.; Oliveira, P.B.d.M. A Multi-Objective Model for the Day-Ahead Energy Resource Scheduling of a Smart Grid with High Penetration of Sensitive Loads. *Appl. Energy* **2016**, *162*, 1074–1088. [CrossRef]
40. Kou, P.; Liang, D.; Gao, L. Stochastic Energy Scheduling in Microgrids Considering the Uncertainties in Both Supply and Demand. *IEEE Syst. J.* **2018**, *12*, 2589–2600. [CrossRef]
41. Karimi, H.; Jadid, S. Optimal Energy Management for Multi-Microgrid Considering Demand Response Programs: A Stochastic Multi-Objective Framework. *Energy* **2020**, *195*, 116992. [CrossRef]
42. Zhang, Y.; Gatsis, N.; Giannakis, G.B. Robust Energy Management for Microgrids With High-Penetration Renewables. *IEEE Trans. Sustain. Energy* **2013**, *4*, 944–953. [CrossRef]
43. Carli, R.; Cavone, G.; Pippia, T.; Schutter, B.D.; Dotoli, M. Robust Optimal Control for Demand Side Management of Multi-Carrier Microgrids. *IEEE Trans. Autom. Sci. Eng.* **2022**, *19*, 1338–1351. [CrossRef]
44. Li, Z.; Zang, C.; Zeng, P.; Yu, H.; Li, H. Two-Stage Stochastic Programming Based Model Predictive Control Strategy for Microgrid Energy Management under Uncertainties. In Proceedings of the International Conference on Probabilistic Methods Applied to Power Systems (PMAPS), Beijing, China, 16–20 October 2016. [CrossRef]
45. Wu, L.; Shahidehpour, M.; Li, Z. Comparison of Scenario-Based and Interval Optimization Approaches to Stochastic SCUC. *IEEE Trans. Power Syst.* **2012**, *27*, 913–921. [CrossRef]
46. Zhu, X.; Yu, Z.; Liu, X. Security Constrained Unit Commitment with Extreme Wind Scenarios. *J. Mod. Power Syst. Clean Energy* **2020**, *8*, 464–472. [CrossRef]
47. Luo, L.; Abdulkareem, S.; Rezvani, A.; RezaMiveh, M.; Samad, S.; Aljojo, N.; Pazhoohesh, M. Optimal Scheduling of a Renewable Based Microgrid Considering Photovoltaic System and Battery Energy Storage under Uncertainty. *J. Energy Storage* **2020**, *28*, 101306. [CrossRef]
48. Zhang, C.; Xu, Y.; Dong, Z.Y.; Ma, J. Robust Operation of Microgrids via Two-Stage Coordinated Energy Storage and Direct Load Control. *IEEE Trans. Power Syst.* **2017**, *32*, 2858–2868. [CrossRef]
49. Duan, Q.; Sheng, W.; Wang, H.; Zhao, C.; Ma, C. A Two-Stage Robust Optimization Method Based on the Expected Scenario for Islanded Microgrid Energy Management. *Discret. Dyn. Nat. Soc.* **2021**, *2021*, 7079296. [CrossRef]
50. Liu, J.; Chen, H.; Zhang, W.; Yurkovich, B.; Rizzoni, G. Energy Management Problems Under Uncertainties for Grid-Connected Microgrids: A Chance Constrained Programming Approach. *IEEE Trans. Smart Grid* **2017**, *8*, 2585–2596. [CrossRef]
51. Prodan, I.; Zio, E. A Model Predictive Control Framework for Reliable Microgrid Energy Management. *Int. J. Electr. Power Energy Syst.* **2014**, *61*, 399–409. [CrossRef]
52. Bordons, C.; Teno, G.; Marquez, J.J.; Ridao, M.A. Effect of the Integration of Disturbances Prediction in Energy Management Systems for Microgrids. In Proceedings of the International Conference on Smart Energy Systems and Technologies (SEST), Porto, Portugal, 9–11 September 2019. [CrossRef]
53. Hu, J.; Shan, Y.; Guerrero, J.M.; Ioinovici, A.; Chan, K.W.; Rodriguez, J. Model Predictive Control of Microgrids—An Overview. *Renew. Sustain. Energy Rev.* **2021**, *136*, 110422. [CrossRef]
54. Raveendran Nair, U.; Costa-Castelló, R. A Model Predictive Control-Based Energy Management Scheme for Hybrid Storage System in Islanded Microgrids. *IEEE Access* **2020**, *8*, 97809–97822. [CrossRef]
55. Patiño, J.; Márquez, A.; Espinosa, J. An Economic MPC Approach for a Microgrid Energy Management System. In Proceedings of the IEEE PES Transmission & Distribution Conference and Exposition-Latin America (PES T&D-LA), Medellin, Colombia, 10–13 September 2014. [CrossRef]
56. Pereira, M.; Limon, D.; Peña, D.M.; de la Valverde, L.; Alamo, T. Periodic Economic Control of a Nonisolated Microgrid. *IEEE Trans. Ind. Electron.* **2015**, *62*, 5247–5255. [CrossRef]
57. Mechleri, E.; Dorneanu, B.; Arellano-Garcia, H. A Model Predictive Control-Based Decision-Making Strategy for Residential Microgrids. *Energies* **2022**, *3*, 100–115. [CrossRef]
58. Incremona, G.P.; Cucuzzella, M.; Magni, L.; Ferrara, A. MPC with Sliding Mode Control for the Energy Management System of Microgrids. *IFAC Pap.* **2017**, *50*, 7397–7402. [CrossRef]
59. Zhang, Y.; Zhang, T.; Wang, R.; Liu, Y.; Guo, B. Optimal Operation of a Smart Residential Microgrid Based on Model Predictive Control by Considering Uncertainties and Storage Impacts. *Sol. Energy* **2015**, *122*, 1052–1065. [CrossRef]
60. Zheng, Y.; Li, S.; Tan, R. Distributed Model Predictive Control for On-Connected Microgrid Power Management. *IEEE Trans. Control. Syst. Technol.* **2017**, *26*, 1028–1039. [CrossRef]
61. Christofides, P.D.; Scattolini, R.; Muñoz de la Peña, D.; Liu, J. Distributed Model Predictive Control: A Tutorial Review and Future Research Directions. *Comput. Chem. Eng.* **2013**, *51*, 21–41. [CrossRef]
62. Carli, R.; Dotoli, M. A Distributed Control Algorithm for Waterfilling of Networked Control Systems via Consensus. *IEEE Control. Syst. Lett.* **2017**, *1*, 334–339. [CrossRef]
63. Ma, Z.; Callaway, D.S.; Hiskens, I.A. Decentralized Charging Control of Large Populations of Plug-in Electric Vehicles. *IEEE Trans. Control. Syst. Technol.* **2013**, *21*, 67–78. [CrossRef]
64. Cominesi, S.R.; Farina, M.; Giulioni, L.; Picasso, B.; Scattolini, R. A Two-Layer Stochastic Model Predictive Control Scheme for Microgrids. *IEEE Trans. Control. Syst. Technol.* **2017**, *26*, 1–13. [CrossRef]

65. Hu, M.C.; Lu, S.Y.; Chen, Y.H. Stochastic Programming and Market Equilibrium Analysis of Microgrids Energy Management Systems. *Energy* **2016**, *113*, 662–670. [CrossRef]
66. Elkazaz, M.; Sumner, M.; Thomas, D. Energy Management System for Hybrid PV-Wind-Battery Microgrid Using Convex Programming, Model Predictive and Rolling Horizon Predictive Control with Experimental Validation. *Int. J. Electr. Power Energy Syst.* **2020**, *115*, 105483. [CrossRef]
67. Kuznetsova, E.; Ruiz, C.; Li, Y.-F.; Zio, E. Analysis of Robust Optimization for Decentralized Microgrid Energy Management under Uncertainty. *Int. J. Electr. Power Energy Syst.* **2015**, *64*, 815–832. [CrossRef]
68. Zhang, H.; Yue, D.; Xie, X. Robust Optimization for Dynamic Economic Dispatch Under Wind Power Uncertainty With Different Levels of Uncertainty Budget. *IEEE Access* **2016**, *4*, 7633–7644. [CrossRef]
69. Yang, J.; Su, C. Robust Optimization of Microgrid Based on Renewable Distributed Power Generation and Load Demand Uncertainty. *Energy* **2021**, *223*, 120043. [CrossRef]
70. Trigkas, D.; Gravanis, G.; Diamantaras, K.; Voutetakis, S.; Papadopoulou, S. Energy Management in Microgrids Using Model Predictive Control Empowered with Artificial Intelligence. *Chem. Eng. Trans.* **2022**, *94*, 961–966. [CrossRef]
71. Feng, C.; Cui, M.; Hodge, B.-M.; Zhang, J. A Data-Driven Multi-Model Methodology with Deep Feature Selection for Short-Term Wind Forecasting. *Appl. Energy* **2017**, *190*, 1245–1257. [CrossRef]

energies

MDPI

Article

Deployment of IoT-Based Smart Demand-Side Management System with an Enhanced Degree of User Comfort at an Educational Institution

S. Charles Raja [1,*], A. C. Vishnu Dharssini [1], J. Jeslin Drusila Nesmalar [2] and T. Karthick [3]

1 Department of Electrical and Electronics Engineering, Thiagarajar College of Engineering, Madurai 625015, Tamil Nadu, India; vishnudharssiniac@student.tce.edu
2 Department of Electrical and Electronics Engineering, Tamilnadu Government Polytechnic College, Madurai 625011, Tamil Nadu, India; jeslineee@kamarajengg.edu.in
3 Quantanics Techserv Pvt. Ltd., Madurai 625016, Tamil Nadu, India; support@quantanics.in
* Correspondence: charlesrajas@tce.edu

Abstract: Nowadays, the Internet of Things (IoT) has a wide impact on many potential applications. The impact of IoT on performing demand-side management (DSM) in an Indian educational institution has not been researched in depth before. In this research work, an IoT-enabled SDSMS (Smart DSM System) has been deployed with the main objective of minimizing electricity tariff and also to tweak the quality of user comfort. It can be feasible by prioritizing available renewable PV solar energy during peak hours in an Indian educational institution. DSM has been performed using day-ahead load shifting and rescheduling the different classes of institutional loads by applying hybrid BPSOGSA (Binary Particle Swarm Optimization and Gravitational Search Algorithm). The BPSOGSA performance on DSM has been evaluated based on electricity tariff, peak demand range, and PAR and compared with the outcomes of both binary conventional algorithms BPSO and BGSA, respectively. The proposed method enhances the degree of user comfort (DUC) by tripping the operation of non-critical institutional loads. Simulation results obtained using MATLAB corroborate that BPSOGSA outperforms both BPSO and BGSA under both DSM scenarios. Before DSM, Peak demand, PAR, and Electricity tariffs were found to be 1855.47 kW, 4.1286, and $2030.67 while after DSM, they reduced to 1502.24 kW, 3.263, and $1314.40 respectively. This indicates a 35.273% reduction in electricity tariff, a 19.037% scale down in peak demand, and a 20.97% reduction in PAR. Finally, the real-time IoT-based SDSMS hardware is implemented at the Renewable energy laboratory for real monitoring of energy consumption via the Blynk application.

Keywords: DSM; BGSA; BPSO; smart grid; load categorizing; DUC; load shifting; Internet of Things

Citation: Charles Raja, S.; Vishnu Dharssini, A.C.; Jeslin Drusila Nesmalar, J.; Karthick, T. Deployment of IoT-Based Smart Demand-Side Management System with an Enhanced Degree of User Comfort at an Educational Institution. *Energies* **2023**, *16*, 1403. https://dx.doi.org/10.3390/en16031403

Academic Editor: Abu-Siada Ahmed

Received: 31 December 2022
Revised: 23 January 2023
Accepted: 27 January 2023
Published: 31 January 2023

1. Introduction

The growing energy demand, the environmental pollution, and the attractive renewable policies lead consumers to adopt renewable energy sources. The emission of greenhouse gases is reduced by prioritizing the utilization of available renewable resource energies rather an the utility grid. Even though the reduction in the emission of greenhouse gases is a major advantage, the interruption of Distributed Generations (DGs) makes the reliability of the grid vulnerable. In the Smart Grid context, DSM technique is extremely promising technique to make the grid stable and dependable, limit consumer energy usage and decrease power generation. The schematic approach of DSM evolves around DR Demand Response programs, energy efficiency programs, and of all energy conservation programs.

It is essential for the DSM framework to have the quality of mitigating certain criteria such as electricity cost minimization, user comfort maximization, and peak demand reduction. This can be achieved using various optimization algorithms. The literature section discusses different optimization algorithms that have been used to implement DSM. Various optimization algorithms such as CS, GWO, PSO, SA, HS, DEA, ACO and GA have been employed. All bio-inspired algorithms have their advantages and limitations. The primitive goal of proposing hybrid forms of algorithms is to achieve better and most promising results by combining appropriate conventional algorithms. BPSOGSA is a proposed hybrid algorithm proposed which is calibrated as both BPSO and GSA for better outcomes. The algorithm offers a decent improvement in the resolution of the intended purpose to minimize electricity costs. The search process in the hybrid BPSOGSA algorithm is supported by agents that replicate the characteristics of BGSA in the research phase and BPSO in the utilization phase. The ultimate aim of hybridizing efficient GSA and PSO is to incorporate the rationale of PSO and the exploration of GSA. Followed by that, distinguished performance of optimization algorithms entirely relies on the tuning algorithm with respect to the system. Generally, it can be achieved by adopting various upgrading methods such as changing the search space, modifying convergence parameters, and lastly, hybridizing the individual algorithms. From the mentioned means of improvisation, hybridizing strengths of individual algorithms results in better performance.

In this extensive study, the hybrid BPSOGSA algorithm is employed in shifting and scheduling the institutional loads, which are classified into two classes, namely, SLs (Shiftable Loads) and NSLs (Non-Shiftable Loads). The hybrid BPSOGSA algorithm optimizes the operating period of each type of load by shifting the load to a period with a lower electricity price. An integrated model that schedules loads using the traditional Ant Colony Optimization (ACO) methodology to minimize overall journey length and unpredictable demand has been used to estimate delivery routes and truck loads [1]. Four heuristic-based strategies for optimum DG location and sizing were proposed by using GSA, PSO, CSA, and GA based on loss minimization criteria [2]. A path for planning and navigation has emerged using GSA, PSO, and Simulated Annealing (SA) algorithms involving autonomous mobile robots [3]. A unique methodology by combining PSO with adaptive GSA for improving FACTS devices to ensure voltage stability in power transmission networks has been established [4]. A unique GSA (Gravitational Search Algorithm) has been evaluated as it is useful in multiple objective problems [5].

In a smart grid, a heuristic search-type controller for energy management has been suggested for the implementation of DSM [6]. Their work examines five different heuristic algorithms, including GA, BFOA, WDO, BPSO, BFOA strategy, and concluded with a suggested hybrid genetic wind-driven (GWD) approach. A familiar Supply Chain Management (SCM) model Vendor Managed Inventory (VMI) model made of a PSO-based algorithm has been developed to find the near optimum [7]. In [7], for defuzzification, the centroid defuzzification method has been used to discover ideal retailer order amounts to minimize overall inventory and transportation costs. A control scheme called PS-GSA has been developed for a double induction wind turbine generator by means of Fuzzy's sliding mode control [8]. A hybrid PSOGSA-based model has been developed to analyze High-Dimensional (HD) data to provide the solution to the convergence problem [9].

A Binary Grey Wolf Optimisation Algorithm (BGWO)-based DSM has been developed to solve the non-linear objective of cost minimization, concerning the electricity tariff of a real-time residential load [10]. Their paper proposes strategic shifting and scheduling of the residential loads optimally deploying the swarm intelligence-based BGWO algorithm. A reinforced DSM approach involving enhanced BGWO has been designed and developed to schedule pre-categorized classes of loads in an educational institution with an aim of deducing peak demand and thereby indirectly minimizing the electricity tariff [11]. A DSM technique based on the PSO algorithm has been developed to diminish the electricity price and schedule different loads in residential, commercial, and industrial areas [12]. A multi-objective stochastic optimization algorithm for scheduling both reserves and energy by allowing simultaneous participation of loads with large wind energy penetration has been proposed [13]. A combination of both the heuristic and randomization algorithms considering both electricity cost and user convenience changes the non-convex method into a convex method [14]. For distributed generation sources, a two-stage optimal scheduler has been formulated [15] and it employs MILP to deduce the bi-objective problem of minimizing net annualized cost TAC and net emission TAE to a single-objective problem. The BPSOGSA algorithm has been utilized in optimally scheduling the available hybrid renewable energy resources for minimizing LCOE and also the probability of power supply failure [16].

The clustering problem has been solved by hybridizing GSA with one of the promising heuristic methods, which is done to improve results gained from GSA [17]. In [18], the convergence speed of GSA is improved by employing position-based learning GSA. In [19], the convergence of GSA is improved by a modified Immune Gravitation Optimization Algorithm (IGOA). The combined quality of both social thinking and individual thinking of PSO is inherited in hybrid PSOGSA which continues the capability of problem solving as GSA aids in parameter identification of hydraulic turbine governing systems [20].

In [21], an algorithm GSA was developed by adopting the concepts of Newton's laws of motion and gravity that form interaction among the search agents [21]. Similarly, the binary version of GSA, namely, BGSA (Binary Gravitational Search Algorithm) was suggested [22]. A DBHS algorithm is developed utilizing pitch variation problems [23]. Whereas, harmony search and pitch adjustments deployed in MBDE have been proposed to solve distinct optimization problems [24]. The discrete binary version of PSO has been developed to deal with discrete space [25] while hybrid binary version of PSOGSA, known as BPSOGSA, has been developed to achieve a better performance with a regard of both exploitation and exploration utilizing positive aspects of both PSO and GSA [26]. A hybrid PSOGSA algorithm suggested to train feedforward neural networks for solving a local convergence problem [27].

A fully distributed and interactive learning approach has been proposed for integrated charging regulators of electric vehicles, to regulate three socio-technical aspects: dependability, discomfort, and fairness [28]. The daily load profiles of residential loads have been assessed by incorporating clustering k-means and self-organizing map algorithms and taking into account the social traits and lifestyles of real-time communities [29]. An IoT-based DSM is proposed using hardware to dynamically optimize the load pattern in actual commercial buildings [30]. The smart socket was designed in real time using the mobile Blynk application for remote monitoring of Power and connected loads management [31]. The socket may be accessed remotely over wireless media. This can be further upgraded by introducing WPC (wireless power communication) for ensuring effective and efficient communication to facilitate user end [32]. Three modes of communication among various access points, co-located power station, and wireless user enhances reliable and fastest means of interaction [33]. For accelerating and promoting user preferences, using an IoT-enabled recommended system is promising [34]. Following this, introducing remote monitoring and control options for data physical fusion technology is essential. Mobile IoT interlinked with appropriate applications paves the way for betterment in performance and also adds research significance [35]. In this case, investigating the smart grid without creating an adverse impact on resiliency and efficiency is the requirement. Hence, IoT and cloud-based detection systems were found to be feasible with integrated AI approaches [36]. All that is necessary for AI is the quality of data for processing. The real-time data retrieved are stored in the cloud/local storage and undergo AI treatment based on the nature of labeling data [37]. The data involved in the Smart grid are of various parameters addressing power quality, operating status, fault detection, and resulting optimal solutions [38].

1.1. Research Gap

From the literature review conducted so far, it can be stated that optimization algorithms were used for different nonlinear problems, in different domains to solve various real-time optimization objectives. Among the previous works, the DSM approaches and techniques were hardly deployed using a real-time educational load profile. This research offers a unique BPSOGSA hybrid optimization algorithm-based demand-side management for KCET—Kamaraj College of Engineering and Technology, an academic higher education institution situated in Southern India.

KCET's infrastructure spans 47.92 acres of land and is scheduled to work from morning 9:00:00 to evening 16:00:00. The institution offers 16 undergraduate and postgraduate programs with 58 highly equipped laboratories of the latest machinery. KCET has a maximum demand of 1000 kVA and is facilitated by a 32 kW solar PV system with battery storage and converter system along with three diesel generator sets, one DG of 500 kVA, and the rest two of 250 kVA. These spacial and temporal factors make the load profile of educational institutions differ from one another. On the whole, the consumption trend of any institutional building completely relies on their working time, the nature of load types used, and the shifting of load operating patterns. The placement of DSM techniques and approaches in educational institutions helps in minimizing the electricity consumption price considerably.

Thus, the multi-objective SDSM system proposed in KCET for reconciling and highlighting following objectives

- Minimize the cost involved in electricity consumption;
- Categorize a wide range of loads for allocating suitable time of operation;
- Accountable reduction in PAR and peak demand;
- To shift and schedule load optimally by considering all the possible constraints;
- Addressing user comfort index (UCI)-based optimization.

1.2. Contributions

The following contributions to the effective implementation of DSM in KCET, an Indian educational institution, have been proposed:

- Internet of Things (IoT)-enabled Smart Demand-Side Management System (SDSMS) has been implemented in real time for energy monitoring and connects the solar power line for the laboratory loads during peak periods;
- The load classification of KCET was obtained based on load shifting with and without interruption;
- The institutional load profile is optimally scheduled by applying the hybrid BPSOGSA algorithm using MATLAB software;
- The load profile of KCET is shifted and scheduled a day ahead effectively to improve the degree of user comfort by utilizing the renewable energy (RE) solar source set up at the institution;
- Next, the real-time hardware implementation of SDSMS was applied to increase the degree of user comfort (DUC) during the peak hours powered by the Solar PV (SPV) resource, and all the determining real-time electrical parameters are monitored with the aid of the Blynk application.

1.3. Novelty of Proposed Work

In this research work, an IoT-based smart demand-side management system is developed using a hybrid BPSOGSA algorithm to achieve the main goal of lowering the cost involved in electricity consumption. It can be made possible by optimally scheduling the different classes of institutional loads considering the operational constraints associated with each class of load. The outcomes of the hybrid BPSOGSA algorithm highlight its performance over other two algorithms—BPSO and BGSA—in electricity tariff reduction, avoiding reach of peak demand, and maintaining PAR. Additionally, the statistical performance of all the three algorithms BPSOGSA, BPSO, and BGSA, has been determined using statistical tests. Holm, Hommel, and Holland tests are considered for this study. Next, the degree of user comfort has been improved by utilizing the solar PV generation located at the institution. Finally, real-time hardware implementation of SDSMS has been deployed at the Renewable Energy lab to improve the degree of user comfort, and the electrical parameters obtained are continuously monitored using the Blynk application.

The paper is organized as follows: The methodology adopted is interpreted in Section 2. Optimization algorithms are outlined in Section 3. The BPSOGSA algorithm is formulated in Section 4, while Section 5 emphases on results and discussions. Section 6 concludes the work with key findings.

2. Proposed Methodology

The hybrid BPSOGSA algorithm is utilized to shift and schedule the two categories of loads classified as Amenable Loads (ALs) and Unamenable Loads (NDLs). Factors such as operating period, starting time, and ending time of each load type are used to divide institutional load data into two classes.

2.1. Design and Development Architecture of SDSMS

The proposed IoT-enabled SDMS is implemented in real time at Renewable Energy laboratory of KCET-Kamaraj College of Engineering and Technology. The deployed real-time power monitor switches the power line either from institutional or solar power. Figure 1 shows the overall implementation of various loads with SDSMS.

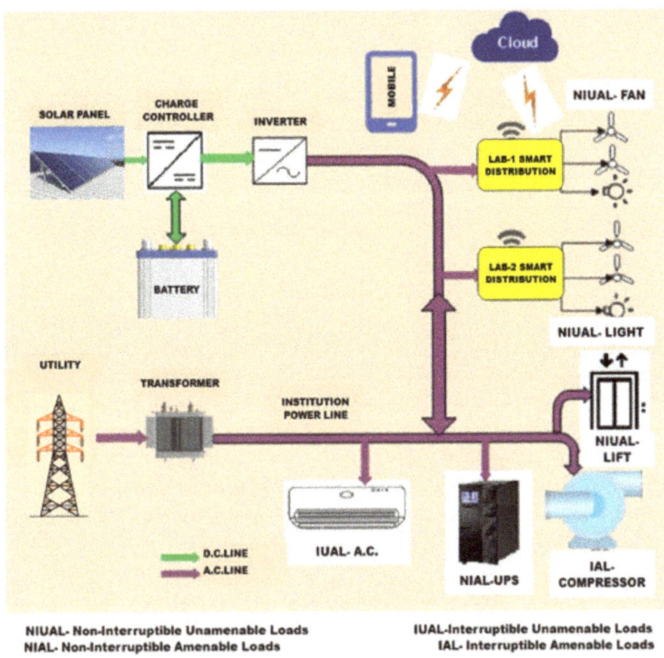

Figure 1. Overall implementation of various loads with SDSMS.

The IoT-based Smart Demand-Side Management System (SDSMS) contains a microcontroller, A.C power measuring module, and relay unit. It successfully reduces the energy consumption cost by supplying solar power to the loads during peak hours (9 a.m. to 12 p.m.) and monitors the electrical parameters of the load. These operations are remotely monitored, validated, and controlled via the Blynk application through open-source TCP/IP protocol. To access the consumers, the electrical parameters are accurately measured and updated with universal timestamps in the cloud server. A power quality analyzer (PQA) is employed to compare the phase voltages and currents obtained from the proposed IoT-based SDMS.

2.2. Classification of KCET Loads

The loads can further be classified based on the functioning of loads as Interruptible and Non-Interruptible loads, as shown in Figure 2. The Amenable loads (AL) have both Interruptible Amenable loads (IALs) and Non-Interruptible Amenable loads (NIALs).

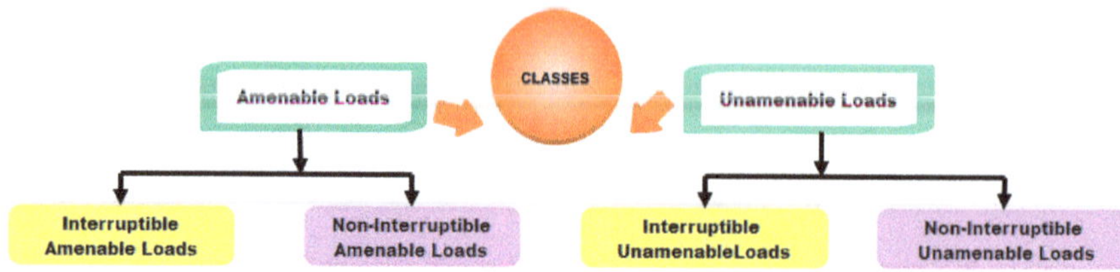

Figure 2. Classification of KCET loads.

Similarly, the Unamenable (UAL) loads are classified into Interruptible Unamenable Loads (IUALs) and Non-Interruptible Unamenable Loads (NIUALs). These loads are shifted and scheduled optimally using the hybrid BPSOGSA algorithm.

2.2.1. Amenable Loads

The loads which have a higher likelihood of shifting and controlling as per the user's comfort within its time of operation are categorized under this category. These schedulable loads have freedom to shift during peak hours for reducing energy consumption in that specific period and electricity tariff considerably. The loads of this type are further classified as IALs and NIALs.

Interruptible Amenable Loads (IALs)

The category of loads which can be both shiftable and schedulable can be operated in its period of usage are classified into Interruptible Amenable loads (IALs). These loads are operated on an intermittent basis within the period of start and completion of the loading process, which corresponds to its duration. Load types such as electric vehicles, pumps, and compressors are classified under this type of load. The period of operation of these load types can be interrupted according to the user's preferences and also with device performance.

Let A be the set of interruptible amenable loads given by $A = \{A_1, A_2, \ldots, A_n\}$. The scheduling vector denotes the Turn ON/Turn OFF status of an IAL in each interval (i = [1, 2, …, t] $\forall i \,\varepsilon\, I$) is depicted as in (1):

$$\xi_a = \left(\xi_a^1, \xi_a^2, \ldots\ldots.., \xi_a^i, \ldots., \xi_a^t\right] \;\forall\, a \in A \tag{1}$$

where n_a is the interruptible amenable loads, i.e., cardinality of A. These Interruptible loads are operated discontinuously without violating total duration of this load operation with discrete starting and ending time. The operating status of an IAL load 'a' in a time interval i is given in Equation (2)

$$\xi_a^i = \begin{cases} 1, ST_a \le i \le ET_a \text{ and } N_a^i \le D_a \text{ and } [ST_a; ET_a] \\ 0,\ i < ST_a \text{ and } i > ET_a \end{cases} = D_a \tag{2}$$

where ST_a = starting time of load 'a',
ET_a = ending time of load 'a',
[ST_a;ET_a] = discontinuous operating period from starting time to ending time,
$N_a{}^i$ = load 'a' operating hours up to ith time period,
D_a = operating period of load 'a' (in total number of hours).

Non-Interruptible Amenable Loads (NIALs)

This type of load is shiftable but only can be switched for continuous operation in the specified time slot finds place under Non-Interruptible Amenable loads (NIALs). These loads are continually operated at their time of commencement and stops within the time to meet their load operation duration. The operation status of these non-interruptible loads is monitored and recorded to schedule the institutional loads optimally. The continuous operation is assumed in these types of loads.

Let B be the set of non-interruptible amenable loads given by $B = \{B_1, B_2, \ldots, B_n\}$. The scheduling vector indicates the Turn ON/Turn OFF status of an NIAL in every interval ($i = [1, 2, \ldots, t]\ \forall i \varepsilon I$) is stated as in Equation (3):

$$\xi_b = (\xi_b^1, \xi_b^2, \ldots\ldots.., \xi_b^i, \ldots., \xi_b^t]\ \forall\ b \in B \quad (3)$$

where n_b is non-interruptible amenable loads, i.e., cardinality of B. These Interruptible loads are operated continuously without violating the total duration of this load operation with specific starting and ending time. The NIAL load 'b' nature in the ith time period is given by Equation (4):

$$\xi_b^i = \begin{cases} 1, ST_b \le i \le ET_b \text{ and } N_b^i \le D_b \text{ and } [ST_b; ET_b] \\ 0,\ i < ST_b \text{ and } i > ET_b \end{cases} = D_b \quad (4)$$

where ST_b = Load 'b' starting time
ET_b = Load 'b' ending time
$[ST_b;ET_b]$ = discontinuous operating period from starting time to ending time
$N_b{}^i$ = load 'b' operating hours up to the ith time period
D_b = load 'b' total number of operating hours.

2.2.2. Unamenable Loads (UALs)

The loads which are non-feasible for scheduling and controlling and can be operated with respect to time are modeled under this category. Those loads are non-shiftable, critical, and primary loads to satisfy the user's comfort.

Interruptible Unamenable Loads (IUALs)

The temperature-dependent loads which have a high consumption range in operating mode whilst demanding low power in standby mode are categorized under IUALs. The temperature-controlled loads may be heating or cooling loads. The cooling type of load air conditioners (ACs) is modeled with respect to a frequency of Turn ON/Turn OFF cycle. In short, the hourly operating frequency of the cycle is accounted for by the quality characteristics of the device. Typical cooling load (air conditioner) considered in this paper switched Turn ON/Turn OFF at a frequency of once in every 15 min. Thus, normal hourly operating frequency of cycle for cooling load is 4, which consumes power above or below this limit during device malfunctions or. For an INAL c, the status of operating c in the interval i is explained as in Equation (5)

$$\xi_c^i = \begin{cases} 1,\ if\ T^c{}_{set} > (T^c{}_{set} + T^c{}_{tot}); (f^i{}_c = 4) \\ 0,\ if\ (T(i)^c{}_{act} < T^c{}_{set});\ (f^i{}_c > 4)||(f^i{}_c < 4)\xi_c^{(i-1)}, if\ T^c{}_{set} \le T(i)^c{}_{act} \le (T^c{}_{set} + T^c{}_{tot}) \end{cases} \quad (5)$$

where T^c_{set} = set point temperature
T^c_{tol} = permissible tolerance limit
f^i_c = operation frequency of cycle during the ith interval
$T(i-1)^c_{act}$ = actual ambient temperature by the $(i-1)$ interval end.

Non-Interruptible Unamenable Loads (NIUALs)

These loads rely entirely on the user's choice of switching Turn ON/Turn OFF in a day. The operating pattern on an hourly basis ensures and prevents the probable reach of sanctioned maximum hourly demand. The load types such as ceiling fans, tube lights, chargers, elevators, LAN communication, and DC fans are classified in NIUALs. The status of these load types is monitored using a separate status controller and the Maximum Power Load Limit (MPLL) is maintained.

2.3. Design of IoT Based SDSMS

The SDSMS illustrated in Figure 3 has a sensor module called PZEM-004T linked with a power supply and an ESP8266 Wi-Fi module that provides details on voltage, power factor, actual power, current, and reactive power data. A relay is developed to change the power supply to the user's convenience load from the institutional power line or solar power source. The low-cost ESP8266 is a Wi-Fi microchip accompanied with full TCP/IP (TCP/IP) stack enhanced with functionality of a microcontroller. This module establishes connectivity with a Wi-Fi network for microcontrollers and easy connections to TCP/IP. The Wi-Fi in turn delivers data to the authorized consumer over a hot spot for sharing system time-series data on the cloud server (Message Queuing Telemetry Transportation MQTT) via Blynk application.

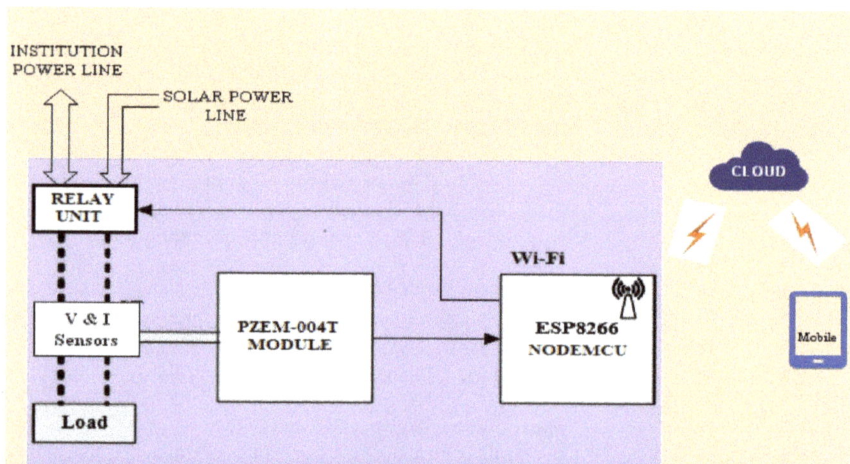

Figure 3. Illustration of the Smart Demand-Side Management System (SDSMS).

In addition, it estimates energy usage and costs so that the user may read the report from anywhere, the report is then sent to the Blynk application. This SDSMS implements the demand side management idea which reduces demand overall and restores the demand profile of the institution. In addition, it boosts the sustainability of the grid by reducing total costs and carbon emissions. The Blynk application offers a great opportunity to receive power measurements and cumulative information over a week or month according to customer requirements at that exact moment in time.

3. Optimization Algorithms

This section details the three optimization algorithms—BPSO Binary Particle Swarm Optimization Algorithm, GSA Gravitational Search Algorithm, and BPSOGSA Binary Particle Swarm Optimization—Gravitational Search Algorithm which are nature-inspired swarm intelligent algorithms. BPSOGSA is a hybrid optimization algorithm.

3.1. Binary Particle Swarm Optimization Algorithm (BPSO)

PSO is a swarm-based optimizer that strives for a rational solution by encoding the fundamental characteristics to interact amongst participants. The particles are first dispersed at random in the solution space, and gradually they start to move around in search of the ideal match for the given optimization problem.

The position vector of PSO is given by Equation (6)

$$X_A = \left(x_A^1, \ldots. x_A^d\right),\ A = 1,2,\ldots,\ N \tag{6}$$

where N = number of search agents, d = dimension of the problem, x_A^d = Ath agent position in the dth dimension. The agent A's velocity vector is given by Equation (7)

$$V_A = \left(v_A^1, \ldots. v_A^d\right),\ A = 1,2,\ldots,\ N \tag{7}$$

Particle movement in the search space is permitted until the maximum possible repetition or the convergence criteria are met. The social and individual thinking capability of the PSO helps in finding the global best solution. The swarming technology with continuous values is adopted to solve the PSO's discrete version.

BPSO algorithm proposed by Kennedy and Eberhart [25] handle discrete optimization problems. The swarming technology with discrete values as decision variables to solve the optimization problem is employed in BPSO.

The fundamental equation to update particle 'A' velocity is in (8) as follows:

$$V_A = V_A + R(p_A - X_A) + R(g_A - X_A) \tag{8}$$

where V_A, X_A, p_A, and g_A are the determining factors of particle, 'A' such as position, velocity, global best and local best position, respectively and R be the generated random positive number within the interval [0, 1].

$$V_A \leq V_{max} \tag{9}$$

The particle 'A' velocity (V_A) is always less than its maximum velocity (V_{max}) in (9). The change in particle position is defined by the following rule (10):

$$if\ rand < S(V_A)\ then\ X_A = 1\ else\ X_A = 0 \tag{10}$$

where the sigmoidal activation function is represented as $S(V_A)$ and the chosen quasi-random number as '*rand*' from a uniform distribution [0, 1]. The following Equation (11) describes the updating of particle position:

$$X_A = X_A + V_A \tag{11}$$

The scheduling vector of BPSO algorithm for each load type is a [1 × 24] vector in (12):

$$S_{BPSO} = (s_1\ s_2\ s_3\ s_4 \ldots\ldots\ldots\ldots\ldots\ldots\ldots\ldots\ s_{24}\,] \tag{12}$$

where S_{BPSO} is the scheduling vector and each variable s_i is either 0 (OFF) or 1 (ON).

3.2. Gravitational Search Algorithm (GSA)

The gravitational force on agent A from agent B at a particular iteration t is expressed as follows in (13):

$$F_{AB}^d(t) = G(t)\frac{M_{pA}(t) * M_{aB}(t)}{R_{AB}(t) + \alpha}\left(X_B^d(t) - X_A^d(t)\right) \tag{13}$$

where M_{pA} is the passive gravitational mass related to agent A, M_{aB} is the active gravitational mass related to agent B, $G(t)$ is gravitational constant at iteration t, α is a constant, X_A is the position of agent A, X_B is the position of agent B and $R_{AB}(t)$ is assumed as the Euclidian gap between agents A and B.

The gravitational constant and the Euclidian gap between agents A and B are determined as in (14) and (15) as follows:

$$G(t) = G_0 \times exp\left(-\alpha * \frac{t}{maxIter}\right) \tag{14}$$

$$R_{AB}(t) = \|X_A(t), X_B(t)\|_2 \tag{15}$$

where $\propto$ is the descending coefficient, G_0 indicates the initial gravitational constant, t is the current iteration, and maxiter shows the maximum limit of iterations.

The total force striking on agent A in a dimension d given by (16):

$$F_A^d(t) = \sum_{B=1,B\neq A}^{N} rand_B F_{AB}^d(t) \tag{16}$$

where $rand_B$ is taken as a random number between [0, 1].

The acceleration of agent A at iteration t as (17):

$$a^d{}_A(t) = \frac{F^d{}_A(t)}{M_{AA}(t)} \tag{17}$$

where M_{AA} is the inertial mass of the Ath agent.

The velocity and position update equations for agent A are given as shown in (18) and (19):

$$V_A^d(t+1) = rand_A \times V_A^d(t) + a_A^d(t) \tag{18}$$

$$X_A^d(t+1) = X_A^d(t) + V_A^d(t+1) \tag{19}$$

where $rand_A$ is a number randomly generated between [0 ,1], $X_A(t)$ is the position of agent A during iteration t, and $V_A(t)$ is the velocity of agent A during iteration t.

The GSA mass movement is estimated based on its fitness weights. Reaching the best global optimum is difficult for GSA due to the slower movement of heavy mass particles. The scheduling vector of BGSA algorithm for each load type is a [1 × 24] vector as in (20):

$$S_{BGSA} = (s_1\ s_2\ s_3\ s_4 \ldots\ldots\ldots\ldots\ldots\ldots\ldots\ldots\ s_{24}] \tag{20}$$

where S_{BGSA} is the scheduling vector and each variable s_i is either 0 (OFF) or 1 (ON).

3.3. Hybrid Binary Particle Swarm Optimization—Gravitational Search Algorithm (BPSOGSA)

The hybrid BPSOGSA optimization algorithm proposed by Mirijalili, Wang, and Coelho [26] combines the social and individual thinking gbest of PSO and the exploration capability of the Gravitational Search Algorithm. GSA suffers from sluggish operations and deteriorates in end iteration and the best solution is used by PSO. The masses with improved fitness values are heavy objects and they converge slowly. Thus, in the final step, the masses with almost the same weights gather around the promising solutions. This finally results in the slow movement of the masses toward the best solution. This behavior of GSA may result in local optimal value rather than the global optimal value. This drawback of GSA and a similar Binary Gravitational Search Algorithm (BGSA) is solved by including the optimal solution by guiding the heavy masses on their way to the global optimal value. Equation (21) combines both PSO and GSA:

$$V_A(t+1) = rand * V_A(t) + c_1' * ac_A(t) + c_2' * (gbest - X_A(t)) \tag{21}$$

where c_1' and c_2' are accelerating factors, $ac_A(t)$ is the acceleration of agent A in iteration t, rand is generated random number using uniform distribution in the interval [0 ,1], $gbest$ is the optimal solution, $X_A(t)$ is the position of agent A in iteration t, and $V_A(t)$ is the velocity of agent A in iteration t.

The position of agents is modified in each iteration using Equation (22):

$$X_A(t+1) = X_A(t) + V_A(t+1) \tag{22}$$

The transfer function must be limited between [0 ,1] and increase as velocity increases. The transfer function is depicted in Equation (23) for relating the velocities of agents with the probabilities of change in their respective positions:

$$S\left(V_{A,B}^d(t)\right) = |\left(tanh\left(V_{A,B}^d(t)\right)\right| \tag{23}$$

After calculating the respective probabilities of agents, their position is updated using the below Equation (24):

$$\begin{aligned} &If\ rand < S\left(V_{A,B}^d(t+1)\right)\ then\ X_{A,B}^d(t+1) = complement\ \left(X_{A,B}^d(t)\right) \\ &else\ X_{A,B}^d(t+1) = X_{A,B}^d(t) \end{aligned} \tag{24}$$

In the gbest variable based on PSO optimisation, the best solution achieved up to now has been stored. The sluggish operation and degeneration of GSA are addressed by the optimum solution. The inclusion of gbest to the speed matrix has hampered the exploration stage as it adds itself to the speed update as a permanent feature. In order to resolve this, a varied adaptive approach is applied. The variables c_1' and c_2' detailed in Equations (25) and (26), respectively, are adaptively decreased or increased, in order that the masses accelerate toward the optimal solution.

$$c_1' = -2\frac{t^3}{T^3} + 2 \tag{25}$$

$$c_2' = 2\frac{t^3}{T^3} + 2 \tag{26}$$

where t is the current iteration and T is the upper limit of iterations.

The scheduling vector of the BPSOGSA algorithm for each load model is a $[1 \times 24]$ matrix vector as in (27):

$$S_{BPSOGSA} = (s_1\ s_2\ s_3\ s_4 \ldots\ldots\ldots\ldots\ldots\ldots\ldots\ldots\ s_{24}\ ] \tag{27}$$

where $S_{BPSOGSA}$ is the scheduling vector and each variable s_i is either 0 (OFF) or 1 (ON).

This adaptive technique is one of the finest ways to permit the transition of an algorithm from the exploration to exploitation stages, given the two phases do not have a distinct limit. The hybrid BPSOGSA optimization algorithm outperforms BPSO and GSA by escalating the movement of heavy masses of GSA to reach the global optimum.

4. Formulation of Binary Particle Swarm Optimization—Gravitational Search Algorithm (Bpsogsa) Based Demand Side Management (DSM)

The classified load is deployed in the proposed BPSOGSA-DSM for scheduling the loads a day ahead optimally. In BPSO, the fitness value updates the gbest value using Equation (11) and in GSA it updates the G value using Equation (14). After updating these values for each agent, the forces and acceleration are determined. The velocity and position values are determined and updated until the stop criterion is met. Finally, the gbest value is returned as the optimal result, using which, the particle's position is determined. The number of particles is the number of load types considered. The flow diagram of the proposed methodology is detailed in Figure 4.

The hybrid BPSOGSA algorithm helps the heavier masses of GSA which become linked at the local optimum and finds it difficult to move towards the global optimum by finding the gbest value. The gbest value of BPSO pushes the heavier masses of GSA to end in global optimum. The institutional load data are classified and the particles are randomly initialized using BPSO. The agents that support positioning the particles and the movement of masses in GSA are initialized. The fitness values of the proposed objective function are determined and the acceleration, forces, velocity, position, and M values are calculated and updated for the particles of KCET load types considered. The gbest and G values of the BPSOGSA assist in shifting and re-scheduling the load types optimally by varying the position of the particles to attain the main objective of electricity tariff minimization portrayed in Figure 5 and Figure 6, respectively. The hybrid BPSOGSA algorithm incorporates the strengths of GSA and BPSO. It pushes the heavy masses of GSA from the local optimum to the global optimum.

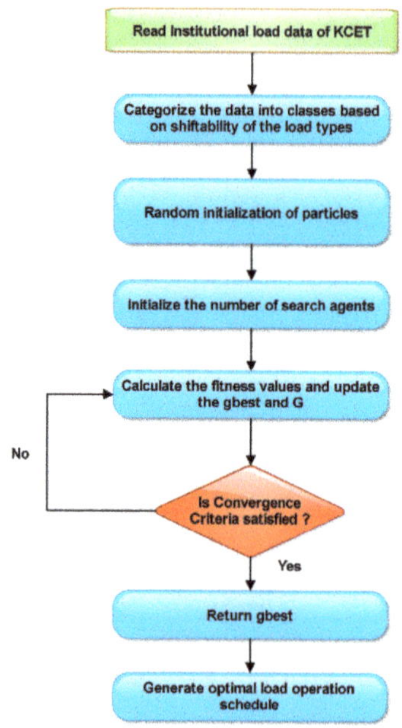

Figure 4. Flow diagram of the proposed methodology.

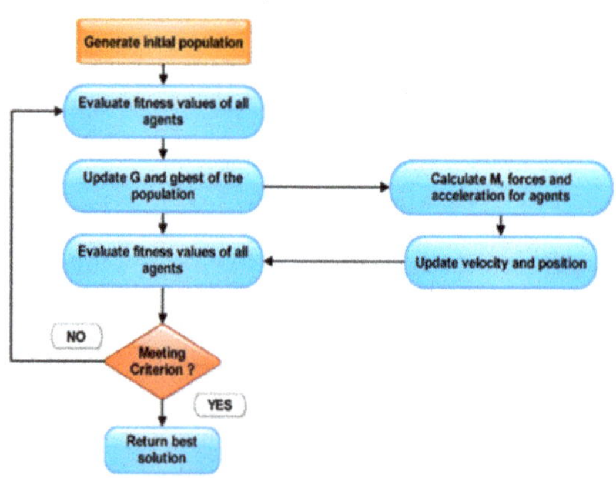

Figure 5. Flowchart of BPSOGSA.

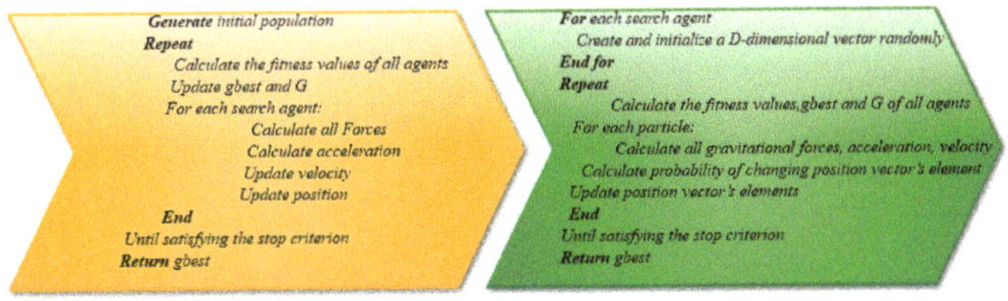

Figure 6. Pseudocode of BPSOGSA.

4.1. Objective Function

A primary objective using the proposed BPSOGSA-DSM strategy is to attain a minimal consumption cost, which allows the institution to optimally shift and schedule loads to recognize user comfort as well. Due to dynamic changes in pricing and power demands, the objective function is formulated as a non-linear function and is given in Equation (28):

$$Min\ :Cost = \sum_{j=1}^{24} \sum_{p=1}^{n} S_p^j * PD_p * Rate^j \quad (28)$$

where cost represents the overall electricity cost consumption, j represents 24 h of a day, n is the type of loads, p is the index for load type, S_p^j= situation of 'p' load at jth hour of operation ('1' as 'ON', '0' as 'OFF'), PD_p = Power demand of pth load, $Rate^j$ = electricity price at the jth period of operation.

4.2. Constraints

4.2.1. Operating Hours of Appliances

The operating hours of all appliances are carefully considered and they are given in Equation (29).

$$subject\ to\ \sum_{j=1}^{24} \sum_{p=1}^{n} S_{1p}^j = D_p \quad (29)$$

where D_p is operating hours of the 'p' load type.

4.2.2. Total Demand

The next important constraint is total demand, which in a day should be equal to before and after what place on the 7-demand side management (DSM). It is given in Equation (30).

$$subject\ to\ \sum_{j=1}^{24} \sum_{p=1}^{n} D_{1p}^j = \sum_{j=1}^{24} \sum_{p=1}^{n} D_{2p}^j \quad (30)$$

where D_{1p}^j is the power demand before DSM for the pth load at the jth hour, D_{2p}^j is the power demand after DSM for the pth load at the jth hour.

4.2.3. Instantaneous Demand

The instantaneous demand should be equal to the maximum power demand always. Equation (31) satisfies this constraint.

$$PD_j \leq PD_{max}\, \forall\, j \in (1,24] \quad (31)$$

where PD_j is jth hour power demand, PD_{max} is maximum demand sanctioned.

4.2.4. Idle Devices

The idle constraint, which is given in Equation (32), is that some appliances remain unused or consume no power except during the operating hours.

$$S_p^t = 0\, \forall\, t < ST, t > ET \text{ and } t\, \epsilon\, (1,24]\ ,\ p\, \epsilon\, (1,n] \quad (32)$$

S_p^t = ON (1) or OFF (0) status of pth load at the tth hour. The Start and End Time are mentioned as ST and ET, respectively.

5. Results and Discussion

The considered test system consists of institutional loads at KCET that have a specific operational pattern. By utilizing BPSO, BGSA, and the hybrid BPSOGSA algorithms, the objective function given in Equation (28) is achieved by implementing DSM in the considered institutional loads using MATLAB as discussed in Section 5.1. Next, the degree of user comfort was enhanced by studying unchanged hours of operation for all load type using the hybrid BPSOGSA algorithm, which was discussed in Section 5.2. Finally, the real-time hardware implementation of the IoT-based Smart Demand Side Management System (SDSMS) for the Renewable Energy laboratory loads in KCET is detailed in Section 5.3.

5.1. Demand before and after Deploying the Hybrid BPSOGSA-DSM

Every day, 9:00 to 16:00 are the working hours of the considered institute KCET, in which 9:00 to 14:00 are considered to be peak hours and before 8:00 and after 15:00 are considered to be the off-peak hours. The contribution of the proposed implementation system is to carry out DSM by moving the possible loads from peak to off-peak hours and reduce peak demand to a maximum level. Figure 7 shows the real-time hourly electricity price, the operating hours are categorized as off-peak, medium, and peak hours based on the power prices in real-time. The price reaches its peak at 10:00 to $27.35/kWh and the lowest price is at 4:00 at $8.1/kWh. Thus, 9:00 a.m. to 14:00 is grouped as peak hour pricing, 5:00 to 8:00 and 14:00 to 19:00 are grouped as medium pricing hours, and 20:00 to 04:00 is grouped as off-peak pricing hours. The electricity tariff at 10:00 is realized as $27.35/kWh, which is noted as the highest electricity pricing and thus, the proposed algorithm shifts the major quantum of loads from 10:00 to off-peak hours. The lowest pricing is $8.1/kWh at the off-peak hour 02:00. Before implementing DSM, the load consumption exceeds about 1855 kW, which contributes to the major rise in total electricity cost. The off-peak load is low, which helps in the possibility of moving the load from peak hours, thereby minimizing the total cost.

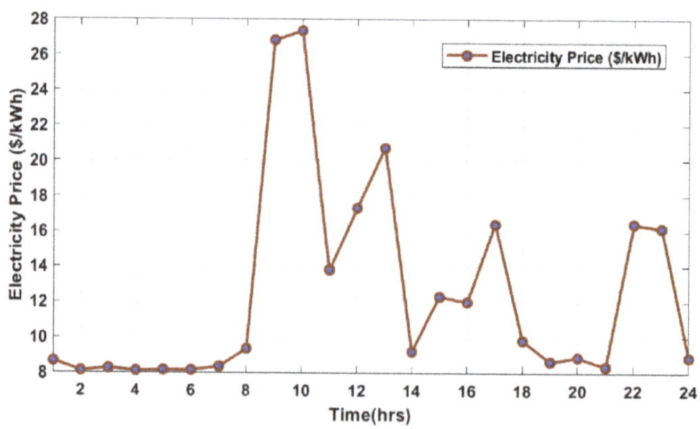

Figure 7. Hourly Electricity Price ($/kWh).

The hybrid BPSOGSA algorithm has the functionalities of both BPSO and GSA. Initially, the gbest of particles of BPSO get randomly distributed to optimally schedule the load types of KCET. The peak load is reduced effectively to 1704.628 kW using the hybrid BPSOGSA optimization algorithm.

Figure 8 compares the demand curves of KCET loads after deploying three different scheduling algorithms, BPSO, BGSA, and BPSOGSA (indicated by yellow, blue, and green colors, respectively), with the demand before deploying BPSOGSA-DSM. The electricity price for each hour is represented by a violet dotted line.

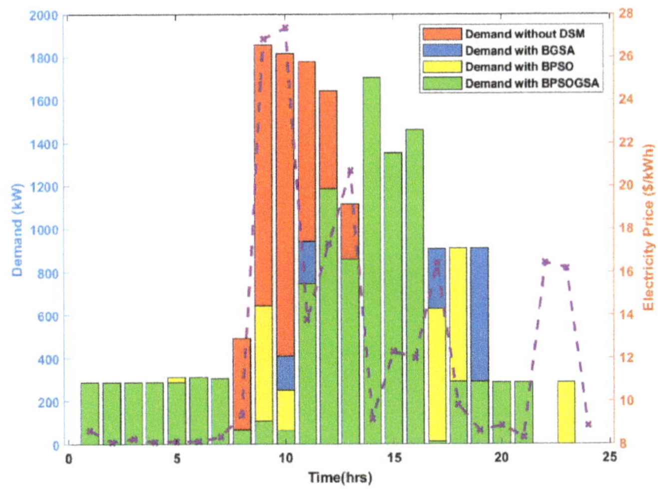

Figure 8. Comparison of demand before and after deployment of DSM with BPSOGSA.

The hybrid BPSOGSA increases the reduction in electricity cost to $716.26, thus proving that the proposed hybrid optimization algorithm yields good results when compared with BPSO and BGSA. The percentage reduction in PAR and Peak demand has improved to 81.295% from 20.98%. Table 1 tabulates the optimized findings of the three optimization algorithms considered along with the decline in PAR, peak demand, and electricity price.

Table 1. Optimized results before and after deployment of DSM techniques.

Criteria	BGSA Indices			BPSO Indices			BPSOGSA Indices		
	PD (kW)	PAR	Cost ($)	PD (kW)	PAR	Cost ($)	PD (kW)	PAR	Cost ($)
Before DSM	1855.4	4.12	2030.6	1855.4	4.12	2030.6	1855.4	4.128	2030.6
After DSM	1756.2	4.012	1532.9	1704.6	3.739	1467.6	1502.2	3.263	1314.4
Reduction	99.24	0.117	497.7	150.84	0.336	562.9	353.2	0.866	716.2
% Reduction	5.349	2.83	24.50	8.129	8.129	27.724	19.03	20.970	35.27

The demand before BPSOGSA-DSM was 1855.47 kW, which is reduced to 1756.23 kW by BGSA, 1704.62 kW by BPSO, and 1502.24 kW by the hybrid BPSOGSA. This makes it clear that it is much more effective in peak demand reduction by shifting 353.23 kW of loads from peak hours of operation. The cost savings achieved by BGSA, BPSO, and BPSOGSA are $497.7, $526.99, and $716.264 respectively. As discussed, the peak hours of operation are from 09:00 to 14:00 and the demand reduction during these hours will boost the reduction in total electricity consumption cost. Figure 9 shows the variation in demand during the peak hours. At 10:00, the demand has fallen from 1855.468 kW to 384.17 kW before and after the deployment of BPSOGSA-DSM. Eventually, this fall will greatly aid a great reduction in electricity consumption costs. Two operating times of the day, 09:00 and 14:00, are considered and the number of load types scheduled at these two hours is analyzed.

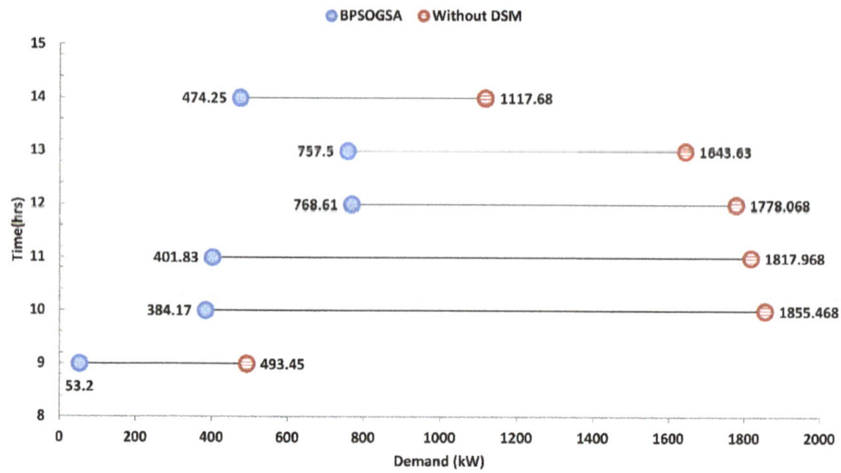

Figure 9. Variations in demand from 9:00 to 14:00 before and after deployment of BPSOGSA-DSM.

Figure 10 depicts the shifting and scheduling of load types for the two pricing times 09:00 and 14:00 before and after deploying the DSM algorithms. The electricity price at 09:00 is 26.82 ($/kWh), which is much higher than that of the price at 14:00 at 9.19 ($/kWh). It is seen that the proposed algorithm shifts the load types from higher pricing to lower pricing. In other words, the individual demand for air conditioner load is 618.35 kW, which is maximum at 09:00 with a pricing of 26.82 ($/kWh) which will increase the overall electricity consumption cost. This load is shifted and scheduled to 14:00 with a lower electricity cost of 9.19 ($/kWh). Thus, the objective of reducing the electricity cost is attained. Each load type is considered as a particle in BPSO, which moves with the force of GSA, thus scheduling loads of KCET optimally. Let us consider the air conditioner of KCET to examine its optimal shifting before and after deploying BPSOGSA-DSM which is shown in Figure 11.

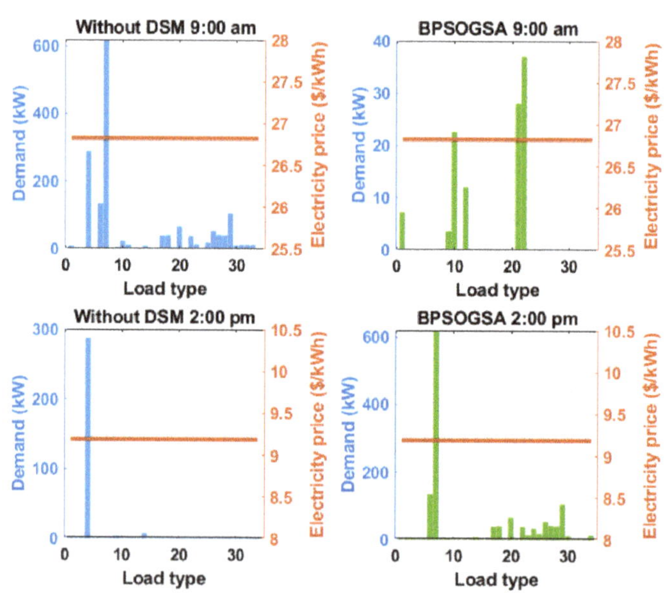

Figure 10. Scheduling of load types before and after BPSOGSA from 09:00 to 14:00.

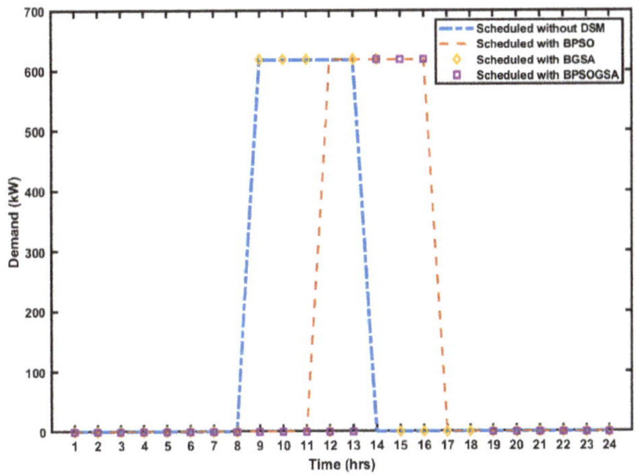

Figure 11. Shifting of 3-phase air conditioner loads before and after deployment of DSM.

Before deploying DSM techniques, the 3-phase air conditioner loads are scheduled for operation at [9–13] hours, where the load is scheduled in peak pricing hours. The BGSA schedules the air conditioner load in [9–11,13,14] which is almost the same as before deploying DSM. This depicts that BGSA does not shift the load effectively to achieve the main objective of electricity cost minimization. After deploying BPSO, the load is shifted to [12–16] a period of operation where the peak pricing hours are neglected. Thus, the total electricity price is reduced considerably. Further, after deploying BPSOGSA, the load is scheduled at [14–18]. The electricity pricing at the 17th and 18th hours is much lower than the 12th and 13th hours. This shifting of loads helps in the reduction of the electricity consumption cost. The performance of algorithms used to plan KCET loads using the proposed DSM method is individually assessed. Table 2 shows the statistical efficiency of BPSO, BGSA, and BPSOGSA algorithms. It can be observed from Table 2 that the statistical performance of BPSOGSA was better than BPSO and BGSA algorithms. The p-values of the BGSA, BPSO, and BPSOGSA algorithms are described in Table 3. The p-values are obtained using Holm, Hommel, and Holland tests with a significance level of $\alpha = 0.05$ for the considered algorithms.

The statistical tests validate the p-value hypothesis of 0.05. Table 4 shows clearly that the BPSOGSA suggested is much superior to the current methods. The statistical details such as the standard deviation of the algorithms were deployed for the test system.

Table 5 shows the load types and their power ratings with the period of operation (in number of hours). The power ratings have been given for the power demand corresponding to each load type. Figure 12 shows the convergence curve of the hybrid BPSOGSA algorithm. The statistical tests performed on these three algorithms depict that the proposed hybrid BPSOGSA is far better than BGSA and BPSO algorithms.

Table 2. Average rankings—KCET load.

Algorithm	Aligned Friedman Test	Friedman Test	Quade Test
BGSA	80.5	3.0	2.9
BPSO	60.5	2.0	2.0
BPSOGSA	30.5	1.0	1.0

Table 3. p-value with significance level α = 0.05—KCET load.

Algorithm	Aligned Friedman Test			Friedman Test			Quade Test		
	Hommel	Holm	Holland	Hommel	Holm	Holland	Hommel	Holm	Holland
BGSA	0.0169	0.016	0.0165	0.0165	0.016	0.0169	0.0165	0.016	0.0169
BPSO	0.025	0.025	0.0252	0.025	0.025	0.0251	0.025	0.025	0.0251
BPSOGSA	0.05	0.05	0.05	0.05	0.05	0.05	0.05	0.05	0.05

Table 4. Statistical details of algorithms.

Algorithm	Best Electricity Cost ($)	Worst Electricity Cost ($)	Mean Electricity Cost ($)	Standard Deviation	Simulation Time (s)
BGSA	1531.54	1535.21	1532.97	1.89	280
BPSO	1465.42	1469.54	1467.68	1.04	242
BPSOGSA	1313.10	1315.22	1314.50	0.82	228

Table 5. Power ratings and period of operation of the load types.

Load Type	Start Time	End Time	Period of Operation (Number of Hours)	Power Demand (kW)
1	9	16	5	0.055
2	8	16	6	0.055
3	10	17	3	0.0385
4	1	23	12	287.09
5	9	20	6	0.013
6	9	16	5	2.5
7	9	16	5	7.45
8	8	16	2	0.003
9	1	23	12	3.5
10	9	16	6	7.5
11	5	10	4	5.5
12	5	10	3	1.5
13	6	10	2	2.2
14	8	17	6	3.7
15	9	16	3	96.42
16	9	16	4	120.8
17	9	16	5	38
18	9	16	5	39
19	9	16	3	3.27
20	9	16	5	64.7
21	9	16	5	28
22	9	16	6	37
23	9	16	4	13
24	9	16	3	32.9
25	9	16	5	17.4
26	9	16	5	51.06
27	9	16	2	39.9
28	9	16	5	39
29	9	16	5	103.99
30	8	15	1	0.5
31	9	16	1	2.5
32	9	16	2	2.5
33	9	16	1	2.5
34	9	16	1	2.5

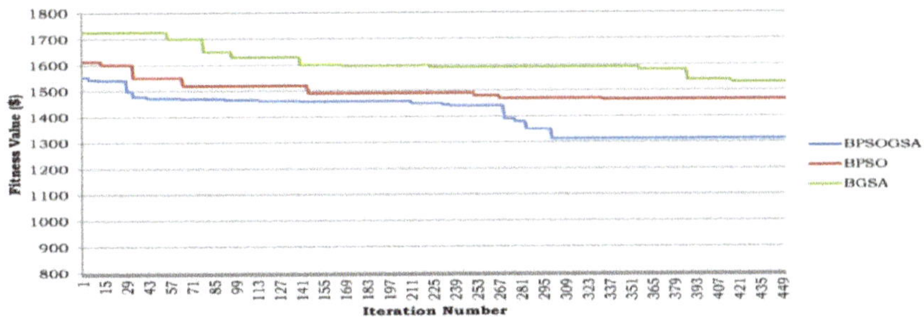

Figure 12. Convergence Curve of the hybrid BPSOGSA.

5.2. Hybrid BPSOGSA-Based Degree of User Comfort (DUC)

The real-time test system where the hybrid BPSOGSA is deployed is set up with 32 kW solar photovoltaic (SPV) generation. Canadian Solar MaxPower CS6U-340 M has been used in KCET. The considered SPV panel has 72 monocrystalline cells with a module efficiency of nearly 17.49%, voltage at open-circuit of 46.2 V, current at short-circuit of 9.48 A, nominal maximum power of 340 W and can be used up to 1500 V DC. Some of the enlisted non-critical institutional loads are categorized as user comfort loads, which have an impact on the satisfaction level of the user to a greater extent. In the proposed method, this level of satisfaction is measured as the DUC. The SPV generation is effectively utilized to increase the DUC with the working of certain user comfort loads.

The hybrid BPSOGSA based DSM technique shifts the operating hours of the load types and optimally schedules them to attain the main objective of electricity cost minimization. This scheduling of loads to different operating hours reduces the user's comfort level. Thus, the SPV generation is utilized to increase the comfort level of the users by applying the hybrid BPSOGSA algorithm. The number of retained hours of operation marks a great impact in increasing the level of satisfaction of the users to the considered load types. Thus, by increasing the number of retained hours of operation and combining the SPV generation, the degree of user comfort (DUC) can be improved.

Table 6 depicts the details of the loads were considered, their operating hours before and after DSM, and the number of non-shifted operational hours before implementing SPV generation. To study and improve the degree of UC, hybrid BPSOGSA is applied along with the SPV generation. The ratio defined here is:

$$Ratio = \left(\frac{Number\ of\ retained\ hours\ of\ operation}{Period\ of\ operation} \right)$$

The DUC increases as the ratio increases (i.e.) DUC is directly related to the number of unchanged operational hours. The period of operation of the load type is constant and so the ratio changes for the number of retained hours of operation. From Figure 13, it can be stated that when the ratio is 0.243, the degree of UC is 13.4. To improve the degree of UC, the generated power from solar energy is utilized.

Table 6. Operating hours of loads before and after DSM.

Load Type	Load Name	Operating Hours before DSM	Operating Hours after DSM	Number of Retained Hours of Operation
2	Ceiling fan 2	[8–13]	[8,11,12,14–16]	3
3	DC fan	[9–11]	[14,16,17]	0
5	LED lamps	[8–13]	[14–16,18–20]	0
6	1 phase air conditioner	[9–13]	[12–16]	2
7	3 phase air conditioner	[9–13]	[14–18]	0
11	Compressor 1	[9–12]	[6–8,10]	1
13	Compressor 2	[6,7]	[7,10]	1
14	Pure water pump	[9–14]	[8,11,14–17]	2

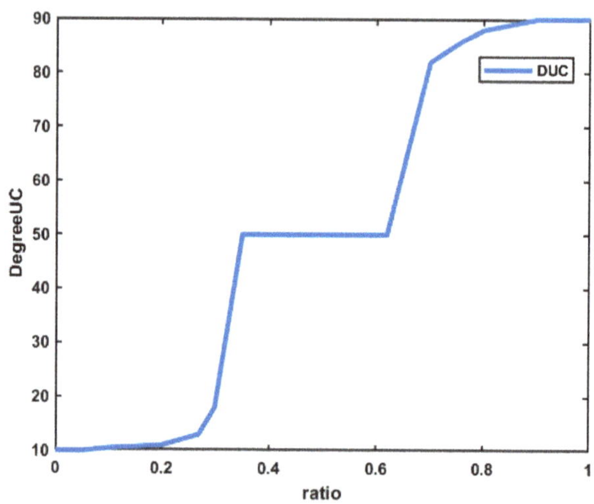

Figure 13. Increasing Degree of UC with the increase in ratio.

5.3. *Real-Time Implementation of SDSMS*

Nowadays, various types of DSM techniques are adopted in the residential sectors as well as institution buildings and commercial building in order to reduce overall electricity consumption cost. The main objective of DSM implementation is to shift the various categories of loads from peak to non-peak hours at the same time without reducing the DUC.

Various DSM approaches with different optimization algorithms were proposed to solve the objective of electricity cost minimization. The hybrid BPSOGSA algorithm optimizes the scheduling of the load types effectively, to attain the main objective of electricity cost minimization at KCET. The SDSMS can decide the power supply to the laboratory loads either from the solar or institutional power line. The measured parameters by SDSMS are compared with the Power Quality Analyser (PQA) is tabulated in Table 7.

Table 7. Comparison of measured parameters of SDSMS with PQA

Model	Phase Voltage (V_{ph})	Phase Current (I_{ph})	Utilization Energy in Peak Hours (kWh)
SDSMS	228.75	7.85	4.84
PQA	229.45	8.25	5.09

The implemented hardware at Renewable Energy Laboratory is shown in Figure 14. The relay unit can be controlled by the microcontroller, based on the changeover timing in the peak period. The PZEM module obtains the electrical parameter from the load power line and sends the data to the microcontroller unit which will send all the information to the cloud server. The consumption details are viewed from any mobile device using the Blynk application.

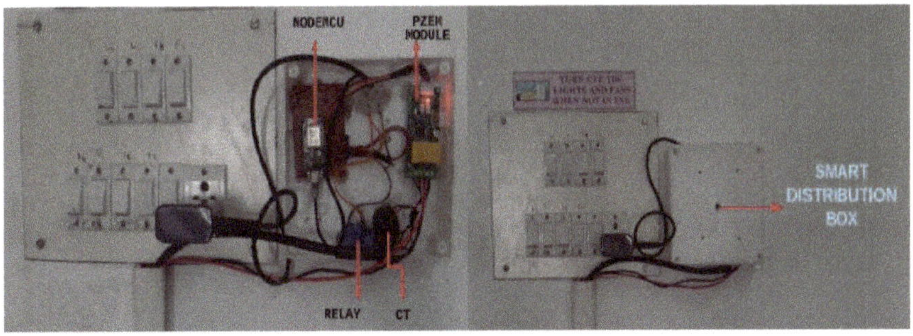

Figure 14. Implemented hardware setup at Renewable Energy Laboratory.

The measured electrical parameters such as energy consumption, voltage, current, and cost using the Blynk application are shown in Figure 15. The electrical parameters and cost are displayed in graph format for viewing the variation at any instant. The changeover timing can be modified according to the user's need through the time scheduler in the Blynk application.

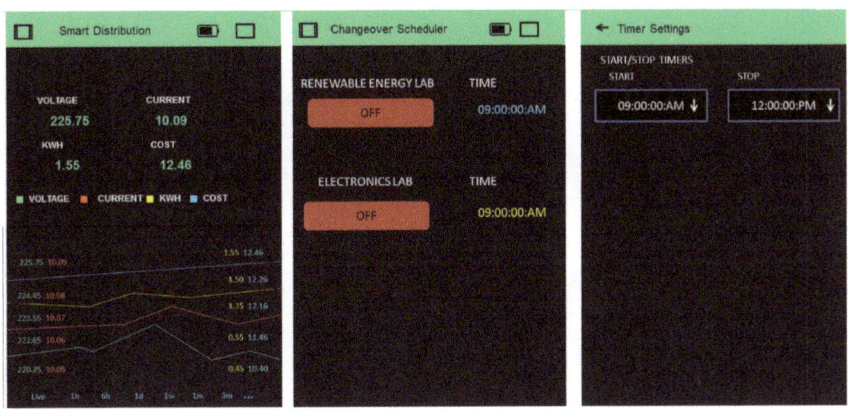

Figure 15. Electrical parameter readings, changeover scheduler, and changeover timing.

6. Conclusions

This paper deploys an Internet of Things (IoT)-based Smart Demand Side Management System (SDSMS) that connects the solar power line for the laboratory loads during peak hours to increase the degree of user comfort (DUC) at an educational institution KCET located at Madurai, Tamilnadu, India. In this paper, DSM for the institutional loads is first implemented using a hybrid BPSOGSA algorithm and the same is integrated as hardware in real time. The KCET loads are categorized into shiftable loads and non-shiftable loads (NSLs) loads. The overall reduction of electricity consumption cost of the institution is achieved by deploying day-ahead load shifting and scheduling in the shiftable loads (SLs). The performance of the hybrid BPSOGSA algorithm has been evaluated in terms of parameters such as PAR and peak demand reduction, and electricity cost minimization by comparing it with individual optimization algorithms BPSO and BGSA using MATLAB. A overall cost reduction is achieved by 35.27% using the proposed hybrid BPSOGSA algorithm, while 27.72% and 24.502% reduction in electricity cost by BPSO, and BGSA respectively. PAR without DSM is 4.12 whereas using BGSA, BPSO, and BPSOGSA, PAR is reduced to 4.01, 3.73, and 3.26 respectively. Furthermore, peak demand without DSM is 1855.47 kW, whereas using BGSA, BPSO, and BPSOGSA, PAR is reduced to 1756.23 kW, 1704.62 kW, and 1502.24 kW, respectively. Hence, BPSOGSA gives the best performance for the minimization of electricity cost, reduction in PAR, and reduction in peak demand. The major highlight of using the hybrid BPSOGSA algorithm-based DSM is that it is effective in eliminating the demerits of both BPSO and GSA algorithms. Further, the degree of user comfort (DUC) has been improved by deploying hybrid BPSOGSA incorporating the Solar PV (SPV) generation at the premises of KCET. The ratio of the number of retained operational hours to the period of operation is directly proportional to the DUC. The use of SPV in the hybrid BPSOGSA algorithm raises the DUC level without ratio dependency. Next, the hardware has been implemented in real time at the Renewable energy laboratory to improve DUC for different load types by optimal scheduling and it reduces the electricity cost during the peak period for the institution, and the corresponding electrical parameters are monitored in real time using the Blynk application.

Author Contributions: Conceptualization, S.C.R.; methodology, S.C.R.; software, A.C.V.D.; validation, S.C.R., J.J.D.N. and T.K.; formal analysis, T.K.; investigation, J.J.D.N.; resources, S.C.R.; data curation, J.J.D.N.; writing—original draft preparation, A.C.V.D.; writing—review and editing, A.C.V.D.; visualization, T.K.; supervision, S.C.R.; project administration, J.J.D.N.; funding acquisition, J.J.D.N. and A.C.V.D. All authors have read and agreed to the published version of the manuscript.

Funding: This research was funded by DST—SERB((File Number: SRG/2019/001946) and Thiagarajar Research Fellowship (TRF)(File Number: TRF/Jul2022/02).

Data Availability Statement: Not applicable.

Acknowledgments: The authors extend their gratitude to the management of their colleges for their constant encouragement and support in doing this research work. The financial assistance under the SRG scheme from DST—SERB is also warmly acknowledged (File Number: SRG/2019/001946). In addition, authors extend their due thanks to the TCE management, Madurai, India for their extensive support and the financial backing from the research excellence strand of the Savitha Project, Thiagarajar Research Fellowship (TRF) scheme (File.no: TRF/Jul2022/02).

Conflicts of Interest: The authors declare that they have no known competing financial interests or personal relationships that could have appeared to influence the work reported in this paper.

Abbreviations

The following abbreviations are used in this manuscript:

ACO	Ant Colony Optimization
BGSA	Binary Gravitational Search Algorithm
BGWO	Binary Grey Wolf Optimization
BPSO	Binary Particle Swarm Optimization
BPSOGSA	Binary Particle Swarm Optimization-Gravitational Search Algorithm
CS	Cuckoo Search Algorithm
DBHS	Discrete Binary Harmony Search algorithm
DEA	Differential Evolution Algorithm
DG	Distributed Generation
DR	Demand Response
DSM	Demand-Side Management
DUC	Degree of user comfort
DUC	Degree of User Comfort
HS	Harmony Search algorithm
IoT	Internet of Things

KCET	Kamaraj College of Engineering and Technology
LCOE	Levelized Cost of Energy
MBDE	Modified-Binary Differential Evolution algorithm
MILP	Mixed Integer Linear Programming
PAR	Peak to average ratio
PAR	Peak to Average Ratio
SA	Simulated Annealing
SDSMS	Smart Demand Side Management System
SDSMS	Smart Demand Side Management System
SPV	Solar Photovoltaic
TAC	Total Annualized Cost
TAE	Total Annualized Emission
WDO	Wind-Driven Optimization algorithm

References

1. Huang, S.H.; Lin, P.C. A modified ant colony optimization algorithm for multi-item inventory routing problems with demand uncertainty. *Transp. Res. Part E Logist. Transp. Rev.* **2010**, *46*, 598–611. [CrossRef]
2. Uniyal, A.; Kumar, A. Comparison of optimal DG placement using CSA, GSA, PSO and GA for minimum real power loss in radial distribution system. In Proceedings of the 2016 IEEE 6th International Conference on Power Systems (ICPS), New Delhi, India, 4–6 March 2016; pp. 1–6.
3. Panigrahi, P.K.; Ghosh, S.; Parhi, D.R. Comparison of GSA, SA and PSO based intelligent controllers for path planning of mobile robot in unknown environment. *Int. J. Electr. Comput. Eng.* **2014**, *8*, 1633–1642.
4. Inkollu, S.R.; Kota, V.R. Optimal setting of FACTS devices for voltage stability improvement using PSO adaptive GSA hybrid algorithm. *Eng. Sci. Technol.* **2016**, *19*, 1166–1176. [CrossRef]
5. Sabri, N.M.; Puteh, M.; Mahmood, M.R. An overview of Gravitational Search Algorithm utilization in optimization problems. In Proceedings of the IEEE 3rd International Conference on System Engineering and Technology, Shah Alam, Malaysia, 19–20 August 2013; pp. 61–66.
6. Javaid, N.; Javaid, S.; Abdul, W.; Ahmed, I.; Almogren, A.; Alamri, A.; Niaz, I.A. A Hybrid Genetic Wind Driven Heuristic Optimization Algorithm for Demand Side Management in Smart Grid. *Energies* **2017**, *10*, 319–319. [CrossRef]
7. Sadeghi, J.; Sadeghi, S.; Niaki, S.T.A. Optimizing a hybrid vendor-managed inventory and transportation problem with fuzzy demand: An improved particle swarm optimization algorithm. *Inf. Sci.* **2014**, *272*, 126–144. [CrossRef]
8. Bounar, N.; Labdai, S.; Boulkroune, A. PSO-GSA based fuzzy sliding mode controller for DFIG-based wind turbine. *ISA Trans.* **2019**, *85*, 177–188. [CrossRef]
9. Sun, S.; Peng, Q. A hybrid PSO-GSA strategy for high-dimensional optimization and microarray data clustering. In Proceedings of the IEEE International Conference on Information and Automation (ICIA), Hailar, China, 28–30 July 2014; pp. 41–46.
10. Hemanth, G.R.; Raja, S.C.; Nesamalar, J.J.D.; Kumar, J. Cost effective energy consumption in a residential building by implementing demand side management in the presence of different classes of power loads. *Adv. Build. Energy Res.* **2020**, *16*, 145–170. [CrossRef]
11. Tamilarasu, K.; Raja, S.C.; Nesamalar, J.J.D.; Rajvikram, M.E.; Mihet-Popa, L. Reinforced Demand Side Management for Educational Institution with Incorporation of User's. *Comfort. Energies* **2021**, *14*, 2855–2855. [CrossRef]
12. Logenthiran, T.; Srinivasan, D.; Phyu, E. Particle swarm optimization for demand side management in smart grid. In Proceedings of the IEEE Innovative Smart Grid Technologies—Asia (ISGT ASIA), Bangkok, Thailand, 3–6 November 2015; pp. 1–6.
13. Zakariazadeh, A.; Jadid, S.; Siano, P. Economic-environmental energy and reserve scheduling of smart distribution systems: A multiobjective mathematical programming approach. . *Energy Convers. Manag.* **2014**, *278*, 151–164. [CrossRef]
14. Park, L.; Jang, Y.; Cho, S.; Kim, J. Residential Demand Response for Renewable Energy Resources in Smart Grid Systems. *IEEE Trans. Ind. Inform.* **2017**, *13*, 3165–3173. [CrossRef]
15. Bhamidi, L.; Sivasubramani, S. Optimal Planning and Operational Strategy of a Residential Microgrid With Demand Side Management. *IEEE Syst. J.* **2020**, *14*, 2624–2632. [CrossRef]
16. Castro, R.C.C. Hybrid Particle Swarm Optimization and Gravitational Search Algorithm for Optimal Sizing of Hybrid Renewable Energy Systems. In Proceedings of the IEEE PES Asia-Pacific Power and Energy Engineering Conference (APPEEC), Macao, China, 1–4 December 2019; pp. 1–6.
17. Hatamlou, A.; Abdullah, S.; Othman, Z. Gravitational search algorithm with heuristic search for clustering problems. In Proceedings of the 2011 3rd Conference on Data Mining and Optimization (DMO), Putrajaya, Malaysia, 28–29 June 2011; pp. 190–193.
18. Shaw, B.; Mukherjee, V.; Ghoshal, S.P. A novel opposition-based gravitational search algorithm for combined economic and emission dispatch problems of power systems. *Int. J. Electr. Power Energy Syst.* **2012**, *35*, 21–33. [CrossRef]
19. Zhang, Y.; Wu, L.; Zhang, Y.; Wang, J. Immune gravitation inspired optimization algorithm. In Proceedings of the 7th international conference on Advanced Intelligent Computing, Zhengzhou, China, 11–14 August 2011; Springer: Berlin/Heidelberg, Germany, 2011; pp. 178–185.

20. Li, C.; Zhou, J. Parameters identification of hydraulic turbine governing system using improved gravitational search algorithm. . *Energy Convers. Manag.* **2011**, *52*, 374–381. [CrossRef]
21. Rashedi, E.; Nezamabadi-Pour, H.; Saryazdi, S. GSA: A Gravitational Search Algorithm. *Inf. Sci.* **2009**, *179*, 2232–2248. [CrossRef]
22. Rashedi, E.; Nezamabadi-Pour, H.; Saryazdi, S. BGSA: Binary gravitational search algorithm. *Nat. Comput.* **2009**, *9*, 727–745. [CrossRef]
23. Wang, L.; Xu, Y.; Mao, Y.; Fei, M. A Discrete Harmony Search Algorithm. In *Life System Modeling and Intelligent Computing*; Springer: Berlin/Heidelberg, Germany, 2010; pp. 37–43.
24. Wang, L.; Fu, X.; Menhas, M.I.; Fei, M. A Modified Binary Differential Evolution Algorithm. In *Life System Modeling and Intelligent Computing*; Springer: Berlin/Heidelberg, Germany, 2010; pp. 49–57.
25. Kennedy, J.; Eberhart, R.C. A discrete binary version of the particle swarm algorithm. In Proceedings of the IEEE International Conference on Systems, Man, and Cybernetics. Computational Cybernetics and Simulation, Orlando, FL, USA, 12–15 October 1997; Volume 5, pp. 4104–4108.
26. Mirjalili, S.; Wang, G.G.; Coelho, L.S. Binary optimization using hybrid particle swarm optimization and gravitational search algorithm. *Neural Comput. Appl.* **2014**, *25*, 1423–1435. [CrossRef]
27. Mirjalili, S.; Hashim, S.Z.M.; Sardroudi, H. Training feedforward neural networks using hybrid particle swarm optimization and gravitational search algorithm. *Appl. Math. Comput.* **2012**, *218*, 11125–11137. [CrossRef]
28. Pournaras, E.; Jung, S.; Yadhunathan, S.; Zhang, H.; Fang, X. Socio-technical smart grid optimization via decentralized charge control of electric vehicles. *Appl. Soft Comput.* **2019**, *82*, 105573–105573. [CrossRef]
29. Llanos, J.; Morales, R.; Núñez, A.; Sáez, D.; Lacalle, M.; Marín, L.G.; Hernandez, R.; Lanas, F. Load estimation for microgrid planning based on a self-organizing map methodology. *Appl. Soft Comput.* **2017**, *53*, 323–335. [CrossRef]
30. Karthick, T.; Raja, S.C.; Nesamalar, J.J.D.; Chandrasekaran, K. Design of IoT based smart compact energy meter for monitoring and controlling the usage of energy and power quality issues with demand side management for a commercial building. *Sustain. Energy, Grids Netw.* **2021**, *26*, 100454–100454.
31. Al-Hassan, E.; Shareef, H.; Islam, M.M.; Wahyudie, A.; Abdrabou, A.A. Improved Smart Power Socket for Monitoring and Controlling Electrical Home Appliances. *IEEE Access* **2018**, *6*, 49292–49305. [CrossRef]
32. Cao, K. Achieving Reliable and Secure Communications in Wireless-Powered NOMA Systems. *IEEE Trans. Veh. Technol.* **1978**, *70*. [CrossRef]
33. Cao, K.; Ding, H.; Li, W.; Lv, L.; Gao, M.; Gong, F.; Wang, B. On the ergodic secrecy capacity of intelligent reflecting surface aided wireless powered communication systems. *IEEE Wirel. Commun. Lett.* **2022**, *11*, 2275–22759. [CrossRef]
34. Yan, S.R.; Pirooznia, S.; Heidari, A.; Navimipour, N.J.; Unal, M. Implementation of a Product-Recommender System in an IoT-Based Smart Shopping Using Fuzzy Logic and Apriori Algorithm. *IEEE Trans. Eng. Manag.* **2022**. [CrossRef]
35. Lv, Z.; Song, H. Mobile Internet of Things Under Data Physical Fusion Technology. *IEEE Internet Things J.* **2020**, *7*, 4616–4624. [CrossRef]
36. Li, J.; Dend, Y.; Sun, W.; Li, W.; Li, R.; Li, Q.; Liu, Z. Resource Orchestration of Cloud-Edge-based Smart Grid Fault Detection. *ACM Trans. Sen. Netw* **2022**, *18*, 46. [CrossRef]
37. Dang, W.; Guo, J.; Liu, M.; Liu, S.; Yang, B. A Semi-Supervised Extreme Learning Machine Algorithm Based on the New Weighted Kernel for Machine Smell. *Appl. Sci.* **2022**, *12*, 9213–9213. [CrossRef]
38. Ge, L.; Li, Y.; Li, Y.; Yan, J.; Sun, Y. Smart Distribution Network Situation Awareness for High-Quality Operation and Maintenance: A Brief Review. *Energies* **2022**, *15*, 828–828. [CrossRef]

energies

MDPI

Article

Power Quality Transient Detection and Characterization Using Deep Learning Techniques

Nuno M. Rodrigues [1,†], Fernando M. Janeiro [2,†] and Pedro M. Ramos [1,*,†]

1 Instituto de Telecomunicações, Instituto Superior Técnico, Universidade de Lisboa, 1049-001 Lisboa, Portugal
2 Instituto de Telecomunicações, Universidade de Évora, 7000-671 Évora, Portugal
* Correspondence: pedro.m.ramos@tecnico.ulisboa.pt; Tel.: +351-218-418-485
† These authors contributed equally to this work.

Abstract: Power quality issues can affect the performance of devices powered by the grid and can, in severe cases, permanently damage connected devices. Events that affect power quality include sags, swells, waveform distortions and transients. Transients are one of the most common power quality disturbances and are caused by lightning strikes or switching activities among power-grid-connected systems and devices. Transients can reach very high magnitudes, and their duration spans from nanoseconds to milliseconds. This study proposed a deep-learning-based technique that was supported by convolutional neural networks and a bidirectional long short-term memory approach in order to detect and characterize power-quality transients. The method was validated (i.e., benchmarked) using an alternative algorithm that had been previously validated according to a digital high-pass filter and a morphological closing operation. The training and performance assessments were carried out using actual power-grid-measured data and events.

Keywords: power quality monitoring; power grid measurements; transient detection; transient characterization; deep learning; convolutional neural networks

Citation: Rodrigues, N.M.; Janeiro, F.M.; Ramos, P.M. Power Quality Transient Detection and Characterization Using Deep Learning Techniques. *Energies* **2023**, *16*, 1915. https://doi.org/10.3390/en16041915

Academic Editor: Miguel Jiménez Carrizosa

Received: 29 December 2022
Revised: 6 February 2023
Accepted: 13 February 2023
Published: 15 February 2023

1. Introduction

Power quality (PQ) monitoring consists of measuring and detecting events that disturb the grid-supplied waveform and may affect the normal behavior of electrical equipment. Power quality events include overvoltages, undervoltages, sags, swells, waveform distortions, flickers and transients [1]. In particular, transients occur more frequently than other events and are short-term disturbances with amplitudes starting at 5–10 % of the rated voltage. Electrical equipment are designed to resist PQ events up to a certain level. If the PQ events exceed certain thresholds, either in duration or amplitude, permanent damage may occur. The acceptable limits are defined by the IEEE and IEC international standards, such as IEEE 1159-2019 [1] and IEC 61000-2-2 [2]. Although measurement and event detection methods have been proposed by these international standards, PQ remains an active research topic, especially in the area of signal processing, with the development and proposal of new algorithms and techniques [3–5].

Artificial neural networks (ANNs) are inspired by the brain network of neurons. The inputs of each neuron are the weighted combinations of the outputs of other neurons, together with a bias component. This combination is applied to an activation function, which provides the non-linearity required by these methods in order to determine the neuron output [6]. ANNs have been used in diverse areas of science, usually for pattern recognition [7] but also in measurements [8] and non-destructive testing [9]. An ANN is trained by examples, using a back-propagation algorithm [6] until it reaches a suitable accuracy. Recent advances in computing power, availability of data and advances on network topologies have led to the development of deep learning (DL) [10]. The main advantages of DL include the ability to use raw data, minimal human intervention and the capacity to search for deeply embedded patterns in the training data. This concept has enabled breakthrough results in many different areas of science and technology, including self-driving cars [11], facial recognition [12], and media content suggestions [13], among many others. The use of DL for power quality event detection and characterization is no exception, with some recent results presented in the literature [14–17]. However, these have been limited only to training sets with simulated/synthetic waveforms and without any real measured data, which cannot capture the full scale of imperfections that characterize real grid-acquired waveforms.

The most frequent PQ events are transients, and many different methods have been used to detect and their amplitude and duration. These include, for example, wavelets [18], time-frequency [19] and morphology methods [20]. The latter can be complex, involving filtering that is followed by dilation and erosion operations, according to a suitable threshold to discard the low amplitude transients. This study proposed the use of deep learning to detect and estimate the amplitude and duration of transients in the electrical grid. To achieve this goal, a two-stage approach was proposed, in which a convolutional neural network (CNN) was used to detect the presence of transients, which fed a second network composed of a convolutional neural network and a bidirectional long short-term memory (BiLSTM) to estimate the transient duration and amplitude. The two networks were trained and validated with actual power grid events that were stored in a publicly available database [21]. The events stored in the database had been detected and characterized by the transient detection algorithm presented in [20]. Therefore, this paper presents the results of the development of a deep learning network, trained and validated with real acquired signals, for the detection and characterization of PQ transient events.

This paper is divided into four sections, including the Introduction and the Conclusions. Section 2 presents a characterization of the measured transients that were used to train, validate and test the proposed DL neural network. Section 3 describes the DL network used for transient detection and the subsequent network that estimated the transient amplitude and duration as well as the main results of this paper.

2. Power Quality Transient Characterization

In Figure 1, four examples of the acquired power grid voltage transients with different characteristics are presented. The amplitudes are shown in per unit (pu) values, which are typical in power quality, (e.g., [1]), to normalize the amplitudes using, in this case, the power grid voltage nominal RMS voltage, which was 230 V in Portugal. The power grid voltages were continuously acquired at 12.5 kS/s using an LEM Hall Effect sensor (LEM LV 25-P) and a 16-bit data acquisition board (DAQ) from National Instruments—NI-USB-9215. The segments were 50 ms long, which corresponded to 2.5 nominal periods at 50 Hz, with 625 samples for each segment.

Figure 1 shows the transients at different power grid voltage phases with different amplitudes, phase locations and durations. Figure 1a shows an oscillatory transient while b and c show negative transients, and the waveform in d includes a positive transient [1]. Notice that other power quality events were not considered here, as the focus was centered only on transient detection and characterization. The represented data showed, in addition to the transient, some harmonic distortion typical in low voltage (LV) power grid voltages.

Transient detection and characterization (e.g., amplitude and duration), can be achieved for example, using wavelet decomposition [22,23] or using a high-pass digital filter followed by a closing morphological operation. The high-pass filter and closing morphological operation method was presented in [24] and implemented in an embedded measurement system described in [20]. This method was used in this study as the reference method (i.e., benchmark) because it was also the process used to detect and measure the real transients in the event database [21] that was used.

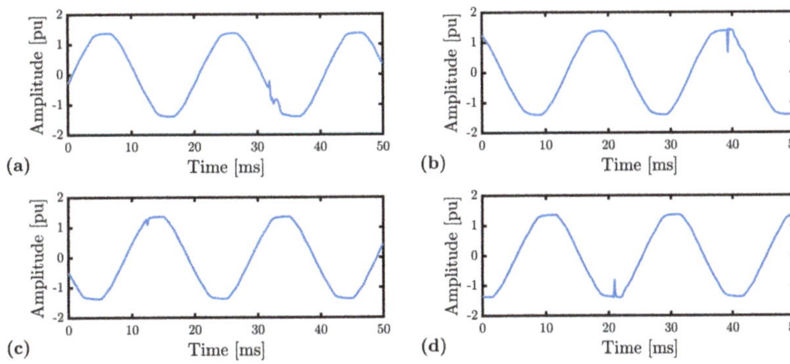

Figure 1. (a–d) Example of four acquired power grid voltage normalized segments. Each segment has 625 acquired samples with 16-bit DAQ at 12.5 kS/s, which corresponds to segments with 50 ms. The four examples highlight the difference in transient amplitudes, durations and relative phase locations.

Figure 2 represents an example of the high-pass filter and morphological algorithm operation used to detect the transients and estimate the transient characteristic parameters (i.e., amplitude and duration). Figure 2a shows an example of an oscillatory transient near 32 ms. The high-pass filter extracts the power grid voltage fundamental component, and the filter-output absolute normalized voltage is represented in Figure 2b. The closing morphological operation, which relied on the dilation operation followed by the erosion, grouped together consecutive oscillations as part of the same transient. Figure 2c shows the result of the morphology closing operation, where the multiple oscillations were joined together to define one unique transient event.

Transient detection was performed by a threshold analysis of the closing operation results (in this paper, a 0.1 pu threshold was considered). If the closing operation results exceeded the pre-defined threshold, a transient was considered detected. The transient amplitude was the maximum of the closing operation, and the transient duration was estimated by the time interval between the first and second interceptions of the closing operation results, with the threshold value using interpolation. For the example in Figure 2, the transient had a 0.119 pu amplitude and 0.72 ms duration. Notice that the harmonic distortions in Figure 2a were present at the filter output, as shown in Figure 2b, but they were not enough to trigger the transient detection process.

The complete dataset used in this work had 9768 transients, and the histogram distribution of the amplitude of these transients is shown in Figure 3. Notice that since the threshold transient detection was set at 0.1 pu, there were no transients with an amplitude below this value. The highest recorded transient in this dataset had a 1.997 pu amplitude.

Figure 4 presents the histogram of the transient durations for the complete dataset. Recall that the duration values were obtained by interpolation from the closing morphological operation interceptions, according to the 0.1 pu threshold, and have, therefore, better resolution than the inverse of the sampling rate (80 µs). The maximum recorded transient duration was 6.6 ms. Notice that there was a significant amount of transients with duration values near 0.2 ms.

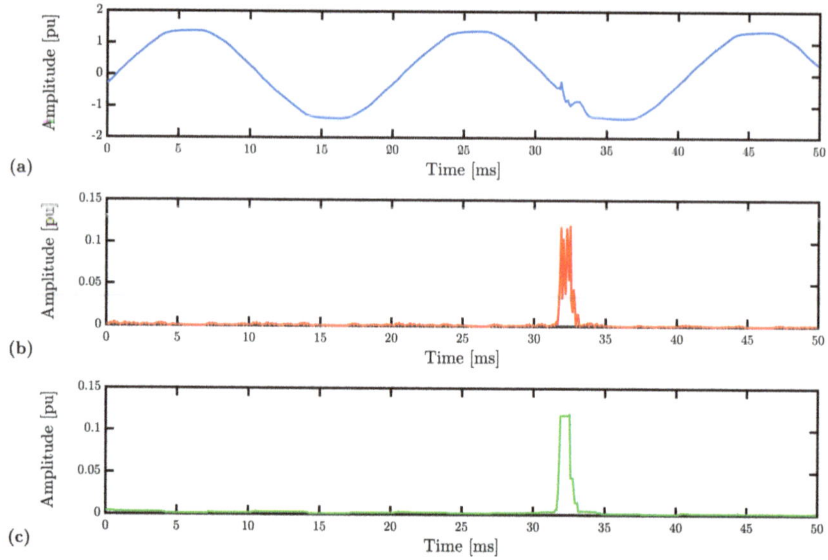

Figure 2. Example of transient detection and characterization using the high-pass filter and closing morphology operation. The input segment is represented in (**a**) while the absolute values of the filter output are presented in (**b**). The result of the closing morphological operation is depicted in (**c**).

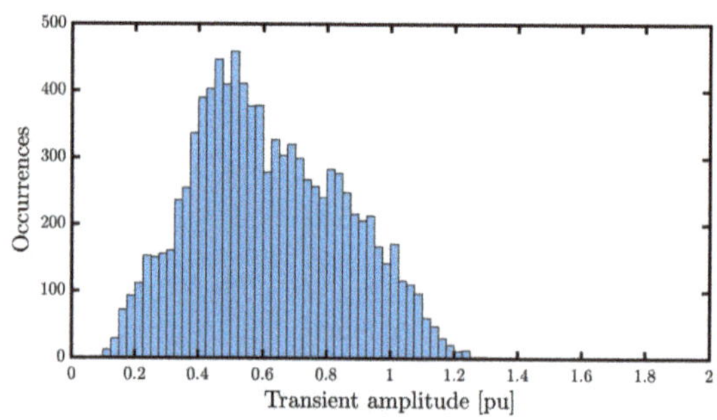

Figure 3. Histogram of the transient amplitudes in the 9768 database segments. Notice that the original transient detection used to build the database ignored events with transient amplitudes below 0.1 pu, and, therefore, there are no events below that amplitude.

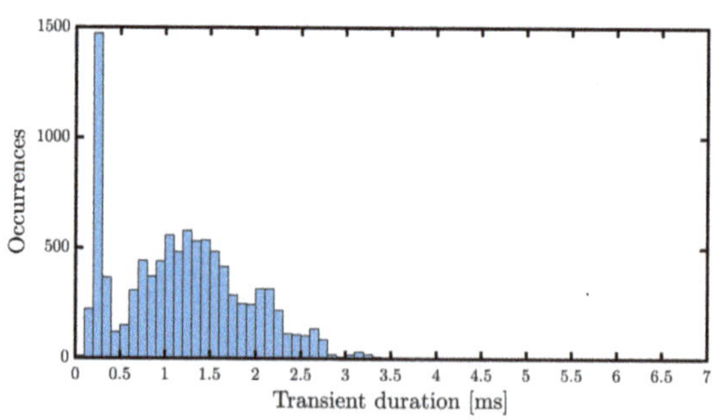

Figure 4. Transient duration histogram of the recorded transient database. The duration was obtained by interpolation of the instances where the closing operation crossed the threshold value.

Finally, Figure 5 depicts the phase location histogram of the recorded transients. Notice that the phase location was not uniform, with more transients occurring near or around 180°.

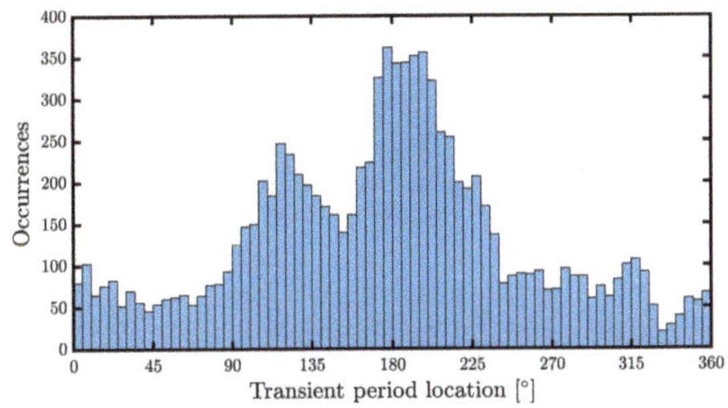

Figure 5. Histogram of the phase location of the transients included in the measured database.

3. Deep Learning for PQ Transient Detection and Characterization

While there are many different architectures for deep-learning networks, the convolutional neural network (CNN) is especially suited for image analysis and classification. These networks have the ability to extract hidden patterns in the provided information by training their internal filters. Therefore, since the problem under study corresponded to the detection and characterization of transients in acquired data segments, the proposed algorithm was based on CNNs. The theory of CNNs is beyond the scope of this work, and a good survey of their theory and applications can be found in [25].

The proposed deep-learning architecture for transient detection and characterization is depicted in Figure 6. It was a two-step process, where the purpose of the first step was to detect the presence of a transient while the second step was only used when a transient was detected, and its objective was to estimate the amplitude and duration of the transient.

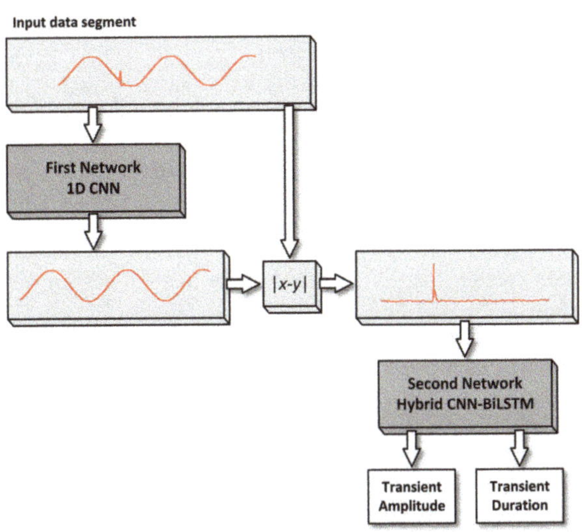

Figure 6. Topology of the proposed deep-learning architecture for the detection and characterization of power quality transients.

The objective of the first step was to extract the input voltage waveforms without transients from the 625-sample input time series. This step was obtained with a 1D CNN, as described in Section 3.1. Power quality transients appeared as spikes in the absolute difference between the input time series and the extracted waveform. Transients were detected by comparison of the absolute difference with a pre-defined threshold.

When a transient was detected, the objective of the second step was to extract information that could be used to estimate the transient amplitude and duration. The network used in the second step was a hybrid CNN-BiLSTM, as described in Section 3.2.

3.1. Transient Detection

The objective of the first network was to obtain the waveform without the transient from the 625-sample acquired and amplitude normalized time series. The topology used in this step is depicted in Figure 7 and includes four layers.

This network was a 1D CNN where the first layer was a convolutional layer, which contains multiple filters with fixed lengths, and its objective is to extract different levels of information from the input time series. Each filter had different coefficients, and their numbers depended on the filter length. The output of each filter was the convolution of the filter with the input time series, along with a stride that was set for the whole layer. Since the objective of this step was to obtain a time series with a length that was the same as the input time series (i.e., 625 samples in this particular application), the stride was set to 1, and padding had to be used to ensure the filter outputs also had the same length. The convolutional layer had 200 filters with lengths of 240 and 200 bias parameters (one for each filter), which resulted in a total of 48 200 optimization parameters. The output of the convolutional layer were 200 time series (the convolution filter outputs), each with 625 samples.

The next three layers were fully connected layers (FCNs), in which the first received the 200×625 data from the convolutional layer and output 100×625, for a total of 20 100 optimization parameters. The second FCN received the 100×625 from the first FCN and output 100×625, for a total of 10 100 optimization parameters. The third and final FCN received the 100×625 from the second FCN and output the final 625 time series, with 101 optimization parameters. Overall, the 1D CNN of this first step had 78 501 parameters that were then optimized during the back-propagation training process.

In DL applications, the hyperparameters that define each layer are defined before the training process. If the desired goal is not achieved, these hyperparameters are adjusted, typically to increase the number of degrees of freedom. The hyperparameters were adjusted using a random search [26] while avoiding any overfitting of the training data. Although other methods for hyperparameter-tuning are available (e.g., grid search or Bayesian search), the use of random search was simple and fast, and it achieved good training results. In this 1D CNN network, the hyperparameters included the number of filters; their size and stride for the convolutional layer; and the number of fully connected layers as well as their distribution to obtain the 625-sample output waveforms without transients.

The training process consisted of inputting the time-series segments of the acquired power grid voltage without PQ events but with the normal voltage grid distortions and noise. Then, we trained the network to estimate the same time series. Back-propagation used the differences between the input time series and the network-estimated time series to adjust the set of optimization parameters of the network, so that the network output was similar to the input (only in the training, validation and test steps, as these datasets did not contain transients). Once the network was trained, whenever a time series with a transient was inserted, the network should output the time series without the event. The difference between the input and output time series indicated if a transient had been detected and included information about the transient for its characterization, namely, the transient amplitude and duration, which was addressed in the second step, as depicted in Figure 6.

For the training process of the 1D CNN, a dataset of 5585 time series, with 625 samples per segment without transients, was used. The validation used a different dataset with 621 segments while the test set had 470 segments. Figure 8 shows four examples of the trained network inputs and outputs, where the blue line corresponds to the network time series input (on the left-side vertical axis) and the red line is the difference between the input and output of the network (right-side vertical axis). In this figure, only one period of the nominal power grid voltage (20 ms) is shown to improve the transient and spike visibility. The example in Figure 8a corresponded to a time series without a transient, and the difference showed the absence of any significant spikes. The difference was always below 0.05 pu, and the harmonic distortions were not registered. The examples in the remaining subplots of Figure 8 correspond to the input time series with transients, and the corresponding spikes are clearly visible and aligned with the time series transients. The situation depicted in Figure 8c had the largest represented amplitude transient, which exceeded 0.5 pu near 47 ms.

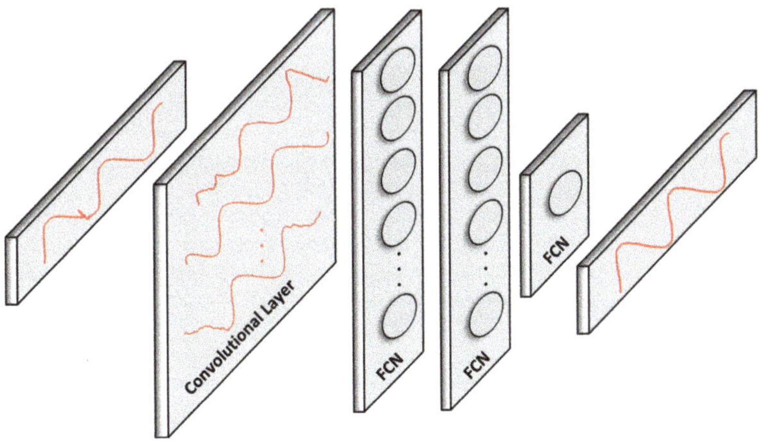

Figure 7. Internal topology of the 1D CNN that aims to extract the power grid voltage waveform without transients. This was then used to detect if a transient occurred by subtracting it from the input time series. The second network used the absolute difference to estimate the transient amplitude and duration.

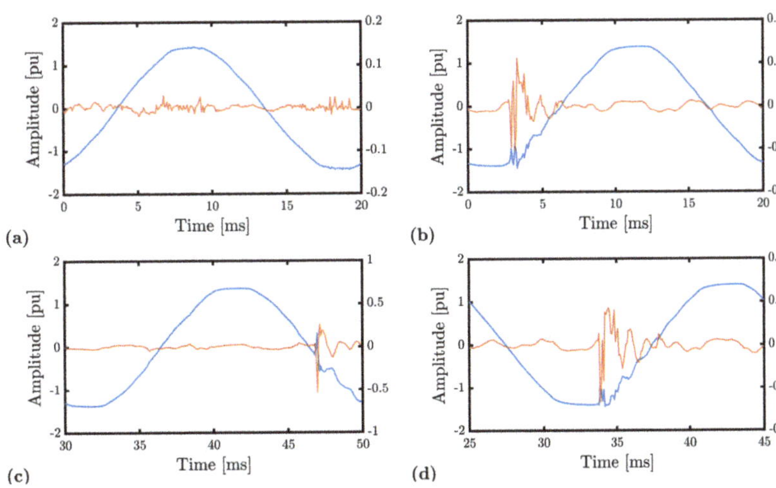

Figure 8. Examples of the operation of the 1D CNN: (**b**–**d**) depict transients while (**a**) is a transient-free segment. The blue lines depict the input time series (left vertical axis) while the red lines represent the difference between the input and output time series (right vertical axis).

3.2. *Transient Parameter Estimation*

The second step of the proposed method began with the absolute difference between the acquired segment and the output segment from the previous step, as shown in Figure 6, and its objective was to estimate the transient amplitude ($\widehat{A_i}$) and duration ($\widehat{D_i}$). The learning process used the transient amplitude (A_i) and duration (D_i) values obtained by the high-pass filter and morphology operation, as described in Section 2. Training was performed using the mean-squared error (MSE) weighted by the average values of the transient amplitudes ($\overline{A}$) and transient durations ($\overline{D}$)

$$\mathrm{MSE}_{\mathrm{weighted}} = \frac{1}{N}\left[\sum_{i=1}^{N}\left(\frac{\widehat{A_i}-A_i}{\overline{A}}\right)^2 + \sum_{i=1}^{N}\left(\frac{\widehat{D_i}-D_i}{\overline{D}}\right)^2\right] \quad (1)$$

where N is the number of data segments used. This weighted MSE was necessary because the two estimated parameters had significantly different orders of magnitude (i.e., the average transient amplitude was $\overline{A} = 0.347\,\mathrm{pu}$, and the average transient duration was $\overline{D} = 0.702\,\mathrm{ms}$). If the same weight were used, the training process would tend to give more relevancy (i.e., weight) to the highest magnitude parameter (in this case, the transient amplitude). In this situation, the lowest magnitude parameter estimation, the transient duration, would become irrelevant in the optimization process, which would result in poor transient duration estimation.

The proposed network topology for this second stage, represented in Figure 9, was a hybrid CNN–BiLSTM. It began with a convolutional layer with 64 filters, with a size of 15, a stride of 1 and no padding, as in this case, there was no need to maintain the size of the time series. The output of this layer was 64×611 with 15×64 weights and 64 biases, for a total of 1024 optimization parameters. The output of the convolutional layer was fed into a bidirectional long short-term memory (BiLSTM) layer with 100 hidden units. The BiLSTM was used because it could learn from the dependencies of the time-series sequence data. The bidirectional version considered the dependencies in both directions by adding one extra LSTM layer with the input data in the reverse order. This BiLSTM layer had 64×800 input weights, 800×100 recurrent weights and 800 output biases, for a total of 132 000 optimization parameters, and it output 200 values. The final layer was a fully connected neural (FCN) network, which output the two estimated parameters (i.e., transient amplitude and duration). This layer had 2×200 weights and 2 biases, for a total of 402 parameters, and these were then optimized during the learning process using the cost function $\mathrm{MSE}_{\mathrm{weighted}}$ from (1). The complete network had a total of 133 426 parameters that were optimized during the back-propagation-based training process.

This network was trained with 7326 acquired data segments with real measured transients that had been pre-processed by the first network. For validation, a set with 1466 similar data segments was used. The testing stage was conducted with a different set with 976 segments.

The results obtained from the test set are shown in Figure 10. Figure 10a depicts the estimated transient amplitudes as a function of the actual transient amplitudes (i.e., benchmarks) while Figure 10b is the corresponding representation for the transient duration. In Figure 10a, the 0.1 pu lower limit of the transient amplitude of the used segments is clearly visible. Figure 10c,d show the estimated transient amplitude and duration error histograms. The proposed method was better at estimating the transient amplitude than the transient duration. This was a reflection of the complexity in estimating the transient duration, which was tied to the threshold that had been used in the baseline algorithm (i.e., high-pass filter and morphological operation) and had never been explicitly used in the DL networks. Specifically, from the absolute difference between the input segment and the clear segment estimated by the first 1D CNN, this two-step hybrid CNN–BiLSTM network must extract the transient duration using only the target values input during the training process as a reference.

The root-mean-square error (RMSE) of the transient amplitude, as obtained using the benchmark values from the filter and closing morphology algorithm, was 0.0197 pu, and the average error was −0.00037 pu. As for the transient duration, the RMSE was 0.2024 ms, and the average error was 0.00039 ms. The estimated amplitude error interval

$[-0.045; 0.034]$ pu was based on 95 % of the situations shown in Figure 10. For the estimated duration error, the corresponding 95 % interval was $[-0.45; 0.33]$ ms. These results again highlighted that the proposed method could achieve better results when estimating the transient amplitude, as compared to the estimated transient duration.

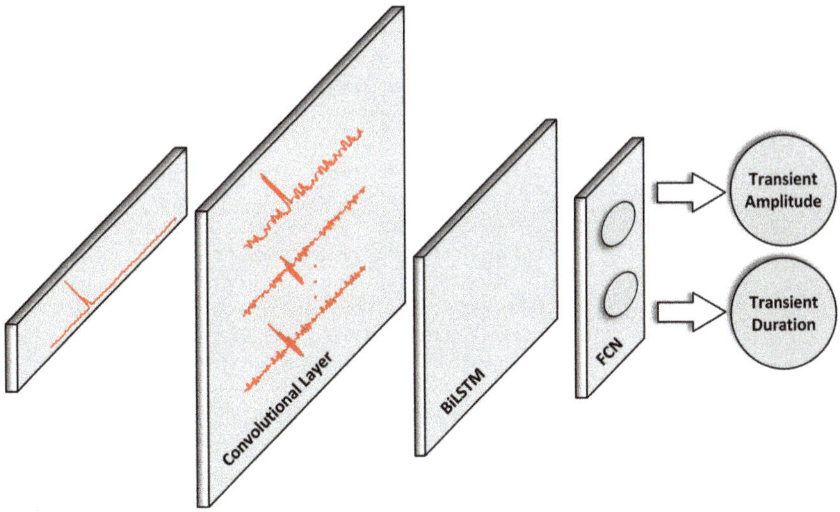

Figure 9. Hybrid CNN–BiLSTM topology for the second step, which receives the absolute difference from the output of the first 1D CNN as well as the acquired data and estimates the transient duration and amplitude.

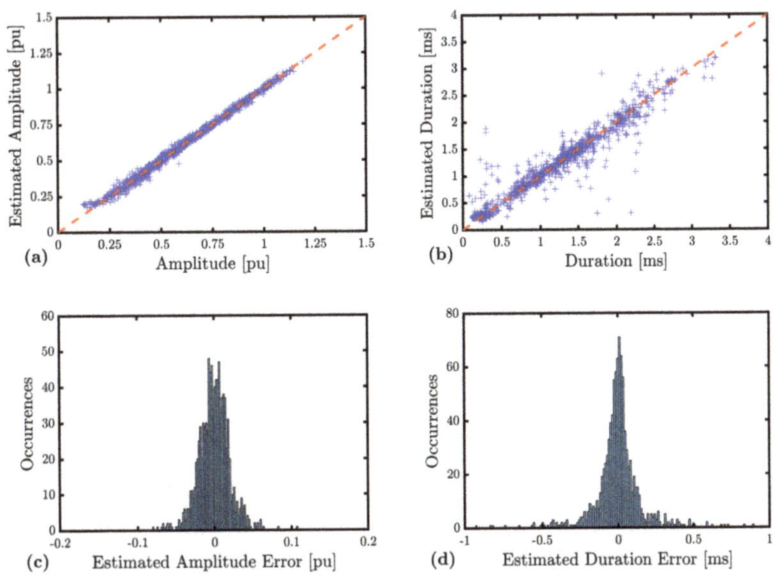

Figure 10. Test set results. In (**a**), the relation between the real transient amplitude (i.e., benchmark) and the proposed method's estimated amplitude is shown while in (**b**), the corresponding relation for the transient duration is plotted. The histogram of the estimated transient amplitude error is shown in (**c**), and the estimated transient duration error is shown in (**d**).

4. Conclusions

This paper addressed the detection and characterization of power quality transients, which are the most common type of disturbances in the power grid. Although widely studied with different approaches, as described in the scientific literature, they usually require technical knowledge and advanced finely tuned signal processing algorithms, such as filtering, morphology operations and wavelets, among others. Previous studies available in the literature have applied deep learning methods to evaluate power quality but only with simulated/synthetic waveforms, which do not include the diverse characteristics of actual grid waveforms. The approach described in this paper used only real measured data to study the performance of deep learning on real grid transients. The strategy used for the deep-learning network divided the problem into two steps: the detection of transients and the characterization of amplitude and duration. A 1D CNN was optimized to extract the voltage waveform without transients, which was not a pure sinewave due to the presence of harmonics and waveform distortions. The second network, a hybrid CNN–BiLSTM, received (as input) the absolute difference between the original acquired signal and the output of the first network, as shown in Figure 6. This network was used to estimate transient parameters and was composed of a convolutional layer, followed by a bidirectional long short-term memory layer, and then completed with a fully connected layer.

The proposed architecture was trained with real acquired data, with and without transients, and the DL test step resulted in a transient RMSE amplitude of 0.0197 pu and 0.2024 ms for the transient duration. This deep-learning approach performed better when estimating the amplitude of the transient, as compared to its duration estimation, as shown in Figure 10. These results validated the use of deep-learning algorithms for PQ transient detection and characterization, using real measured data for training and validation.

Future work will include the implementation of the trained networks in an embedded measurement system (e.g., in an STM32Cube.AI) for the continuous detection and characterization of real-time power grid transient events. The assessment of the performance of the measurement system will be compared with previously developed embedded measurement systems that use a high-pass filter and morphological operations.

Author Contributions: Conceptualization and methodology, N.M.R., F.M.J. and P.M.R.; software, N.M.R.; writing, review and editing, N.M.R., F.M.J. and P.M.R. All authors have read and agreed to the published version of the manuscript.

Funding: This work was developed under the PhD program of the Fundaç ao para a Ciência e a Tecnologia (FCT) reference SFRH/BD/130327/2017 and is funded by FCT/MCTES through national funds and, when applicable, co-funded EU funds, under the project UIDB/EEA/50008/2020.

Data Availability Statement: Not applicable.

Conflicts of Interest: The authors declare no conflict of interest.

References

1. *IEEE 1159-2019*; (Revision of IEEE Std 1159-2009). IEEE International Standard—IEEE Recommended Practice for Monitoring Electric Power Quality. IEEE: New York, NY, USA, 2019. [CrossRef]
2. *IEC 61000-2-2*; IEC International Standard—Electromagnetic Compatibility (EMC)—Environment—Compatibility Levels for Low-Frequency Conducted Disturbances and Signaling in Public Low-Voltage Power Supply Systems, 2nd ed. IEC: Geneva, Switzerland, 2018.
3. Rodrigues, N.; Janeiro, F.; Ramos, P. Implementation of Goertzel-based frequency estimation for power quality monitoring in embedded measurement systems. *Metrol. Meas. Syst.* **2022**, *29*, 455–468. [CrossRef]
4. Bandla, P.; Vairavasundaram, I.; Teekaraman, Y.; Kuppusamy, R.; Nikolovski, S. Real time sustainable power quality analysis of non-linear load under symmetrical conditions. *Energies* **2022**, *29*, 57. [CrossRef]
5. Han, Y.; Feng, Y.; Yang, P.; Xu, L.; Xu, Y.; Blaabjerg, F. Cause, classification of voltage sag, and voltage sag emulators and applications: A comprehensive overview. *IEEE Access* **2020**, *8*, 1922–1934. [CrossRef]
6. Haykin, S. *Neural Networks and Learning Machines*, 3rd ed.; Prentice-Hall: Upper Saddle, NJ, USA, 2009.

7. Abiodun, O.; Jantan, A.; Omolara, A.E.; Dada, K.V.; Umar, A.M.; Linus, O.U.; Arshad, H.; Kazaure, A.A.; Gana, U.; Kiru, M.U. Comprehensive review of artificial neural network applications to pattern recognition. *IEEE Access* **2019**, *7*, 158820–158846. [CrossRef]
8. Daponte, P.; Grimaldi, D. Artificial neural networks in measurements. *Measurement* **1998**, *23*, 93–115. [CrossRef]
9. Rosado, L.; Janeiro, F.; Ramos, P.; Piedade, M. Defect characterization with eddy current testing using nonlinear-regression feature extraction and artificial neural networks. *IEEE Trans. Instrumen. Meas.* **2013**, *62*, 1207–1214. [CrossRef]
10. Pouyanfar, S.; Sadiq, S.; Yan, Y.; Tian, H.; Tao, Y.; Reyes, M.; Shyu, M.-L.; Chen, S.-C.; Iyengar, S. A survey on deep learning: Algorithms, techniques, and application. *ACM Comput. Surv.* **2018**, *51*, 1–36. [CrossRef]
11. Tabernik, D.; Skočaj, D. Deep learning for large-scale traffic-sign detection and recognition. *IEEE Trans. Intell. Transp. Syst.* **2020**, *21*, 1427–1440. [CrossRef]
12. Guo, G.; Zhang, N. A survey on deep learning based face recognition. *Comput. Vis. Image Underst.* **2019**, *189*, 102805. [CrossRef]
13. Nguyen, H.; Wistuba, M.; Grabocka, J.; Drumond, L.; Schmidt-Thieme, L. Personalized deep learning for tag recommendation. *Lect. Notes Comput. Sci.* **2017**, *10234*, 186–197. [CrossRef]
14. Wang, S.; Chen, H. A novel deep learning method for the classification of power quality disturbances using deep convolutional neural network. *Appl. Energy* **2019**, *235*, 1126–1140. [CrossRef]
15. Ma, J.; Zhang, J.; Xiao, L.; Chen, K.; Wu, J. Classification of Power Quality Disturbances via Deep Learning. *IETE Tech. Rev.* **2017**, *34*, 408–415. [CrossRef]
16. Oliveira, R.; Bollen, M. Deep learning for power quality. *Electr. Power Syst. Res.* **2023**, *214*, 108887. [CrossRef]
17. Salles, R.; Ribeiro, P. The use of deep learning and 2-D wavelet scalograms for power quality disturbances classification. *Electr. Power Syst. Res.* **2023**, *214*, 108834. [CrossRef]
18. Gaing, Z.-L. Wavelet-based neural network for power quality disturbance recognition and classification. *IEEE Trans. Power Deliv.* **2004**, *19*, 1560–1568. [CrossRef]
19. Wang, M.; Rowe, G.; Manishev, A. Classification of power quality events using optimal time-frequency representations, theory and application. *IEEE Trans. Power Deliv.* **2004**, *19*, 1496–1503. [CrossRef]
20. Radil, T.; Ramos, P.; Janeiro, F.; Serra, A. PQ monitoring system for real-time detection and classification of disturbances in a single-phase power system. *IEEE Trans. Instrumen. Meas.* **2008**, *57*, 1725–1733. [CrossRef]
21. IT-GIM Group. Smart-Monitor Power Quality Events Database. Available online: http://smartmonitor.lx.it.pt/ (accessed on 30 January 2023).
22. Santoso, S.; Powers, E.J.; Grady, W.M.; Hofmann, P. Power quality assessment via wavelet transform analysis. *IEEE Trans. Power Deliv.* **1996**, *11*, 924–930. [CrossRef]
23. Gaouda, A.M.; Salama, M.M.A.; Sultan, M.R.; Chikhani, A.Y. Power quality detection and classification using wavelet-multiresolution signal decomposition. *IEEE Trans. Power Deliv.* **1999**, *1476*, 1469–1476. [CrossRef]
24. Radil, T.; Ramos, P.; Janeiro, F.; Serra, A. An efficient approach to detect and classify power quality disturbances. *COMPEL-Int. J. Comp. Math. Electr. Electron. Eng.* **2008**, *27*, 1178–1191. [CrossRef]
25. Kiranyaz, S.; Avci, O.; Abdeljaber, O.; Ince, T.; Gabbouj, M.; Inman, D. 1D convolutional neural networks and applications: A survey. *Mech. Syst. Signal Proc.* **2021**, *151*, 107398. [CrossRef]
26. Feurer, M.; Hutter, F. Hyperparameter Optimization. In *Automated Machine Learning*; Feurer, M., Hutter, F., Kotthoff, L., Vanschoren, J., Eds.; Springer: Berlin/Heidelberg, Germany, 2019; pp. 3–33. [CrossRef]

energies

Article

Integration of Local and Central Control Empowers Cooperation among Prosumers and Distributors towards Safe, Efficient, and Cost-Effective Operation of Microgrids

Paolo Tenti [1] and Tommaso Caldognetto [2,*]

1 Department of Information Engineering, University of Padova, 35131 Padova, Italy
2 Department of Management and Engineering, University of Padova, 36100 Vicenza, Italy
* Correspondence: tommaso.caldognetto@unipd.it

Abstract: The advent of energy communities will revolutionize the energy market. However, exploiting their full potential requires innovations in the structure and management of low-voltage grids. End users shall be aggregated within microgrids, where their physical interaction is possible and coordinated operation of power sources and energy storage systems can be achieved. Moreover, meshed network topologies will enable multiple paths for the power flow. The combination of smart control and meshed networks can dramatically improve microgrid performance in terms of power quality, efficiency, and resilience to transients and faults. Ubiquitous control of the power flow becomes possible, as well as active fault clearing and isolation of subgrids without tripping circuit breakers. This paper proposes a control approach that pursues such goals without requiring modification of control and communication hardware implemented in commercial inverters. Instead, a revision of control firmware, integrated with local measurements, allows retrofitting existing plants to improve microgrid operation. Further improvements may derive from the installation of community power sources and energy storage systems, which can extend microgrid operation to pursue demand response and islanding. The potential of the proposed control methods is demonstrated by simulation considering a standard microgrid under different operating conditions.

Keywords: energy community; microgrid; local control

Citation: Tenti, P.; Caldognetto, T. Integration of Local and Central Control Empowers Cooperation among Prosumers and Distributors towards Safe, Efficient, and Cost-Effective Operation of Microgrids. *Energies* **2023**, *16*, 2620. https://doi.org/10.3390/en16052320

Academic Editor: Miguel Jiménez Carrizosa

Received: 3 February 2023
Revised: 21 February 2023
Accepted: 23 February 2023
Published: 28 February 2023

1. Introduction

The increasing penetration of renewable energy sources (RES) and energy storage systems (ESS) in low-voltage networks makes more and more interesting and feasible the aggregation of end users to form energy communities (EC). According to the EU directive "Clean energy for all Europeans" [1], these aggregations will become prime actors in the electrical market and will play a fundamental role to improve grid efficiency and flexibility and, more generally, to support a safe, reliable, and cost-effective operation of the electrical network.

From a technical point of view, the most effective type of aggregation consists in the creation of microgrids connecting neighboring users fed by the same distribution grid [2,3]. This has the potential to allow full exploitation of any local energy sources and control abilities, resulting in improved quality and efficiency of operation, nearly ubiquitous control of the power flow, and aggregated demand response [4]. The performances can further be improved by the inclusion of community energy storage systems (CESS), which can support the microgrid operation during transients and allow temporary storage or release of energy, e.g., for demand response, servicing, fault clearing, or even islanding [5,6].

While the installation of energy sources and storage systems is merely an economical issue, the coordination of their operation within the microgrid is not trivial. A huge literature is available on this subject, proposing a variety of solutions based on centralized, distributed or hybrid control, which make use of various types of powerline or wireless [7] communication or can be even communication-less. The practical implementation of such control techniques is often limited by the incompatibility of communication protocols as well as control features implemented in the power electronic converters interfacing the sources with the mains, which often rely on proprietary standards [8].

The most widely applicable control approach implements droop techniques [9], which do not require communication among units, and automatically adjust the active and reactive power fed by every inverter based on local voltage and frequency measurements. However, problems can arise under transient conditions, when power oscillations may occur due to the interaction of distributed control units [10]. Moreover, droop techniques hardly include reactive power compensation and load unbalance correction, which would help to increase both power quality at user terminals and distribution efficiency.

In this paper, we analyze control solutions that can be implemented locally, at the level of each single inverter, and permit significant improvement of microgrid operation without requiring modification of inverter structure and rating, but a revision of control firmware. In particular, the implementation of reactive power and load unbalance compensation are proposed at the local level, according the Steinmetz compensation approach [11,12], which is currently applied in high-power SVCs (static VAR compensators), e.g., for single-phase AC railways and arc furnaces. The same principle can further be applied to groups of users, or to entire subgrids, owing to the flexibility of the approach and the potential benefits in terms of overall performance.

All proposed control approaches are verified by simulation in a low-voltage benchmark network proposed by CIGRE, to evaluate the control performance in a standard operation environment [13].

2. Principles of Local Control

Currently, most commercial inverters interfacing distributed energy sources (RES and ESS) with the mains operate at unity power factor, i.e., they feed AC currents in phase with the line voltages with an amplitude determined by the balance between the source power fed into the DC link and that transferred to the AC grid.

However, some recent standards (e.g., Italian CEI 021-2022) prescribe that PV inverters shall be able to regulate the power factor down to 0.95 in the inductive or capacitive region, either by local control or in response to external commands. This may require some overrating of the inverter, but does not exhaust its control capacity. Actually, pulse-width-modulated (PWM) inverters can control the waveform of the currents fed to the grid under the constraint to meet the DC-to-AC power balance. Such constraint can be alleviated by the storage capacitor connected to the inverter DC terminals [14], which can accommodate the energy fluctuations corresponding to temporary power unbalance.

In the following, we will show how the wide control ability of PWM inverters can be used to compensate, by proper control of the AC currents fed to the grid, the reactive power and phase unbalance generated by the loads nearby.

2.1. Reactive Power Control and Phase Unbalance Compensation

The compensation principle is the same proposed by prominent scientist Charles Proteus Steinmetz in early 20th century [11,12,15,16]. However, while he solved the problem by connecting line-to-line a proper set of reactive components, we may approach the solution by separately controlling the positive, negative, and zero sequence components of the AC inverter currents. In particular:

- *In-phase positive-sequence currents determine the active power*; that is, the average power flowing through the inverter. They can be adjusted to achieve the desired DC-to-AC power flow;
- *In-quadrature positive-sequence currents determine the reactive power* fed by the inverter to the grid, which can be adjusted to compensate for that absorbed by the load and/or to provide a desired reactive power flow at the inverter terminals.

While the above terms can be defined for single-phase systems too, the following apply to three-phase systems only:

- *Negative-sequence currents* can be adjusted to compensate for the unbalanced active and reactive power absorption of the loads. Notably, under sinusoidal voltage supply such current terms do not imply any average DC-to-AC power transfer; rather, they correspond to a fluctuating power, which must be accommodated by the filter capacitor in the DC link of the inverter;
- *Zero-sequence currents* may compensate for any homopolar currents absorbed by the loads in four-wire distribution grids.

Obviously, the implementation of the compensation duties in the inverter control requires: (i) the measurement or estimation of load currents, and (ii) the computation of sequence components.

The theme of current estimation is relevant for implementing any type of compensation techniques, and it is addressed by a wide literature [17,18]. Thus, it will not be discussed herein.

2.2. Determination of Sequence Components in the Time Domain

The sequence components can be determined in the frequency domain, according to the method proposed by another prominent scientist, Charles LeGeyt Fortescue, in 1918 [19]. More conveniently, in three-phase inverters, the sequence components can be determined directly in the time domain according to the method proposed in [20], which can be implemented by a simple computation algorithm in the inverter control.

Let $f_1(t)$, $f_2(t)$, and $f_3(t)$ be any triplet of variables in the time domain (currents or voltages) measured in phase 1, 2, and 3 of a three-phase network. We may define their sequence components as follows.

- *Zero-sequence* component. The reference term is:

$$f_o(t) = \frac{1}{3}(f_1(t) + f_2(t) + f_3(t)) \tag{1a}$$

The corresponding phase components are:

$$\begin{bmatrix} f_{o1}(t) \\ f_{o2}(t) \\ f_{o3}(t) \end{bmatrix} = \begin{bmatrix} f_o(t) \\ f_o(t) \\ f_o(t) \end{bmatrix} \tag{1b}$$

- Positive sequence components. Let T be the line voltage period; the reference term is:

$$f_p(t) = \frac{1}{3}\begin{pmatrix} f_1(t) - f_o(t) + f_2(t+T/3) - f_o(t+T/3) \\ +f_3(t+2T/3) - f_o(t+2T/3) \end{pmatrix} \tag{2a}$$

The corresponding phase components are:

$$\begin{bmatrix} f_{p1}(t) \\ f_{p2}(t) \\ f_{p3}(t) \end{bmatrix} = \begin{bmatrix} f_p(t) \\ f_p(t-T/3) \\ f_p(t-2T/3) \end{bmatrix} \tag{2b}$$

- Negative sequence components. The reference term is:

$$f_n(t) = \frac{1}{3}\begin{pmatrix} f_1(t) - f_o(t) + f_2(t-T/3) - f_o(t-T/3) \\ +f_3(t-2T/3) - f_o(t-2T/3) \end{pmatrix} \tag{3a}$$

The corresponding phase components are:

$$\begin{bmatrix} f_{n1}(t) \\ f_{n2}(t) \\ f_{n3}(t) \end{bmatrix} = \begin{bmatrix} f_n(t) \\ f_n(t+T/3) \\ f_n(t+2T/3) \end{bmatrix} \tag{3b}$$

Notably, the sequence components result by simply adding time-shifted terms, and can be adjourned at the beginning of each period on the basis of the values stored in the previous cycle. The result is a fast and simple computation algorithm, easily implemented in the control firmware.

2.3. *Types of Local Control*

Based on the previous definitions, we may identify various types of local control, which are characterized by different types of measurements and communication among neighbor units. In the following section, the impact of such control techniques on microgrid performance will be analyzed by simulation on a standard benchmark network proposed by the CIGRE [21].

2.3.1. Autonomous Control

With this type of control, each inverter operates autonomously and no communication is required among them. Local load currents are measured or estimated and their reactive and/or sequence components are determined and fed to the inverter control to perform the desired compensation. Each type of compensation must carefully be analyzed for its impact on the ratings of the inverter and the filter capacitor. In fact, single-phase reactive compensation causes a large power fluctuation in the DC link of the inverter, close to the rated load power, while three-phase reactive compensation causes negligible power fluctuation. Moreover, load unbalance compensation can involve considerable power fluctuations associated to negative-sequence currents (up to 150% of rated load power), while the compensation of zero-sequence currents does not involve power fluctuations at all.

This type of control can be extended to the case of a single inverter compensating multiple loads nearby. Obviously, the currents of all targeted loads must be measured and fed to the inverter control, to allow an aggregate compensation.

2.3.2. Cooperative Control

This type of control applies when groups of inverters share the compensation duty. This requires slow communication within each group, to exchange data about load currents and implement a power-sharing criterion. Actually, this criterion can be determined offline if the compensation duty is shared according to nominal quantities; for example, the rated kVA of inverters, the distance among neighbor nodes, the characteristics of the distribution wires, etc.

In general, cooperative control methods based on Steinmetz compensation do not necessarily require knowledge of grid parameters, or sophisticated system modeling, or advanced estimation algorithms. Therefore, they can be implemented in the control firmware of commercial inverters with limited additional computation burden, and without requiring a revision of the control hardware.

This type of solution can effectively be applied in energy communities, where the energy resources and the inverters partly belong to end users and partly to the community as a whole. Actually, the synergistic operation of private and collective resources can significantly improve the network operation without requiring additional equipment or infrastructure.

2.3.3. Integration of Local Control with Centralized Control

The above local control techniques can be integrated with a centralized control, if any, either to comply with commands issued at a higher level of the control hierarchy (e.g., related to demand response, power quality, islanding) or to implement electrical safety procedures within the microgrid (e.g., active fault clearing, isolation of faulty lines, separation of subgrids for maintenance).

3. Benchmark Microgrid, Daily Operation, and Performance Factors

3.1. Microgrid Architecture

To analyze the operation and impact of the local control method proposed in the previous section, we consider the low-voltage microgrid shown in Figure 1, which corresponds to a European low-voltage benchmark network proposed by CIGRE [21]. This network has been already considered in a previous paper [13] and its configuration and parameters are recalled here for completeness.

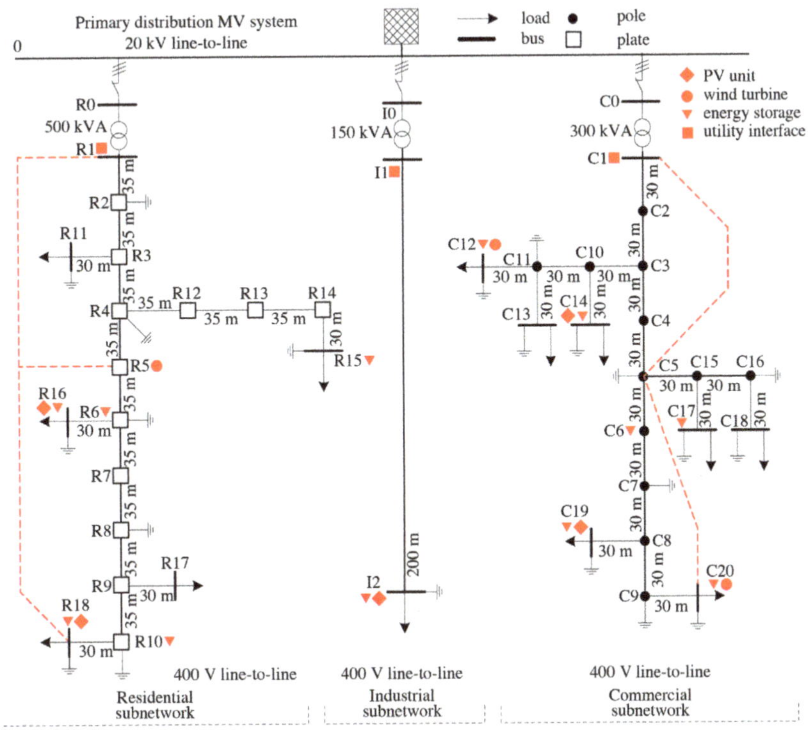

Figure 1. Considered low-voltage European testbench. Details on network parameters are reported in Appendix A [13].

The benchmark network includes three subnetworks: residential, commercial, and industrial. The characteristics and parameters of loads, sources, and distribution feeders are detailed in [13] and are reported in Appendix A. In particular, Table A1 specifies the type of feeders; Table A2 shows the connections among grid nodes, the characteristics of the loads fed by the grid, and the parameters of the MV/LV transformers feeding the three subnetworks; Table A3 specifies the characteristics of distributed energy sources.

As compared to the CIGRE benchmark network, wind turbines (WT, indicated by circular marker), photovoltaic systems (PV, diamond marker), and distributed energy storage systems (ESS, triangular marker) similar to those proposed for the residential subnetwork were added in the commercial and industrial sections, too. Moreover, energy storage units have been added as indicated in Figure 1 to allow the implementation of electrical safety procedures and extended local control. All distributed power sources, either RES (renewable energy sources) or ESS, interface to the grid by an electronic power converter that performs as a controllable current source. A special function is assigned to the utility interfaces (UI, square marker) located next to the points of common coupling

(PCC) of each subgrid. They include ES units interfaced with the grid by voltage-controlled inverters and perform as voltage-tracking units during on-grid operation, and as voltage-forming units during off-grid operation. UIs play a fundamental role in controlling the power flow from and to the mains and are the basic elements to manage islanding and demand response and to handle transient situations due to faults or fast variations of network behavior.

The ESS located at nodes R6 and C6 are community units and support the UIs during fast transients. Under normal operation, they play as reactive and unbalance compensators according to the control methods described in the following sections.

To complement the CIGRE network, we also included the connecting lines shown in red color, which transform the original radial topology of each subnetwork to a meshed one. In practice, this corresponds to double the main power cable in the cable duct.

As discussed in a previous paper [13], choosing a meshed topology offers considerable advantages in the case of low-voltage microgrids: (i) the voltage stability at grid nodes improves; (ii) the power loss in distribution feeders reduces; (iii) faulty lines can be isolated without shutting down entire subgrids. Moreover, if a faulty area is surrounded by active nodes equipped with grid-tied inverters, the faulty currents can also be cleared by a proper control of the inverters, without opening the circuit breakers (electronic fault clearing). The faulty lines can then be isolated by opening disconnectors or by removing fuses. The same approach can be followed to isolate subgrids for maintenance.

3.2. Daily Power Profiles

Figure 2 shows typical daily power profiles of the three subgrids of Figure 1, according to CIGRE standards for residential, commercial, and industrial microgrids in Europe.

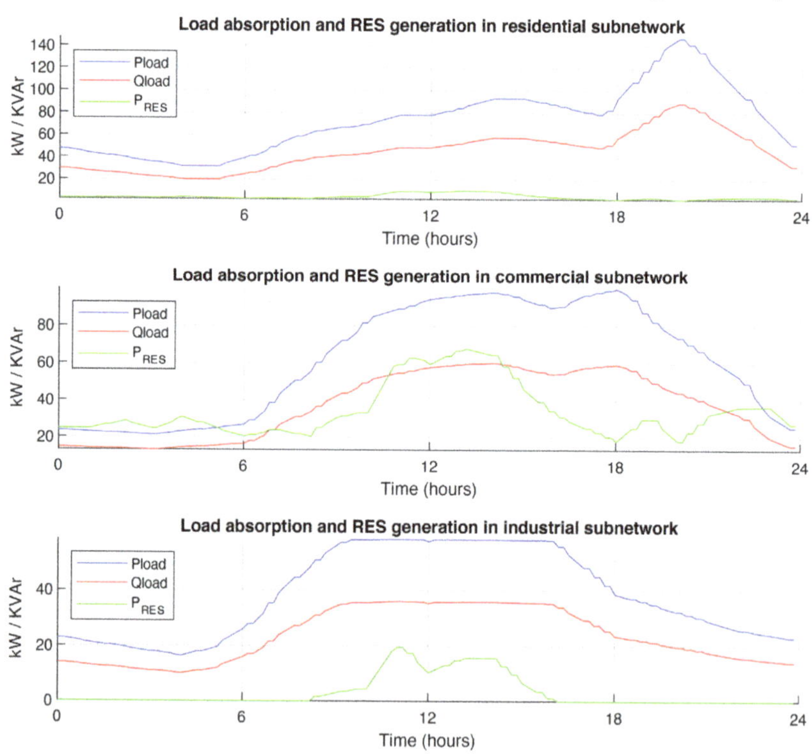

Figure 2. Typical 24 h power profiles of subgrids in Figure 1.

The upper diagram shows the daily behavior of residential loads (active and reactive power absorption) together with the generation profile of renewable energy sources. The central and lower diagrams show the same quantities for the commercial and industrial subgrids, respectively. While the residential part is characterized by higher load power absorption, renewable energy generation is more relevant in commercial and industrial parts.

3.3. Mathematical Definition and Computation of Performance Factors

Figure 3 shows the daily behavior of three specific performance factors measured at the points of common coupling (PCCs) between each subgrid and the mains. Before discussing the behavior of such factors, let us introduce their mathematical definition and physical meaning in general terms.

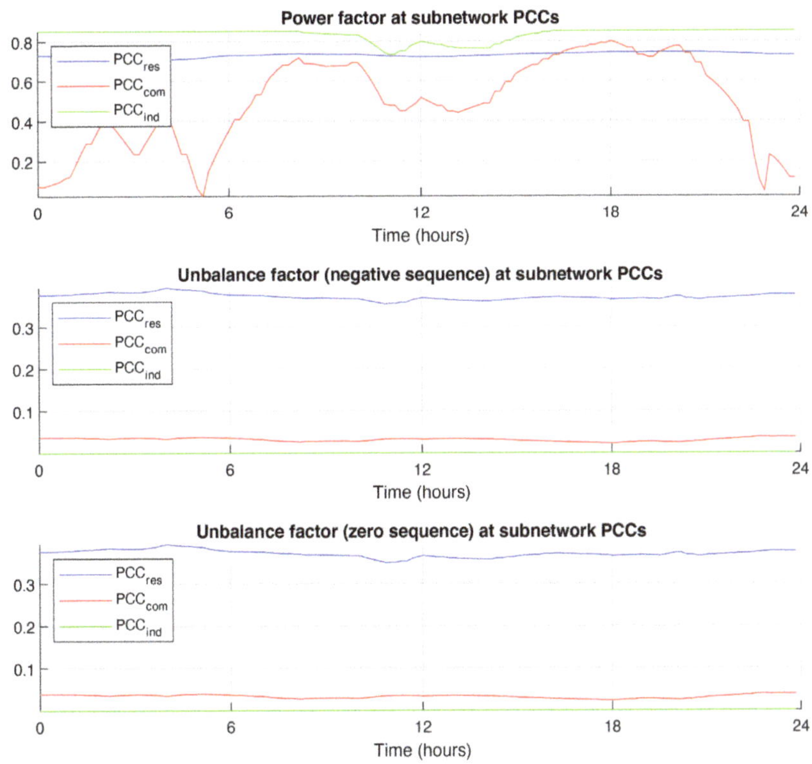

Figure 3. Behavior of three specific performance factors measured at the points of common coupling (PCCs) between each subgrid and the mains.

In three-phase systems under periodic operation, the power factor (PF) is generally defined as:

$$PF = \frac{P}{A} = \frac{\langle \boldsymbol{v}, \boldsymbol{i} \rangle}{\|\boldsymbol{v}\| \cdot \|\boldsymbol{i}\|} \tag{4a}$$

where $\boldsymbol{v}$ and $\boldsymbol{i}$ are the vectors of phase voltages and currents at a given network port, P is the active power flowing through that port, and A is the corresponding apparent power. The mathematical operator $\langle \cdot \rangle$ indicates internal product, while $\|\cdot\|$ refers to the square norm (i.e., L^2 norm). In the time domain, for three-phase systems the power factor can be computed as:

$$PF = \frac{P}{A} = \frac{\sum_{k=1}^{3} \int_0^T v_k \, i_k dt}{\sqrt{\sum_{k=1}^{3} \int_0^T v_k^2 \, dt} \cdot \sqrt{\sum_{k=1}^{3} \int_0^T i_k^2 \, dt}} \tag{4b}$$

where T is the line period, and $(v_k,\ i_k)$ indicate instantaneous phase quantities. Under sinusoidal operation, the PF can be computed as a function of RMS voltages and current as:

$$PF = \frac{P}{A} = \frac{\sum_{k=1}^{3}(V_k \cdot I_k \cdot \cos\varphi_k)}{\sqrt{\sum_{k=1}^{3} V_k^2} \cdot \sqrt{\sum_{k=1}^{3} I_k^2}} \tag{4c}$$

where $(V_k,\ I_k)$ are the RMS voltage and current in phase k, and φ_k is their angular shift.

Equation (4c) can also be rewritten as a function of the sequence components of phase voltages (V_p, V_n, V_o) and currents (I_p, I_n, I_o). Owing to the orthogonality of the sequence terms defined by Equations (1)–(3), from (4c) we derive:

$$\begin{aligned} PF = \tfrac{P}{A} &= \frac{\sum_{k=1}^{3}(V_k \cdot I_k \cdot \cos\varphi_k)}{3\sqrt{V_p^2+V_n^2+V_o^2} \cdot \sqrt{I_p^2+I_n^2+I_o^2}} \\ &\approx \frac{\sum_{k=1}^{3}(V_k \cdot I_k \cdot \cos\varphi_k)}{3V_p \cdot \sqrt{I_p^2+I_n^2+I_o^2}} \end{aligned} \tag{4d}$$

Equation (4d) clearly shows the impact on the power factor caused by reactive power absorption, which affects angular shifts φ_k, and the asymmetrical behavior of phase currents, generated by the load unbalance, which may cause non-negligible negative and zero sequence terms I_n, I_o to appear. In contrast, the impact of voltage asymmetry is usually minimal, since the negative- and zero-sequence voltage components at PCCs are negligible as compared to the positive sequence term.

The other two performance factors of Figure 3 describe the current asymmetry (power unbalance) measured at the PCCs, and are defined as:

$$\text{Unbalance factor (negative sequence)}: \ \chi_n = \frac{I_n}{I_p} \tag{5a}$$

$$\text{Unbalance factor (zero sequence)}: \ \chi_o = \frac{I_o}{I_p} \tag{5b}$$

- χ_n reflects the presence of unbalanced power absorption in the supply cables at the PCCs, and it is mainly due to asymmetrical single-phase loads connected line-to-line or line-to-neutral.
- χ_o reflects the presence of current circulation in the neutral wire, mainly due to single-phase loads connected line-to-neutral.

As mentioned before, the impact of voltage asymmetry is negligible and is not considered in our computation.

3.4. *Performance Factors at PCC*

Figure 3 shows the behavior of the above performance factors measured at the PCCs for the power profiles shown in Figure 2.

The upper diagram shows the daily behavior of the power factor, separately for each subgrid, while the central and lower diagrams depict the behavior of unbalance factors.

- The PF behavior of the residential subgrid is nearly constant along the 24 h interval. This regular profile reflects, on one side, the limited impact of renewable energy generation as compared to load power absorption and, on the other side, the fact that the power factor of residential loads keeps nearly constant at all times. Consequently, the ratio between active power P and apparent power A at residential PCC shows limited variations. Its relative low value (nearly 0.7) is mostly determined by the significant load unbalance, which derives from single-phase loads connected line-to-neutral and causes relevant unbalance factors (nearly 40% for both negative and zero sequence terms).

- The PF behavior of the commercial subgrid shows considerable variations, which are mainly determined by the irregular load power absorption, very low during the night, and by the impactful generation of renewable energy during daytime (both factors affect the ratio P/A). In contrast, in this subgrid the effect of load unbalance is limited, thanks to the significant amount of power absorbed by three-phase symmetrical loads.
- Finally, the industrial subgrid reveals a relatively high PF at all times, due to the coincidence of high-power absorption and generation during daytime, and low-power absorption and generation during nighttime. The unbalance factors are always negligible since almost every load is symmetrical in the three phases.

4. Local Control Methods and Their Impact on Performance Factors

In this section, we discuss two types of control approaches.

The first approach, called *autonomous control*, is applicable to each inverter tied to the grid without requiring any communication with other units or the grid manager. It only requires a modification of the inverter control firmware and the measurement (or estimation) of the active and reactive power absorbed by local loads (connected to the same grid terminals of the inverter), with a possible extension to surrounding loads.

The second approach, called *cooperative control*, implies the coordination of a group of inverters acting in the same subgrid. This method requires: (i) Collection and sharing of data on the overall active and reactive power absorbed by the loads in the subgrid, separately for each phase; this can be done by load power measurement, or even by day-ahead forecast, that should be sufficiently precise and reliable for this type of control, which is inherently stable and does not significantly degrade the grid operation even in case of data inaccuracy; (ii) Defining a suitable power sharing criterion among the inverters; for this purpose, sharing coefficients based on the rated kVA of each inverter can be determined offline, with the double advantage of avoiding real-time computations and ensuring uniform exploitation of inverters' capacity. Even this control method requires updating the control firmware of the inverters. Moreover, some communication ability is needed, which is, however, consistent with mandatory requirements of modern grid codes.

4.1. Autonomous Control

The reference operation mode of the microgrid (*Mode 0*) is that depicted in Figures 2 and 3 and refers to the case when RES inverters feed to the AC grid just the active power produced by energy sources, without compensating the load reactive power or the power unbalance. Moreover, all energy storage units remain off.

We now consider three types of autonomous control, which correspond to different operating modes of the microgrid.

- *Mode 1—Local reactive power compensation*. In this case, each inverter compensates the reactive power absorbed by the local loads fed at the same grid terminals. This type of compensation can be implemented in single- and three-phase inverters.
- *Mode 2—Local reactive power and unbalance compensation*. In this case, each inverter compensates for the reactive power absorbed by local loads and for the negative- and zero-sequence currents generated by the asymmetry of these loads. While negative-sequence compensation is applicable to any type of three-phase inverter, zero-sequence compensation requires a connection to the neutral wire. With this type of control, every grid node equipped with a three-phase inverter can perform at unity power factor.
- *Mode 3—Reactive power and phase unbalance compensation of vicinity loads*. The previous control methods do not allow reactive and unbalance compensation of loads connected to passive grid nodes, i.e., not equipped with inverters. However, Mode 2 can be extended to the case where inverters receive information on the power absorption of vicinity loads and develop an aggregate compensating action. For example, in the network of Figure 1, we aggregate nodes R10 and R17, so that the inverter in R10 compensates for all loads at nodes R10 and R17. Similarly, we aggregate neighbor nodes R11 and R15, C12 and C13, C17 and C18.

4.2. Cooperative Control

Following the proposal introduced in Section 2.3.2 and further discussed at the beginning of this section, we may apply cooperative control to all inverters of each subgrid. The resulting operation mode is:

- Mode 4—Cooperative compensation in subgrids. In this case, we exploit the control ability of the utility interfaces, located next to each PCC, to strengthen the compensation action performed by the other inverters connected to the same subgrid.

4.3. Active Node Structure and Control Implementation

4.3.1. Active Node Structure

Figure 4a depicts the general structure of an active node, equipped with an AC/DC pulse-width-modulated (PWM) inverter interfacing the AC grid with the DC circuit. This latter generally includes a power source and/or energy storage unit in addition to filter capacitor C_d. AC loads can also be connected at the same grid terminals of the inverter. In Figure 4a we define:

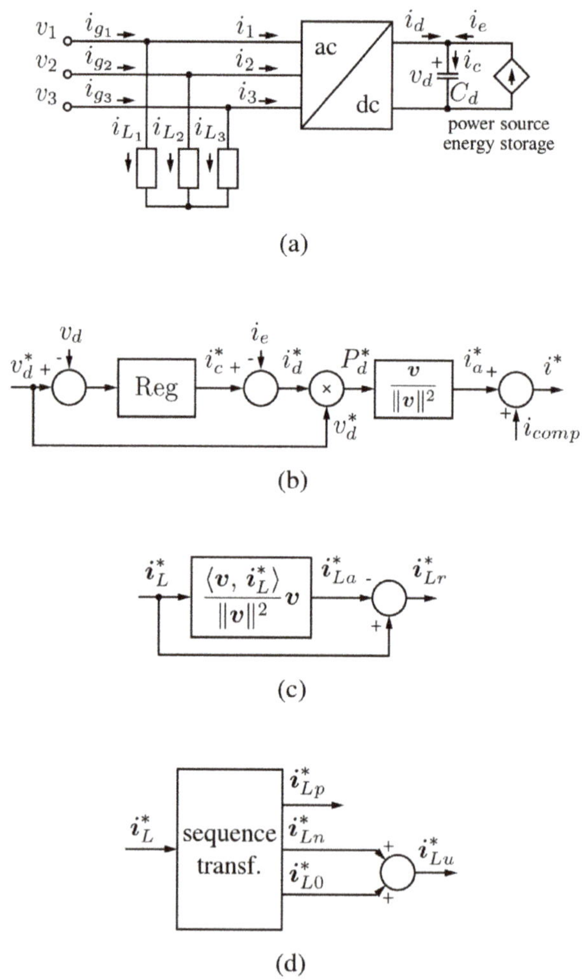

Figure 4. Control quantities at active node (i.e., inverter) level. (**a**) general structure of active nodes; (**b**) DC voltage control and AC current references; (**c**) determination of active and reactive load currents; (**d**) determination of unbalance current references.

- $\boldsymbol{i}_g$, $\boldsymbol{i}_L$, and $\boldsymbol{i}$ as respectively the vectors of the total currents drawn from the AC grid, absorbed by the AC loads, and entering the inverter AC terminals;
- v_d and i_d as respectively the DC voltage and the current fed by the inverter to the DC link;
- i_c and i_e as respectively the current absorbed by DC filter capacitor C_d and that fed by the DC source, either power source or energy storage unit.

4.3.2. General Control Principle

As known, PWM inverters allow a high flexibility to control the waveform of AC currents, under the constraint of input/output power balance and within the limits set by kVA power rating. All proposed control methods suit these boundaries, while taking full advantage of the control flexibility to impress the desired reactive and unbalance compensation currents. The general control principle is depicted in Figure 4b.

The main control loop refers to DC voltage v_d, since keeping v_d at the reference value ensures the balance between inverter input and output power. Actually, any mismatch between these power terms would cause the energy in capacitor C_d to increase or decrease, thus causing a variation of v_d. Fort this purpose, the voltage control loop adjusts the value of DC current i_d which, in turn, reflects on the amplitude of the active currents entering the inverter.

This is graphically represented in Figure 4b, where the control amplifier *Reg* adjust capacitor current reference i_c^* to keep voltage v_d at desired reference v_d^*. Inverter DC current reference i_d^* is then determined by detracting source current i_e from i_c^*. By multiplying DC voltage reference v_d^* times DC current reference i_d^* the control estimates DC power reference P_d^* which, under the assumption of input/output power balance, also coincides with the AC active power reference P_a^*. Active ac current references $\boldsymbol{i}_a^*$ are then determined as those quantities proportional to AC voltages $\boldsymbol{v}$ which comply with power reference $P_a^* = P_d^*$. Finally, inverter AC current references $\boldsymbol{i}^*$ are determined by adding compensation terms $\boldsymbol{i}_{comp}$ to active current references $\boldsymbol{i}_a^*$. Since reactive and unbalance compensation currents do not alter the active power flow, the inverter will eventually absorb the required active power while performing the compensation action required by the selected control method.

4.3.3. Determination of Reactive and Unbalance Compensation Currents

According to the above control methods, each inverter is required to compensate totally or partially the reactive and unbalance currents of one or more vicinity loads. The compensation duty assigned to each inverter depends on the selected control method and the power sharing criterion.

Assuming that $\boldsymbol{i}_L^*$ is the vector of the equivalent load currents to be compensated by the inverter of Figure 4a, the control must firstly separate the active and reactive components of such currents. The separation procedure is shown in Figure 4c. The active components $\boldsymbol{i}_{La}^*$ are first determined as those currents proportional to line voltages $\boldsymbol{v}$, which convey the same active power of currents $\boldsymbol{i}_L^*$. The reactive components $\boldsymbol{i}_{Lr}^*$ are then determined by difference. If only a reactive compensation is required, compensation terms $\boldsymbol{i}_{comp}$ of Figure 4b simply coincide with the opposite of equivalent load reactive currents $\boldsymbol{i}_{Lr}^*$.

If unbalance compensation is also required, the control determines the sequence component of currents $\boldsymbol{i}_L^*$ according to the transformation equations of Section 2.2. The procedure is depicted in Figure 4d, where negative-sequence term $\boldsymbol{i}_{Ln}^*$ and zero-sequence term $\boldsymbol{i}_{Lo}^*$ are added to determine total unbalance currents $\boldsymbol{i}_{Lu}^*$. These latter terms are then accounted for in the computation of compensation term $\boldsymbol{i}_{comp}$.

In the most general case of reactive and unbalance compensation, we set:

$$\boldsymbol{i}_{comp} = -\boldsymbol{i}_{Lr}^* - \boldsymbol{i}_{Lu}^* \tag{6}$$

Remarkably, the control relies on basic control functions at the inverter level. These consist of output active and reactive output power, which are foreseen in current grid standards [22] too, and inverse sequence current injection, which is a basic functionality widely discussed in the literature [23].

4.4. Impact of Control Methods on Grid Performance

- In this section, we analyze separately the performances of residential, commercial, and industrial subgrids, since their different characteristics facilitate the discussion on the impact of the various control techniques on overall microgrid operation.

4.4.1. Impact on the Power Factor of Commercial Subgrids

First, we analyze the effect of the above control methods on the power factor measured at the PCC of the commercial subgrid. We refer to this subgrid because of the large variations of its power factor shown in Figure 3.

The results of the analysis are summarized in Figure 5.

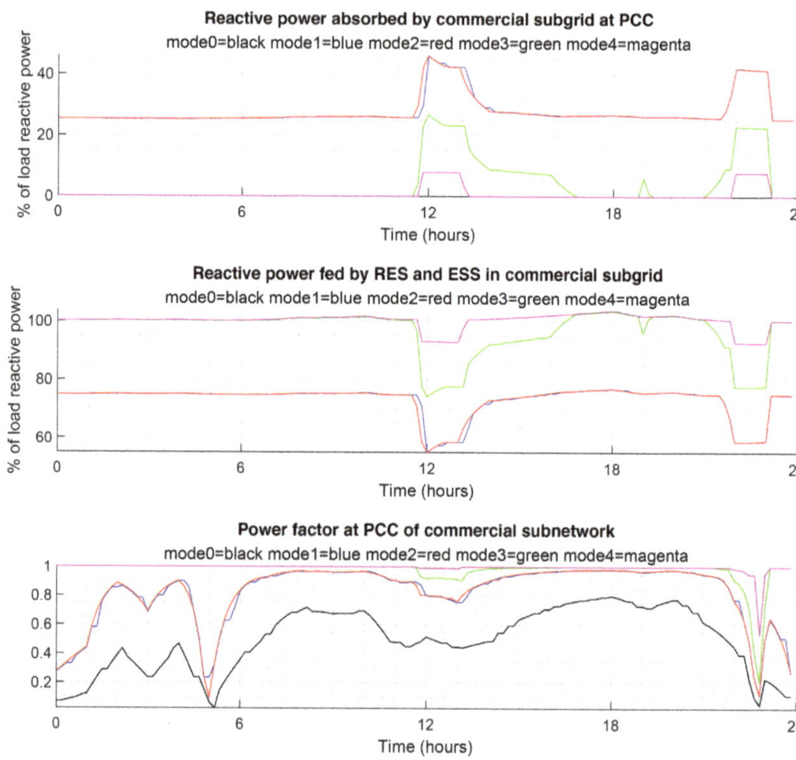

Figure 5. Effect of control methods on the power factor at the PCC.

- The upper diagram shows the time behavior of the reactive power fed by the mains at PCC for the various control methods. The values are expressed in percentage of the total reactive power absorption of the loads in the subgrid. In every operating mode, the reduction of the reactive power fed by the mains is impressive. In mode 1, the maximum value does not exceed 40% of load power, with an average of about 25%, and the same happens in mode 2. The reason for the similar behavior in modes 1 and 2 is that the reactive power fed by the inverters is the same in both cases (it coincides with the reactive power of local loads), the main difference being related to load unbalance compensation, which is a minor issue in commercial subgrids. Instead, modes 3 and 4 provide a greater reduction of reactive power absorption, since the aggregation of vicinity loads (mode 3) or the cooperative control within the entire subgrid (mode 4) allow much better compensation of the overall reactive power taken by the loads. Note that the compensation is less effective in the central hours of the day, where renewable energy generation is high, and in late afternoon, where load power absorption is maximum. In both cases the compensating action is limited by the kVA ratings of the inverters.

- The central diagram shows the corresponding behavior of the reactive power fed by the inverters within the subgrid, whose action is increasingly effective when adopting more sophisticated control methods. This diagram further clarifies that a full compensation of load reactive power is not possible even for mode 4 (cooperative control in the entire subgrid) in the central hours of the day. The reason is that in those hours the generation of renewable energy is maximum, thus the inverters are primarily committed to feed the active power produced by the sources and their residual capacity for load reactive power compensation is limited by the kVA ratings. In presence of an energy storage system, the flexibility of operation could obviously be extended, e.g., by temporarily storing the energy produced by the power sources to recover the full reactive power compensation ability.
- The lower diagram shows the effect of the various control methods on the power factor measured at the PCC. As compared to the reference situation of mode 0, the benefit is appreciable in modes 1 and 2, and is nearly ideal (unity power factor) in modes 3 and 4.

Overall, all proposed control methods allow significant improvements over the reference operating mode. In particular, the compensation of vicinity loads (mode 3) provides excellent performance without requiring the implementation of collective measurements or power sharing among inverters.

4.4.2. Impact on the Unbalance Factors of Residential Subgrid

We now analyze the impact of the various control methods on the unbalance factors in the residential subgrid, which mostly suffers the effects of load unbalance. The results are summarized in Figure 6.

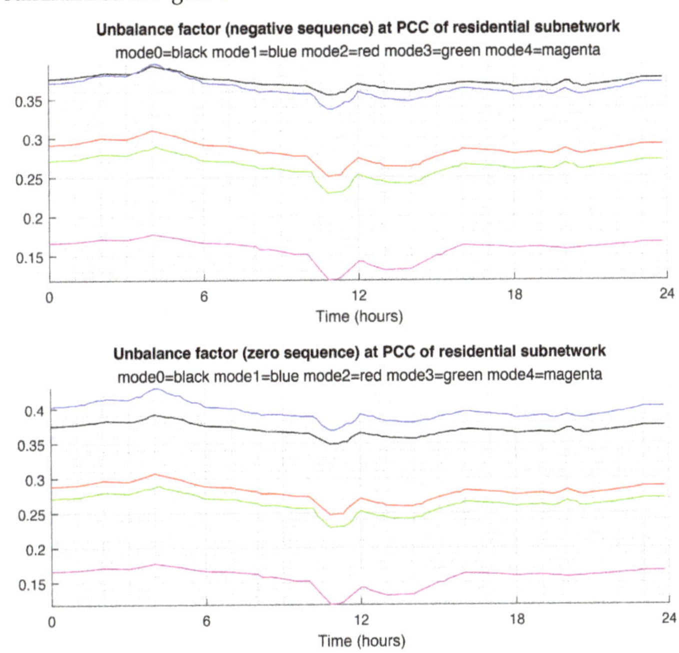

Figure 6. Effect of the control methods on the unbalance factor of the residential subgrid.

- The upper diagram refers to the behavior of the unbalance factor related to the presence of negative-sequence currents at PCC. As compared to the reference operation of mode 0, the pure reactive power compensation method of mode 1 does not reduce the unbalance factor. In fact, in this mode the inverters feed symmetrical reactive currents and do not substantially contribute to the elimination of negative-sequence components. The other operating modes perform significantly better, and the cooperative control of mode 4 reduces the unbalance factor to nearly one-third.
- The lower diagram refers to the behavior of the unbalance factor related to the presence of zero-sequence currents at PCC. As compared to the reference operation of mode 0, the pure reactive power compensation method of mode 1 slightly increases the unbalance factor. The reason lies in the symmetrical operation of the inverters, which increases positive-sequence currents without impacting on zero-sequence load unbalance. The other operating modes ensure increasingly appreciable reductions of the unbalance factors when switching from mode 2 to mode 3 to mode 4.

4.4.3. Impact on Overall Efficiency and Total Voltage Deviation

The overall efficiency is defined as:

$$\eta = \frac{P_{load}}{P_{fed}} = 1 - \frac{P_{loss}}{P_{fed}} = 1 - \frac{P_{transf} + P_{wires} + P_{inv}}{P_{PCC} + P_{gen} + P_{ESS}} \tag{7}$$

where P_{load} is total power absorbed by the loads, and P_{fed} is the total of the power fed by the mains at PCC (P_{PCC}) and that generated by the sources in the microgrid (P_{gen}, generated by any type of local power sources, and P_{ESS} fed by ESS). P_{loss} is total power loss, i.e., the sum of transformer power loss (P_{transf}), power loss in the distribution wires (P_{wires}), and inverter loss P_{inv} (which accounts for internal losses of energy sources too).

The total voltage deviation is defined as:

$$\Delta_v = \sqrt{\sum_{k=1}^{N} (V_k - V_{nom})^2} \tag{8}$$

where N is the total number of grid nodes, V_k is the RMS value of the voltage measured at node k, and V_{nom} is the rated RMS grid voltage. Term Δ_v is a root-mean-square factor and gives a general indication on the steadiness of the voltage profiles across all nodes of the microgrid. As an alternative indicator, the minimum voltage across the entire microgrid can be considered.

Figure 7 shows the time behavior of the overall efficiency η and total voltage deviation Δ_v in the various operating modes. In all cases, the improvement over the reference situation of *mode 0* is appreciable.

- The upper diagram shows that the efficiency is generally higher for modes 2 and 3, due to the local elimination of useless current terms (reactive currents and negative/zero-sequence components), which do not flow through the grid wires toward the PCC. In mode 1 this happens partially, since only the reactive power absorbed by the loads is compensated locally. Instead, mode 4 is characterized by additional losses since the circulation of compensating currents happens through longer paths than for modes 2 and 3. As compared to reference operation *mode 0*, where inverters do not perform any compensation actions, all proposed control methods provide considerable efficiency improvements along the entire time period.
- The lower diagram shows that all control methods ensure a reduction of the total voltage deviation as compared to the reference situation. This happens because the local compensation of reactive and unbalance power prevents circulation of useless reactive and unbalance current terms, thus reducing the voltage drop across the grid wires.

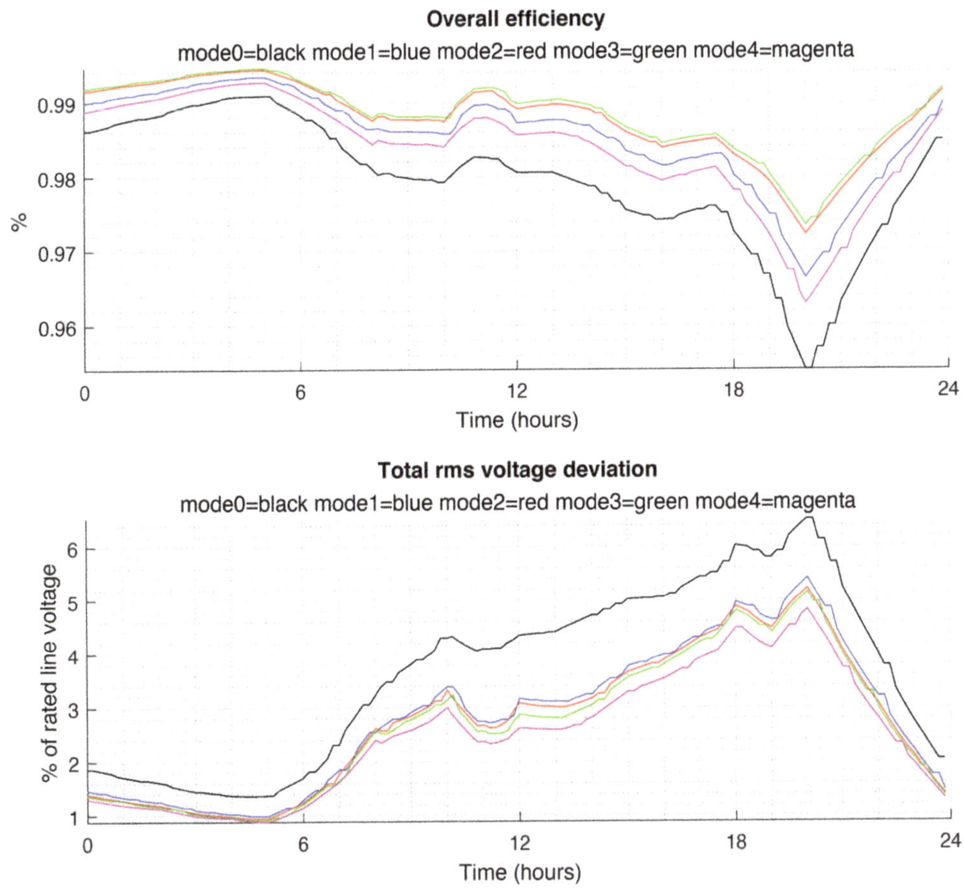

Figure 7. Effect of control methods on the overall efficiency and RMS voltage deviations.

It is also interesting to see how the distribution and inverter losses change during the time for the various control methods. This is shown in Figure 8, where the upper diagram refers to distribution losses, which essentially track the behavior of load power and slightly increase in case of cooperative control, where the compensation currents flow though longer paths. The lower diagram refers to conversion losses, which increase in the central hours of the day, where the power fed by renewable sources is maximum. Note that the adoption of cooperative control methods helps to reduce the inverter losses, thanks to power sharing. It must also be observed that the diagram was drawn assuming inverters with an efficiency above 94%, which is consistent with, for example, [24].

Finally, Figure 9 shows the minimum node voltage measured across the entire microgrid for the various control methods. Noticeably, while local compensation of vicinity loads helps the steadiness of voltage profiles, the requirement of unity power factor at PCCs affects it, since reactive and unbalance currents flow through the microgrid, causing additional voltage drops along the distribution lines.

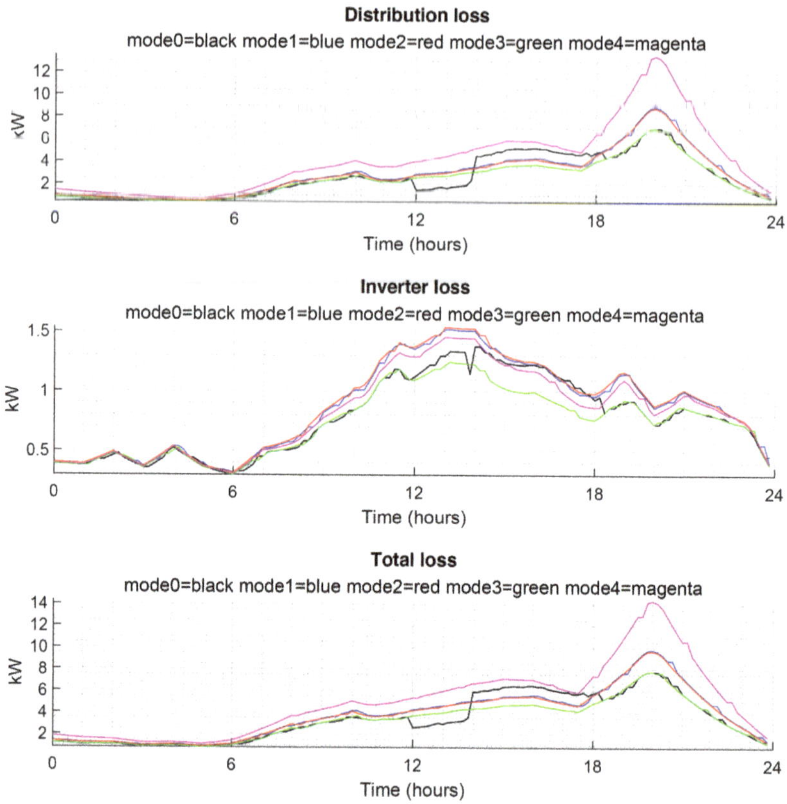

Figure 8. Loss analysis reporting, from top to bottom, distribution loss, inverters loss, and total loss.

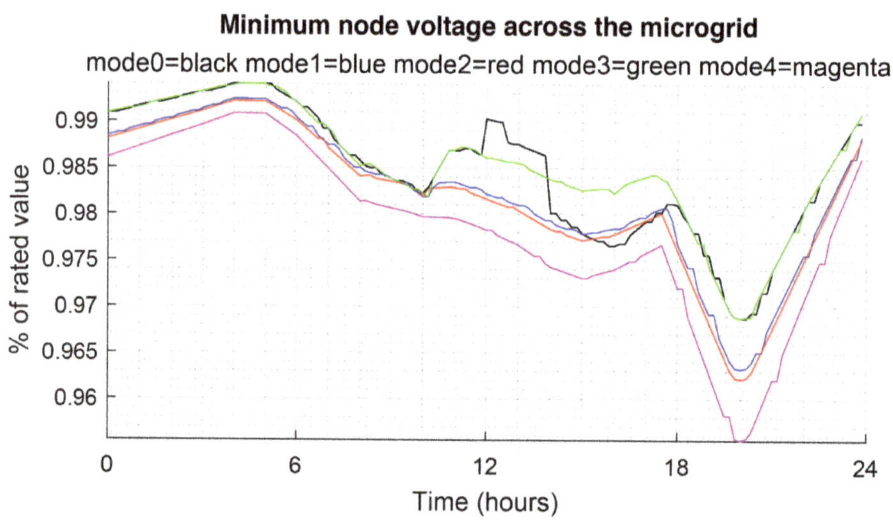

Figure 9. Minimum node voltage measured across the entire microgrid for the various control methods.

Overall, the proposed local control modes show the ability to reduce the power loss and stabilize the voltage profiles within the entire microgrid.

5. Integrating Local Control and Central Control

As mentioned in Section 2.3.3, the above local control methods can profitably be integrated with centralized control to extend the operational limits of the microgrid. Here we consider an integration approach that does not require any modifications of the above reactive and unbalance compensation techniques conducted at a local level.

Rather, we extend the same local control approach to include active power too, assuming that local energy storage units can be driven to temporarily compensate (e.g., to accomplish islanding or demand response) for the active power absorbed by their vicinity loads. This operation mode is toggled on or off by a logical command issued by the central controller, without any additional data exchange with local units.

The only units requiring bidirectional communication with the central controller are the utility interfaces, which interact with the grid manager to implement additional functions such as demand response or islanding.

5.1. Integration of Local and Central Control for Reactive Power and Load Unbalance Compensation

For the sake of demonstration, let us assume that the compensation functions described in the previous chapter are integrated with a power factor control action done by the UIs under the guidance of central control. For this purpose, the information on the reactive and unbalance currents absorbed at the PCC are fed by the central controller to the UIs, which adapt their operation to provide the residual compensation needed to complement that performed by local control methods. The result is a unity power factor operation at all PCCs irrespective of the type of local control used by the inverters in the microgrid. This means that at any time the currents fed by the mains are purely active, symmetrical, and in phase with the line voltages. This means that all reactive and unbalance currents taken by the loads are compensated within the microgrid, without affecting the mains.

To verify that the integrated control does not affect the overall efficiency and voltage stability at grid nodes, Figure 10 shows the same quantities of Figure 7 when the UIs are driven by central control. Noticeably, the voltage deviation factor is substantially unaffected, while the overall efficiency is further improved. It is also remarkable that the highest efficiency is achieved by implementing locally the simplest control technique (mode 2), where the inverters compensate for local loads only, without requiring data exchange with other units in the same subgrid.

This demonstrates that driving the UIs by central control may ensure nearly ideal operation at the PCCs, high efficiency, and steady voltage profiles without requiring data sharing among the inverters in the same subgrid.

5.2. Integration of Local and Central Control for Active Power Management in Case of Islanding

The integration of local and central control has the potential to allow a precise management of the power flowing at the PCCs, making possible the implementation of demand response, or even islanding of entire subnetworks. For the sake of demonstration, we consider the case when, in the time interval from 12 a.m. to 2 p.m., the residential subgrid is driven in the islanded mode, while in the remaining time the operation remains in mode 4.

The corresponding time behavior of some relevant quantities of the residential subgrid is shown in Figure 11.

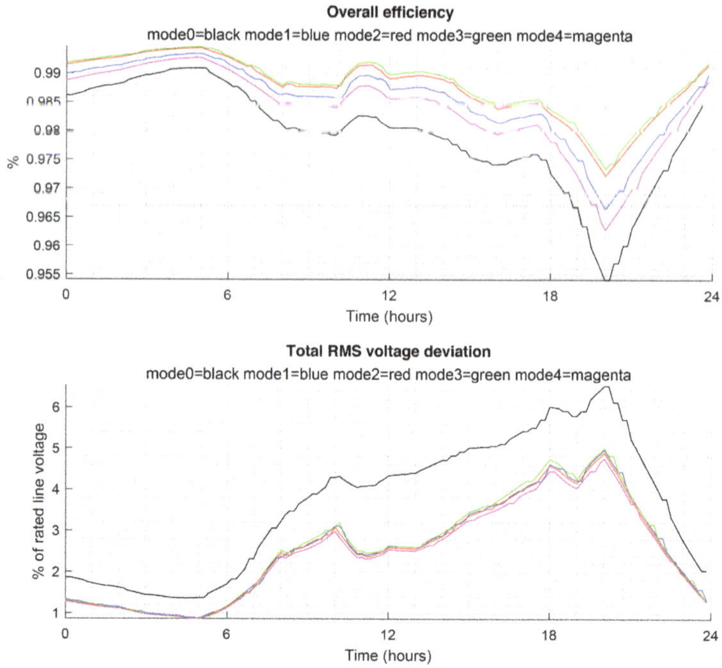

Figure 10. Effect of control methods on the overall efficiency and RMS voltage deviations, as in Figure 7, but with the UI driven by central control. Voltage deviation factor is unaffected, overall efficiency is improved.

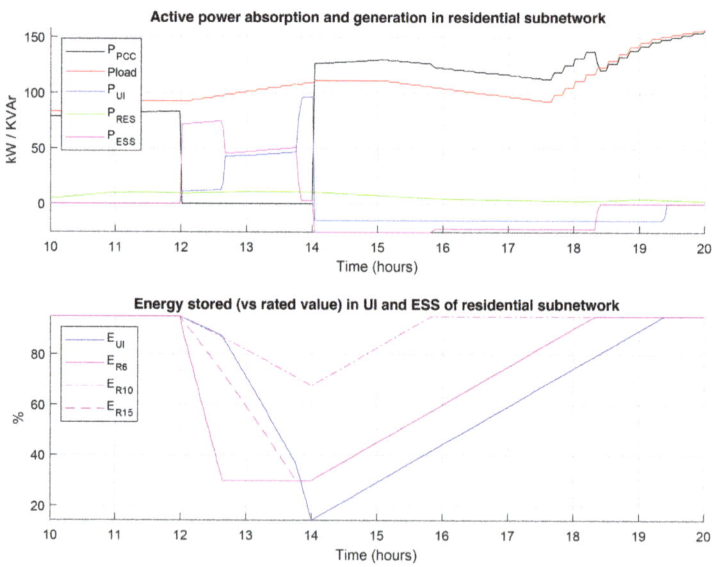

Figure 11. Integrating local and central control to implement islanded operation in residential subgrid.

At 12 a.m. the active power compensation command is issued to all nodes equipped with energy storage (R6, R10, R15), and each unit suddenly begins to feed the active power requested by the associated loads (loads R16 and R18 for R6; load R17 for R10; loads R11 and R15 for R15). The UI at node R1 receives data measured at PCC and arranges its output currents to compensate for the residual active and reactive power and unbalance factors.

At 2 p.m. the islanding condition is removed, and the energy storage units enter a charge recovery phase at constant current. At 7.30 p.m. this phase also ends and the operation returns to regular mode 4.

- The upper diagram shows the active power: fed by the mains at PCC (black line); absorbed by all loads in the subgrid (red); fed by the utility interface at node R1 (blue); generated by the renewable energy sources in the microgrid (green); fed by the energy storage units at nodes R6, R10, and R15 (magenta).
- The lower diagram shows the behavior of the energy stored in the ES units at node R1 (UI, blue line), R6 (magenta, continuous), R10 (magenta, dash-dotted), R15 (magenta, dashed).

We observe that before 12 a.m. the load power is fed by the mains at PCC and by RES only. After the islanding command is issued at 12 a.m., the power fed at PCC vanishes, and the corresponding power gap is quickly filled by UI and ES units. Most of the power is initially fed by ES units, until at 12.30 a.m. the unit at R6 reaches the minimum charge limit (30% of rated value) and stops feeding power. The UI steps up suddenly to fill the gap, and further increases its power contribution at 1.30 p.m., when the ES in R10 also reaches the lower charge limit. At 2 p.m. the UI's energy is nearly exhausted (15% of rated value) and the islanding phase ends. In the following hours the storage units recover the full charge at constant current, and the mains feeds the load power plus the energy recovery power of ES units. At 7.30 p.m. the energy recovery phase is completed and the system returns to mode 4 (unity PF at PCC).

Note that, even for this type of operation, the local units operate autonomously based on local power data. The required active power control can be implemented by upgrading the control firmware to include the on/off toggle for load power compensation.

6. Discussion

In this paper, we proposed a control approach that can be applied for retrofitting existing grid-tied inverters with the aim of improving their individual operation as well as the performance of the surrounding grid.

In particular, three types of control were discussed, which are characterized by different performances but require only handling of the power and current data measured locally:

1. With the simplest type of control, the inverters are driven to compensate the total reactive power absorbed by the loads fed at the same grid terminals of the inverter. This only requires a revision of the control firmware and local reactive power measurements. This type of control prevents useless reactive power terms flowing through the distribution feeders, resulting in reduced distribution loss and limited fluctuations of the voltages at grid nodes.
2. In the case of three-phase inverters, the above approach can be extended to the compensation of the power unbalance generated by local loads. This requires a more sophisticated control algorithm in the inverter control firmware; moreover, local phase power and currents must be measured to allow computation of negative- and zero-sequence terms. The result is an increase in the computational burden of each inverter which, however, keeps well within the capability of commercial controllers. The corresponding performance improvement is very appreciable in terms of power factor at the point of common coupling since the symmetry and phase shift of the currents fed by the mains significantly improve. The overall efficiency also increases, since most useless current terms are prevented from flowing in the distribution feeders. For the same reason also, the voltage profiles improve since the asymmetrical voltage drops caused by the circulation of negative- and zero-sequence currents are attenuated.

3. A further extension is achieved if each inverter performs reactive and unbalance compensation for a set of vicinity loads. The control algorithm is the same as in the previous case; however, the range of measurement extends to cover neighbor loads.

Two additional control methods have been considered, which further improve the performances but require data sharing among the inverters and/or interaction between the central control and the utility interfaces:

4. The first method coordinates all inverters of a subgrid to compensate for the total reactive power absorption and load unbalance generated within the subgrid. While providing excellent performances, this method requires data exchange within the subgrid and the implementation of suitable power sharing criteria among the inverters. Correspondingly, its implementation is more complex and demanding than for the previous cases.

The second solution requires the implementation of the simple control method mode 2 in all inverters but the utility interfaces. These latter receive data on total power absorption from the central controller. Correspondingly, each UI computes and compensates for any residual reactive and unbalance power terms in the subgrid. The results are unity power factor at the PCCs, high efficiency and voltage stability, without overstressing the UI inverters, which are supported by any other inverters in the same subgrid.

Overall, the simplicity of control is appreciable as compared to that of solutions based on centralized or distributed optimum or consensus-based control techniques, which may require dynamic and adaptive control algorithms and high-frequency communication of ubiquitous data collected by fast meters. Even though the considered methods are presented referring to steady-state considerations, they are also effective with typical load and generation dynamics, as shown by the time-varying 24 h profiles considered in the presented case studies. In addition, the response in case of the occurrence of fast transients of similar control algorithms has been already analyzed considering realistic conditions; for example, in [25], showing that no significant impact occurs on the main microgrid performances.

Notably, the control methods presented in this paper do not suffer from potential instability, since the interaction among inverters is minimal; in fact, the limited voltage variations occurring during normal operation of the grid do not substantially affect the control methods presented here.

Finally, we observe that the proposed control performs properly even in case of measurement inaccuracy or estimation error, since moderate control errors by local inverters do not affect significantly the overall efficiency and voltage stability of the network. In any case, the performances at PCCs can be adjusted by corrective actions done by the utility interfaces.

7. Conclusions

The establishment of energy communities as active players in the energy market requires the capacity to handle the energy resources in a coordinate way. This is the basis for energy trading, whose reward may justify collective efforts to strengthen the facility and improve the effectiveness and resilience of energy management.

Microgrids represent the most viable solution to aggregate neighboring end users who share their resources for mutual benefit. Exploiting the full potential of such aggregations, however, requires smart control and coordination of energy resources, which may be difficult if the inverters tied to the distribution grid are not equipped with flexible control and communication.

To circumvent this limit, we presented some methods of local control which do not require communication and can be implemented even in existing inverters by updating their control firmware.

We also discussed more advanced solutions, which rely on community energy storage units, equipped with flexible control and communication interfaces, and permit control of the power flow at the points of common coupling with the mains.

The application of the proposed control methods to a benchmark network demonstrated that even pure local control techniques can significantly improve the quality of operation of the microgrid from several perspectives, e.g., energy efficiency, voltage steadiness, and power unbalance. It was also shown that integrating local and central control makes possible more sophisticated and rewarding interaction with the mains, including demand response and islanding.

Overall, it seems that several planned benefits of energy communities can be achieved without substantial renovation of existing infrastructure and equipment. In fact, retrofitting existing plants and establishing community energy storage systems would provide a wide set of operational tools to control and trade energy in the wholesale market.

Author Contributions: Conceptualization, P.T.; methodology, P.T.; software, P.T.; validation, P.T.; data curation, P.T. and T.C.; writing—original draft, P.T.; writing—review & editing, T.C.; visualization, T.C.; project administration, P.T. All authors have read and agreed to the published version of the manuscript.

Funding: This research was partially funded by the Department of Management and Engineering (DTG), University of Padova, research project ADPE.

Data Availability Statement: The data presented in this study are available on request from the corresponding author.

Conflicts of Interest: The authors declare no conflict of interest.

Appendix A

This appendix reports the parameters of the testbench grid displayed in Figure 1. Table A1 specifies the type of feeders, Table A2 shows the connections among grid nodes, the characteristics of the loads tied to the grid, and the parameters of MV/LV transformers feeding the three subnetworks, and Table A3 specifies the characteristics of distributed energy sources.

Table A1. Parameters of the cables used in the application example in Figure 1.

Cable Type	Effective Area (mm^2)	R Phase Wire (Ω/km)	L Phase Wire (mH/km)	R Neutral Wire (Ω/km)	L Neutral Wire (mH/km)	Rated Current (A_{rms})
UG1	240	1.63×10^{-1}	4.33×10^{-1}	4.90×10^{-1}	1.50	430
UG2	150	2.66×10^{-1}	4.81×10^{-1}	7.33×10^{-1}	1.81	325
UG3	120	3.26×10^{-1}	5.03×10^{-1}	8.60×10^{-1}	2.01	290
UG4	25	1.54	6.56×10^{-1}	2.33	4.63	120
UG5	35	1.11	6.21×10^{-1}	1.93	4.03	145
UG6	70	5.69×10^{-1}	5.54×10^{-1}	1.29	2.75	215
OH1	50	3.87×10^{-1}	9.39×10^{-1}	6.89×10^{-1}	1.50	172
OH2	35	5.24×10^{-1}	9.77×10^{-1}	8.38×10^{-1}	1.56	145
OH3	16	1.15	1.05	1.84	1.69	93
I1	150	2.66×10^{-1}	4.81×10^{-1}	7.33×10^{-1}	1.81	325

Table A2. Details about the connections among grid nodes, the characteristics of the loads tied to the grid, and the parameters of MV/LV transformers feeding the three subnetworks considered in the application example in Figure 1.

Residential Subnetwork					Commercial Subnetwork					MV/LV Transformers (Yn)				
Line	Node from	Node to	Length (m)	Cable Type	Line	Node from	Node to	Length (m)	Cable Type	Name	Node from	Node to	kVA	Vcc
R1	R1	R2	35	UG1	C1	C1	C2	30	OH1	RT	R0	R1	500	5%
R2	R2	R3	35	UG1	C2	C2	C3	30	OH1	CT	C0	C1	300	6%
R3	R3	R4	35	UG1	C3	C3	C4	30	OH1	IT	I0	I1	150	2%

Table A2. *Cont.*

Residential Subnetwork					Commercial Subnetwork					MV/LV Transformers (Yn)				
Line	Node from	Node to	Length (m)	Cable Type	Line	Node from	Node to	Length (m)	Cable Type	Name	Node from	Node to	kVA	Vcc
R4	R4	R5	35	UG1	C4	C4	C5	30	OH1	Residential loads				
R5	R5	R6	35	UG1	C5	C5	C6	30	UG6	Node	kW	kVAr	Conn	Random term vs. CIGRE profile
R6	R6	R7	35	UG1	C6	C6	C7	30	OH2	R11	13	8	Yn	±40%
R7	R7	R8	35	UG1	C7	C7	C8	30	OH2	R15	60	38	Yn	±40%
R8	R8	R9	35	UG1	C8	C8	C9	30	OH2	R16	46	30	Yn	±40%
R9	R9	R10	35	UG1	C9	C3	C10	30	OH2	R17	4.25	2.65	3n	±40%
R10	R3	R11	35	UG4	C10	C10	C11	30	OH2	R18	40	25	Yn	±40%
R11	R4	R12	35	UG2	C11	C11	C12	30	OH3	Commercial loads				
R12	R12	R13	35	UG2	C12	C11	C13	30	OH3	C12	17	10.5	Yn	±30%
R13	R13	R14	35	UG2	C13	C10	C14	30	OH3	C13	6.8	4.2	Yn	±30%
R14	R14	R15	30	UG3	C14	C5	C15	30	OH3	C14	21.5	13	Yn	±30%
R15	R6	R16	30	UG6	C15	C15	C16	30	OH3	C17	13.5	8.5	Yn	±30%
R16	R9	R17	30	UG4	C16	C15	C17	30	OH3	C18	6.8	4.2	Yn	±30%
R17	R10	R18	30	UG5	C17	C16	C18	30	OH3	C19	21.5	13	Yn	±30%
Residential subnetwork meshing					C18	C8	C19	30	OH3	C20	17	10.5	Yn	±30%
R18	R1	R5	140	UG1	C19	C9	C20	30	UG4	Industrial loads				
R19	R5	R18	200	UG1	Commercial subnetwork meshing					I2	60	37	Yn	20%
Industrial subnetwork					C20	C1	C5	120	UG1					
I1	I1	I2	200	UG2	C21	C5	C20	150	UG1					

Table A3. Characteristics of the distributed energy resources considered in the application example in Figure 1.

Type	Node	Inverter Rating (kVA)	Rated/Max Power (kW)	Rated Energy (kWh)	Random Term vs. CIGRE Profile
PV unit	R16	55	4		±20%
	R18	50	3		±20%
	C14	25	20		±20%
	C19	25	20		±20%
	I2	80	20		±20%
Wind turbine	R5	7	5.5		±30%
	C12	25	20		±20%
	C20	25	20		±20%
Energy storage	R6	35	15/30	30	
	R10	25	10/20	20	
	R15	70	60/120	120	
	R16	55	50/100	100	
	R18	50	40/80	80	
	C6	35	15/30	30	
	C12	20	20/40	40	
	C14	25	20/40	40	
	C17	15	15/30	30	
	C19	25	20/40	40	
	C20	20	17.5/35	35	
	I2	80	60/120	120	
Utility interface	R1	120	50/100	100	
	C1	120	50/100	100	
	I1	80	35/70	70	

References

1. Publications Office of the European Union. *Clean Energy for All Europeans*; Publications Office of the European Union: Luxemburg, 2019; p. 24.
2. Saha, D.; Bazmohammadi, N.; Vasquez, J.C.; Guerrero, J.M. Multiple Microgrids: A Review of Architectures and Operation and Control Strategies. *Energies* **2023**, *16*, 600. [CrossRef]
3. Abbasi, M.; Abbasi, E.; Li, L.; Aguilera, R.P.; Lu, D.; Wang, F. Review on the Microgrid Concept, Structures, Components, Communication Systems, and Control Methods. *Energies* **2023**, *16*, 484. [CrossRef]
4. Salehi, N.; Martinez-Garcia, H.; Velasco-Quesada, G.; Guerrero, J.M. A Comprehensive Review of Control Strategies and Optimization Methods for Individual and Community Microgrids. *IEEE Access* **2022**, *10*, 15935–15955. [CrossRef]
5. Kandari, R.; Neeraj, N.; Micallef, A. Review on Recent Strategies for Integrating Energy Storage Systems in Microgrids. *Energies* **2022**, *16*, 317. [CrossRef]
6. Tenti, P.; Caldognetto, T. On Microgrid Evolution to Local Area Energy Network (E-LAN). *IEEE Trans. Smart Grid* **2019**, *10*, 1567–1576. [CrossRef]
7. Dambrauskas, P.; Syed, M.H.; Blair, S.M.; Irvine, J.M.; Abdulhadi, I.F.; Burt, G.M.; Bondy, D.E.M. Impact of Realistic Communications for Fast-Acting Demand Side Management. *CIRED—Open Access Proc. J.* **2017**, *2017*, 1813–1817. [CrossRef]
8. Marafao, F.P.; Alonso, A.M.d.S.; Goncalves, F.A.S.; Brandao, D.I.; Martins, A.C.G.; Morales Paredes, H.K. Trends and Constraints on Brazilian Photovoltaic Industry: Energy Policies, Interconnection Codes, and Equipment Certification. *IEEE Trans. Ind. Appl.* **2018**, *54*, 4017–4027. [CrossRef]
9. Tayab, U.B.; Roslan, M.A.B.; Hwai, L.J.; Kashif, M. A Review of Droop Control Techniques for Microgrid. *Renew. Sustain. Energy Rev.* **2017**, *76*, 717–727. [CrossRef]
10. Crivellaro, A.; Tayyebi, A.; Gavriluta, C.; Gross, D.; Anta, A.; Kupzog, F.; Dorfler, F. Beyond low-inertia systems: Massive integration of grid-forming power converters in transmission grids. In Proceedings of the IEEE Power & Energy Society General Meeting (PESGM), Montreal, QC, Canada, 2–6 August 2020; pp. 1–5.
11. Qingzhu, W.; Mingli, W.; Jianye, C.; Guiping, Z. Model for optimal balancing single-phase traction load based on Steinmetz's method. In Proceedings of the IEEE Energy Conversion Congress and Exposition, Atlanta, GA, USA, 12–16 September 2010; pp. 1565–1569.
12. Steinmetz, C.P. *Theory and Calculations of Electrical Apparatus*; Mc-Graw Hill: New York, NY, USA, 1917.
13. Tenti, P.; Caldognetto, T. Generalized Control of the Power Flow in Local Area Energy Networks. *Energies* **2022**, *15*, 1416. [CrossRef]
14. Gavriluta, C.; Candela, I.; Rocabert, J.; Etxeberria-Otadui, I.; Rodriguez, P. Storage system requirements for grid supporting PV-power plants. In Proceedings of the IEEE Energy Conversion Congress and Exposition (ECCE), Pittsburgh, PA, USA, 14–18 September 2014; pp. 5323–5330.
15. Todeschini, G.; Emanuel, A.E.; Ferrero, A.; Morando, A.P. A Poynting Vector Approach to the Study of the Steinmetz Compensator. *IEEE Trans. Power Deliv.* **2007**, *22*, 1830–1833. [CrossRef]
16. Czarnecki, L.S. From Steinmetz to currents' physical components (CPC): History of power theory development. In Proceedings of the International Conference on Applied and Theoretical Electricity (ICATE), Craiova, Romania, 6–8 October 2016; pp. 1–10.
17. Pau, M.; Ponci, F.; Monti, A.; Muscas, C.; Pegoraro, P.A. Distributed State Estimation for Multi-Feeder Distribution Grids. *IEEE Open J. Instrum. Meas.* **2022**, *1*, 1–12. [CrossRef]
18. Radhoush, S.; Bahramipanah, M.; Nehrir, H.; Shahooei, Z. A Review on State Estimation Techniques in Active Distribution Networks: Existing Practices and Their Challenges. *Sustainability* **2022**, *14*, 2520. [CrossRef]
19. Fortescue, C.L. Method of Symmetrical Co-Ordinates Applied to the Solution of Polyphase Networks. *Trans. Am. Inst. Electr. Eng.* **1918**, *37*, 1027–1140. [CrossRef]
20. Tenti, P.; Willems, J.L.; Mattavelli, P.; Tedeschi, E. Generalized Symmetrical Components for Periodic Non-Sinusoidal Three-Phase Signals. *J. Electr. Power Qual. Util.* **2007**, *XIII*, 9–15.
21. CIGRE Task Force. *C6.04.02 Benchmark Systems for Network Integration of Renewable and Distributed Energy Resources*; International Council on Large Electric Systems: Paris, France, 2014.
22. *IEEE Std 1547-2018*; IEEE Standard for Interconnection and Interoperability of Distributed Energy Resources with Associated Electric Power Systems Interfaces (Revision of IEEE Std 1547-2003). Institute of Electrical and Electronics Engineers: Piscataway, NJ, USA, 2018; pp. 1–138.
23. Li, Y.R.; Nejabatkhah, F.; Tian, H. *Smart Hybrid AC/DC Microgrids: Power Management, Energy Management, and Power Quality Control*; John Wiley & Sons: Hoboken, NJ, USA; IEEE Press: Chichester, UK, 2023; ISBN 9781119598398.
24. Vlahinić, S.; Franković, D.; Komen, V.; Antonić, A. Reactive Power Compensation with PV Inverters for System Loss Reduction. *Energies* **2019**, *12*, 4062. [CrossRef]
25. Abedini, H.; Caldognetto, T.; Mattavelli, P.; Tenti, P. Real-Time Validation of Power Flow Control Method for Enhanced Operation of Microgrids. *Energies* **2020**, *13*, 5959. [CrossRef]